D0835480

# Cell Clones:
# Manual of Mammalian Cell Techniques

# Cell Clones: Manual of Mammalian Cell Techniques

EDITED BY

**C. S. Potten**
Head of Epithelial Kinetics Section, Paterson Laboratories, Christie Hospital and Holt Radium Institute, Manchester, UK

**J. H. Hendry**
Head of Radiobiology Section, Paterson Laboratories, Christie Hospital and Holt Radium Institute, Manchester, UK

CHURCHILL LIVINGSTONE
EDINBURGH LONDON MELBOURNE AND NEW YORK 1985

CHURCHILL LIVINGSTONE
Medical Division of Longman Group Limited

Distributed in the United States of America by Churchill Livingstone Inc., 1560 Broadway, New York, N.Y. 10036, and by associated companies, branches and representatives throughout the world.



First published 1985

ISBN 0 443 02805 2

British Library Cataloguing in Publication Data
Cell clones: manual of mammalian cell techniques.
1. Cell proliferation
2. Clone cells
3. Mammals — Cytology
I. Potten, C. S. II. Hendry, J. H.
599.08′7 QH605

Library of Congress Cataloging in Publication Data
Cell clones: manual of mammalian cell techniques.
Includes index.
1. Mammals — Cytology — Technique. 2. Clone cells.
3. Cell culture. I. Potten, C. S., 1940– .
II. Hendry. J. H. [DNLM: 1. Cells, Cultured — physiology. 2. Cell Division. 3. Regeneration. 4. Histological Technics. QS 525 M265]
QL739.15.M36 1984 599′.087 84-11416

Produced by Longman Group (FE) Ltd
Printed in Hong Kong

# Contributors

**S. E. Al-Barwari**
Department of Anatomy and Histology, College of Medicine, University of Salahaddin, Erbit, Iraq

**G. H. Buisman**
SSDZ Hospitals, Delft, Holland

**K. H. Clifton**
Department of Human Oncology and Radiology, Wisconsin Clinical Cancer Center, University of Wisconsin Medical School, Madison, Wisconsin, USA

**A. M. Connor**
Department of Therapeutic Radiology, University of Louisville School of Medicine, Louisville, Kentucky, USA

**V. D. Courtenay**
Division of Biophysics, Radiotherapy Research Unit, Institute of Cancer Research, Sutton, Surrey, UK

**R. Cox**
MRC Radiobiology Unit, Harwell, Oxford, UK

**J. Denekamp**
Section of Radiobiology Applied to Therapy, Gray Laboratory of the Cancer Research Campaign, Mount Vernon Hospital, Middlesex, UK

**P. J. Deschavanne**
Laboratoire de Radiobiologie des Cellules Normales et Cancéreuses, Unité Inserm 247, Institut Gustave-Roussy, Villejuif, France

**R. Dover**
Department of Histopathology, Royal Postgraduate Medical School, Hammersmith Hospital, London, UK

**M. Fox**
Section of Experimental Chemotherapy, Paterson Laboratories, Christie Hospital and Holt Radium Institute, Manchester, UK

**M. N. Gould**
Department of Human Oncology and Radiology, Wisconsin Clinical Cancer Center, University of Wisconsin Medical School, Madison, Wisconsin, USA

**M. Guichard**
Laboratorie de Radiobiologie des Cellules Normales et Cancéreuses, Unité Inserm 247, Institut Gustave-Roussy, Villejuif, France

**J. H. Hendry**
Head of Radiobiology Section, Paterson Laboratories, Christie Hospital and Holt Radium Institute, Manchester, UK

**R. P. Hill**
Physics Division, University of Toronto, The Ontario Cancer Institute, Toronto, Canada

**J. W. Hopewell**
CRC Normal Tissue Radiobiology Group, Research Institute (University of Oxford), Churchill Hospital, Oxford, UK

**S. Hornsey**
MRC Cyclotron Unit, Hammersmith Hospital, London, UK

**R. L. Jirtle**
Duke University Medical Center, Durham, North Carolina, USA

**N. F. Kember**
Physics Department, The Medical College of St Bartholomews Hospital, London, UK

**J. Kummermehr**
Institut für Biologie, Abteilung Strahlenbiologie und

Biophysik, Gesellschaft fur Strahlen-und Umweltforschung MBH, Neuherberg, West Germany

**B. I. Lord**

Section of Experimental Haematology, Paterson Laboratories, Christie Hospital and Holt Radium Institute, Manchester, UK

**E. P. Malaise**

Laboratoire de Radiobiologie des Cellules Normales et Cancéreuses, Unité Inserm 247, Institut Gustave-Roussy, Villejuif, Cedex, France

**K. A. Mason**

Section of Experimental Radiation Oncology, Center for Health Sciences, The University of California, Los Angeles, California, USA

**W. K. Masson**

MRC Radiobiology Unit, Harwell, Didcot, Oxford, UK

**G. Michalopoulos**

Duke University Medical Center, Durham, North Carolina, USA

**C. Mothersill**

St Lukes Hospital, Rathgar, Dublin, Ireland

**C. S. Potten**

Head of Epithelial Kinetics Section, Paterson Laboratories, Christie Hospital and Holt Radium Institute, Manchester, UK

**H. S. Reinhold**

Radiobiological Institute, REP — Institutes of the Organisation for Health Research TNO, Rijswijk, The Netherlands

**R. Schofield**

Section of Experimental Haematology, Paterson Laboratories, Christie Hospital and Holt Radium Institute, Manchester, UK

**C. P. Sigdestad**

Department of Therapeutic Radiology, Louisville School of Medicine, University of Kentucky, Louisville, Kentucky, USA

**N. G. Testa**

Section of Experimental Haematology, Paterson Laboratories, Christie Hospital and Holt Radium Institute, Manchester, UK

**H. R. Withers**

Department of Experimental Radiation Oncology, Center for Health Sciences, The University of California, Los Angeles, California, USA

# Contents

# Introduction

The ability of radiation to abolish the capacity of cells to divide repeatedly has long been recognised as its major effect in renewing tissues. A culture technique for assessing this ability for mammalian cells was described by Puck & Marcus in 1956. When a dilute suspension of single cells was seeded in nutrient agar in a Petri dish, the cells divided repeatedly and formed isolated clusters. These clusters are called clones when it is known that all cells in the cluster are formed from a single ancestral cell. The clusters are called colonies when the number of ancestors per cluster is one or more than one.

Radiation reduces the ability of cells to form clones or colonies, and this is called variously a reduction in viability, or in reproductive integrity, or in survival. The first assay for assessing the proliferative ability of single cells in vivo was developed by Hewitt & Wilson in 1959, using the ability of single leukaemic cells to multiply, after transplantation into a recipient mouse, and induce leukaemia. Since then there has been a continued increase in the range of tissues where cells can be persuaded to produce regenerative foci. Such techniques have found wide application in the fields of radiobiology and pharmacology (with clinical implications for radiotherapy and chemotherapy), and also in biological studies of specific tissues or cell types. A clear example of this is the dramatic increase in our understanding of the cellular basis of haemopoiesis which arose from the introduction of the spleen colony assay for bone marrow stem cells in 1961 by Till & McCulloch. A similar increase in our knowledge followed the development of ingenious assays for epithelial colony-forming cells by Withers from 1967 onwards, for epidermis, gastro-intestinal mucosa, testis and kidney. Some techniques have now been used extensively to study a variety of responses, while others are still in their infancy and have so far been used for little more than the determination of a radiation dose-response curve. This is commonly one of the initial steps undertaken as a technique is being developed for a new tissue. Most of the contributors to this volume illustrate the use of their techniques by presenting such a radiation survival curve.

We, the editors, thought it would be of value to collect together all the available types of colony regeneration techniques. These techniques could be useful in several fields aside from radiobiology. It is sometimes difficult to find out particular technical details about assays in tissues with which one has little direct experience. We selected authors, all of whom had some first hand experience of using the techniques, even if they were not the originators or developers of the technique. Sometimes the pioneers of the system have long since moved into other fields of interest. We asked all authors to describe the methods involved in such a way that a novice could perform the techniques and also to provide information on the types of data which can be generated. We also asked the authors to discuss the question of what type of cell is being assessed. This is a difficult question to answer in some instances since to some extent the ability to answer it is simply related to the time since the technique was first introduced, i.e. to the amount of work which has been undertaken using the procedure.

The system using cells in culture measures cell survival directly. The survival curves generated are usually true cell survival curves, uncomplicated by unknown levels of cellular multiplicity — i.e. each colony arises from an isolated single cell and not from a pair or a clump of cells. Each colony thus represents a 'clone' of cells derived from a single ancestral cell, and this has been demonstrated with some colonies in vivo by using chromosome markers (spleen colonies) or enzyme markers (liver colonies). In other cases the single cell origin of colonies is less clear but can be reasonably assumed because of the low level of cell survival after high dose irradiation: here the chances of two surviving clonogenic cells lying so close together that their clones fuse to make one colony is very remote. This line of reasoning can be applied, for example, to the macrocolony assays in skin or gut. In principle, there may be a genetic (and hence phenotypic) heterogeneity 'built into' the original DNA so that the descendants in the clone appear dissimilar. Also, the descendants may develop some genetic heterogeneity during subsequent successive divisions, or may develop phenotypic heterogeneity as the consequence of changes in the repression or depression of parts of the DNA, i.e. adoption of different differentiation options.

In yet other systems the colonies observed are the result of cells surviving from organ subunits or 'proliferative units', e.g. tubules in testis or kidney, and intestinal crypts, each of which may contain many colony-forming cells. When the number of regenerated units is only slightly less than the number in untreated animals, the surviving units should be regarded as colonies, and not clones, since most will have regenerated from more than one clonogenic cell. However, after a certain radiation dose the survival curves for the units become true cell survival curves since each colony becomes increasingly likely to be a true clone derived from a single surviving cell in each structure. The survival curve measured by counting colonies can be transformed into a survival curve for clonogenic cells by applying a factor calculated using Poisson statistics, which describe the random exponential distribution of surviving clonogenic cells after irradiation. An example of this is the crypt microcolony assay. The fact that the macrocolony assay in gut can be seen to represent an extension to higher doses of the crypt survival curve further supports the conclusions that the colonies being measured at high doses in the microcolony assay are indeed clones derived from single surviving cells.

It requires many divisions for a true surviving cell to make a clone, conventionally 5 to 6 or more divisions are considered necessary for cells in culture, probably 7 to 8 or more for microcolonies in the gut, and many more, probably 10 to 13, for macrocolonies. It should be noted that if there is insufficient stimulus for division of the colony-forming cells, small colonies may be produced, resulting from very few divisions of true colony-formers. Other small colonies may result from colony-formers which were adequately stimulated, but were reproductively incompetent. At one extreme the colonies hypothetically could contain only two cells. In this case the test to which the cells are being put is whether or not they can be induced to enter a single round of DNA synthesis and mitosis. This is clearly a far less stringent test than the test of whether or not they can divide many times. The number of rounds of cell division required in the assay will tend to determine the apparent radiosensitivity of the system because some genetic injury affecting division is expressed only in later divisions. Thus cells which undergo only one round of cell division will appear more radioresistant than those undergoing several rounds. Generally speaking the radiosensitivity increases as the stringency of the test, in terms of the number of cell divisions needed, is increased up to a point and then remains roughly constant. The 'break-point' where further stringency makes little difference is often 5 or 6 cell divisions. This forms the basis to the choice of the common 'cut-off limit' of 50 cells for counting colonies in vitro. With sectioning techniques in vivo, the number of cells appearing in a cluster in the section and scored as a colony is for example 10 with the intestinal microcolony technique, but only 2 with kidney tubules.

For many systems the true clonal nature of the colonies has not been proved using marker techniques, but reasonably can be assumed as discussed above. However, even then there are some possible doubts. For example, with culture systems, single cells are plated. These attach to the surface and colonies are eventually observed. These most probably result from the clonal expansion of a single clonogenic cell, but they could in some cases require the local accumulation of several migratory cells before a colony is produced. If only one of the coalesced group subsequently forms a colony, i.e. the others act merely as 'helpers' in some way, the majority of cells in the colony can still be regarded as a clone. This may apply to fibroblastic colonies derived from mouse bone marrow which also contain endothelioid cells and macrophages. It is perhaps worth noting that some systems have a particular range of cell densities which is optimum for growth, i.e. if cells are plated too sparsely they do not form colonies, either because they lack long-ranging growth factors or perhaps because they are too far apart to coalesce and perform their helper function. We have called the book *Cell Clones: Manual of Mammalian Cell Techniques* in spite of the several assumptions of the clonal origin of the colonies grown using many of the techniques. We have not edited the text, or titles, of many chapters where the authors have used the term 'clonal' since it seemed that in many cases although the clonal nature of the colonies has not been proved it is indeed a reasonable assumption.

A point that should be considered is the nature and normal function of the cells that are assayed by these techniques. The methods using established cell lines assay the reproductive integrity of all proliferative cells. Other techniques using primary cultures assay colony-forming cells which may include both maturing proliferative cells as well as their progenitors, the stem cells. Macrocolony assays in vivo measure the response of clonogenic cells and it is often assumed that these constitute the stem-cell population. However, the ability to undergo clonal growth itself is not a feature solely of stem cells. Maturing proliferative cells may well be capable of many cell divisions and could produce a clone. For example, granulocytic precursor cells derived from haemopoietic stem cells are capable of up to about 6 divisions, and differentiating epidermal basal cells and maturing intestinal crypt cells normally may be capable of 3–4 divisions. It is possible that these capabilities could be expanded under certain conditions. However, these clones *eventually* differentiate and mature, and exhaust their division potential. Also, these clonogenic cells would be incapable of reproducing further similar clonogenic cells. Therefore the true criterion for identifying stem-cell clones is clonal growth together with the concomitant production of further new clonogenic (stem)cells. These clonogenic cells possess a long-term proliferative potential with the probable production of one or more differentiated transitory cell populations. The spleen colony assay satisfies these requirements because the clones have been shown by chromosome marker techniques to be derived from a single cell, and they contain maturing erythroid and granuloid cells as well as further spleen colony-forming cells. Others systems also may be classed in the same category, because for example epithelial macrocolonies contain maturing cells which re-epithelialise the denuded epithelial surface and hence demonstrate the permanence expected when they contain further clonogenic cells.

In hierarchical tissues, the stem cells are required for the initiation of the amplifying series of cell divisions needed for the continual replenishment of the functional mature-cell population. Although stem-cell colonies assay this maintenance ability, there are other contributory factors to continued tissue function after injury which should not be forgotten, but which are not the subject of this volume, namely the environmental support for the regenerating lineages and the correct spatial reconstruction of these lineages in the tissue.

We thank the authors for their contributions, their forebearance with our editorial suggestions and their helpful comments on this introduction.

# 1

*J.H. Hendry*

# Mathematical aspects of colony growth, transplantation kinetics and cell survival

## COLONY GROWTH

### Definitions

The ability of cells to divide repeatedly has long been considered to be the most important parameter to be quantitated in renewing tissues exposed to cytotoxic agents. Cells with this ability are commonly called *viable cells*. (It should be noted that this term is also sometimes used to describe whether a cell is morphologically and often functionally intact.) When radiation is the cytotoxic agent, the major expression of injury is the reduction in viability i.e. the proportion of cells which retains *reproductive integrity* or the *surviving fraction* of cells according to this criterion.

When cells divide repeatedly and form discrete isolated groups of progeny, these groups are called *colonies*. In cases where colonies have been shown to arise from divisions of a single parental cell, for example, by radiation-induced chromosomal markers in the case of spleen colonies (Becker et al, 1963), the originator of a colony should be called a *clonogenic cell* (*clonogen*) or *colony-forming cell* (CFC). The cells in the clone would be genetically identical, although they may have differentiated and be phenotypically different. When colonies contain several cell types of unknown origin, the originators should be called a colony-forming unit (CFU).

Many authors use the term clonogenic cell or a similar connotation, e.g. microcolony-forming cell, in situations where clonality has not been demonstrated but is inferred, for example by agreement with mathematical predictions of single cell responses e.g. exponential survival or transplantation kinetics. However, there are some examples in vitro of a curvilinear relationship between the number of cells plated and the number of colonies produced (see Xu et al, 1983 and Ch. 3), and of an unexpected similar relationship in vivo between the number of cells injected and the number of tumours produced (Porter et al, 1973). In such situations more colonies are produced than would be expected from the higher number of cells seeded, and cell-cooperation, a density-dependent effect of helper cells or factors, or an immune mechanism in vivo acting preferentially on single cells rather than on cells in groups, can be invoked. Furthermore, exponential survival data, although predicted from single-cell considerations, cannot in some instances exclude the necessity of several cells for survival (discussed further below). Hence the term CFU (or micro-CFU), which can embody these effects, is preferable in these cases to the term clonogenic cell.

When colonies arise from single cells, and in the simplest case where all the daughter cells are also capable of division, the number of cells in the colony will increase from 1 to 2, 4, 8, 16, 32 etc. This is exponential growth, and after 'n' divisions, there will be $2^{(T/Tc)}$ cells in the colony. If the time between divisions is constant (the cell cycle time, Tc), then after time T there will have been (T/Tc) divisions. Hence after time T there will be $2^{(T/Tc)}$ cells in the colony. For example, if Tc is 12 hours, then after 7 days (a common assay time in vitro) the colonies should contain $2^{14}$ or 16 000 cells. If there is an initial lag of one cycle time before growth commences at this rate, the final number of cells would be less by a factor of 2, and if 2 cycles, by a factor of 4. Other combinations can be calculated similarly. Colonies of 1000 cells are often grown and counted in assays in vitro, but in assays in vivo the number of cells per colony can vary markedly, e.g. about $10^6$ for spleen colonies (Ch. 2) and about 50 cells for liver follicles in fat pads (Ch. 16). Colonies are defined generally as those containing more than 50 cells. Those containing less than 50 cells are sometimes called small colonies or clusters. The number 50 is relatively arbitrary, but it indicates that cells have undergone more than 5 divisions. It is common experience that most cells capable of undergoing 5 divisions are capable of many more, so that these can be classified as the true survivors.

### Self-renewal versus differentiation of clonogenic cells

Most colonies of normal cells differentiate into one or more lineages of maturing cells, and the presence of mature non-dividing cells will lead to a slowing of colony growth in terms of total cell number. This may also occur to some extent if the maturing dividing cells have longer cycle times than their precursors. Small colonies or clusters can arise from maturing cells which still possess a limited division potential, e.g. BFU-E (Ch. 3). However, these will be temporary, disappearing at the end of the life-span of the mature cells, as the colony will contain no further progenitor cells. Larger colonies can originate from stem cells, defined as those cells which can renew themselves as well as differentiate into all types of maturing progeny characteristic of that cell hierarchy. The content of stem cells at any time can be calculated in some situations, as follows.

When a stem cell divides it could give rise on average to '2p' stem cells and 2(1-p) differentiating cells, where p is the self-renewal probability (Till et al, 1964). If p is constant throughout colony growth, the total number of stem cells after n divisions would be $(2p)^n$. Clearly, if p = 1 there is only self-renewal and no differentiation, and if p = 0.5 the number of stem cells will remain constant. The latter is analogous to the steady-state situation in vivo, where cell production balances cell loss. As the total number of cells after n divisions will be $(2)^n$ if all the differentiating cells are dividing, then the fraction of cells which are stem cells after n divisions will be $(2p)^n/(2)^n = (p)^n$. Also, after time T, the number of stem cells present will be $(2p)^{T/Tc}$, where Tc is the cell cycle time, and the time for the stem cells to double in number, the doubling time (TD), will be $Tc.\frac{\ln(2)}{\ln(2p)}$. Hence, if there is a fixed average probability of differentiation at each cycle, the stem cell population will grow more slowly than the total cell number, which is doubling at a rate given directly by Tc. In the later stages of colony growth the maturing cells may well have longer cycle times, or emigrate from the colony, and hence the fraction of total cells which is stem cells could be higher than predicted. The above considerations are based on a constant value of Tc among stem cells. If there is a distribution of cycles times, which is likely, then the number of stem cells present at any time (T) will be different from the simple relationship above. An example of this effect, which is probably the other extreme but which is amenable to calculation, is where there is an exponential distribution of cycle times. In this case the average number of stem cells (M) present at any time (T) is given by:

$$\ln M = [0.693.(2p-1).\frac{T}{Tc}]$$

(Schofield et al, 1980).

The degree of spread of the stem-cell content between colonies depends on the value of p. This has been discussed with reference to spleen colonies and can be expressed mathematically as:

$$V^2 = \frac{(2-2p)}{(2p-1)} + \frac{1}{M}$$

(Vogel et al, 1968, 1969), where V = coefficient of variation of CFU-S per colony = standard deviation/mean CFU-S per colony; p = probability of self-renewal; M = mean CFU-S per colony. Hence p can be measured when the distribution of stem cells among colonies can be assessed. Values of p for normal CFU-S range between about 0.62 (Vogel et al, 1968) and 0.68 (Schofield et al, 1980). As the doubling time of CFU-S in spleen colonies is about 20 hours, a value for p of 0.65 would give Tc = 7.5 hours.

The number of CFU-S per colony can be changed by various cytotoxic treatments to the original graft, e.g. after IMS (Schofield & Lajtha, 1973). The reduced content of CFU-S in colonies of a given age is commonly used as a measure of an induced qualitative defect in the surviving stem-cell population, e.g. Botnick et al (1981). However, this should not be interpreted automatically in terms of a reduction in the value of p, because if this were so, the doubling time of CFU-S in the colony would change assuming that the cycle time is unchanged. Most examples where this has been measured show a *similar* rate of growth of the CFU-S population, but the growth curve is shifted in time due to a lag or some other mechanism (Schofield & Lajtha, 1983).

A parameter related to p is the extinction probability '$\omega$' (Vogel et al, 1968, 1969). If p is constant throughout colony growth, then a fraction (1-p) of stem cells will differentiate at the first division and will not form a colony. A similar fraction should differentiate at the second and subsequent divisions, and this effect will reduce the initial number of CFC which finally produce colonies. The probability of 'extinction' reaches an asymptotic value of [(1-p)/p] after about 5 generations (Vogel et al, 1968), and this is about 0.63 for normal CFU-S. Thus, only $(1-\omega)$ = 37 per cent of CFC would produce colonies, and hence the expected number of potential CFC would be greater than the measured number of CFC by a factor (1/0.37) = 2.7.

These techniques have not yet been applied to other colony assays, but the advent of grafting techniques for producing colonies derived from cells in other tissues makes their application possible.

In contrast, a cell loss factor ($\emptyset$) was introduced for the growth of tumours (Steel, 1968). This is the rate of cell loss i.e. cells lost per unit time expressed as a fraction of the rate of cell production. If cell 'loss' is taken to denote solely loss of self-renewal ability, then $\emptyset$ = [(1-p)/p], which is the same expression as for $\omega$. However, cell 'loss' in tumours refers to the physical removal of cells, not solely their loss of self-renewal ability. A discussion of 'p' in differentiating cell populations in tumours can be found elsewhere (Mackillop et al, 1983).

## SAMPLING TECHNIQUES

### The Poisson distribution

All the assays described in this book are based on techniques using estimated sample sizes. When the total number of cells (N) plated in vitro or at risk in vivo can be counted accurately, rather than estimated from a count of a small aliquot, binomial statistics should be applied as the surviving number ranges between N and zero. When N is estimated from a random sample, i.e. when aliquots of a suspension are plated out, Poisson statistics can be used as an approximation because N could range between zero and very large (almost infinite) values. When the mean number of cells is small, the Poisson distribution will be skew, as many of the counts will be zero by chance. When the mean is large (above about 10), the distribution becomes more symmetrical and tends towards a normal distribution.

The Poisson distribution is very common in radiobiology, not only because of the sampling techniques employed but also because of the random nature of deposition of energy, and hence the production of biological events, by radiation. The probability of any given count can be calculated from: $f = \frac{e^{-m}.m^n}{n!}$ where m is the mean count, n is the count which occurs with probability f, and $n! = [n \times (n-1) \times (n-2) x.....x1]$. Common examples of its applications are given below.

**Intracellular events (distribution of hits)**

If cells receive on average 1 lethal hit, the fraction of cells escaping i.e. not being hit, will be $e^{-1} = 0.37$ or 37 per cent. Another fraction $(e^{-1} \times 1^1)/1 = e^{-1} = 37$ per cent of cells will receive 1 lethal hit, $(e^{-1} \times 1^2)/(2 \times 1) = (e^{-1})/2 = 18.5$ per cent of cells will receive 2 hits, and $(e^{-1} \times 1^3)/(3 \times 2 \times 1) = (e^{-1})/6 = 6$ per cent will receive 3 hits. With 2 lethal hits on average per cell, $e^{-2} = 13.5$ per cent of cells will escape, 27 per cent will receive 1 hit, and 27 per cent will receive 2 hits, etc.

Clearly, the fraction of cells escaping 'x' lethal hits is $e^{-x}$, and when x is linear with the dose delivered this forms the basis of a simple exponential survival curve (see below).

**Extracellular events (groups of cells)**

In many of the assays for survival, the clonogenic cells are grouped into structures containing similar numbers of these cells. After radiation there will be a distribution of the number of surviving cells per structure which approximates to a Poisson distribution when the initial number of clonogenic cells per structure is large and the number of survivors is small, so that the range in numbers about the mean can be from zero to this large initial value. When colonies can grow from 1 or more clonogenic cells, the fraction of ablated structures (F), can be used to calculate the corresponding mean number of surviving clonogenic cells per structure (m). $F = e^{-m}$, and hence $m = -\ln F$. Values of m can be plotted on the logarithmic scale of semi-log graph paper versus dose on the linear scale to give a conventional cell survival curve. However, this method can be used only at relatively high doses where the number of surviving cells per structure has been reduced on average to near unity or below so that some structures are ablated (see Chs 5, 12, 17). If the standard error on F is f, then the standard error on $m = f/F$. Also, it should be recognised that the number of surviving cells per surviving structure $= [m/(1-F)] = [-\ln F/(1-F)]$. The validity of using these calculations has been confirmed in two assays by measuring the distribution of colonies between different areas of epidermis (Withers, 1967; Hendry, 1984) or intestine (Ch. 5), where the distribution was not significantly different from a Poisson distribution.

Two related extensions of these principles are (1) when the possibility is considered that one clonogenic cell cannot form a colony, and co-operation of two or more is required, as noted in Chapter 17, and (2) when a larger number of clonogenic cells is required to rescue a tissue or an animal, where each cell has a small but finite probability of rescue. These effects are described below in the section on cell survival curves and multicellular structures.

**Transplantation kinetics**

Another similar and common situation is where colonies arise from one or more CFU after injecting serially-diluted inocula of a cell suspension. This applies to assays in fat pads (Chs 15, 16), and also to $TD_{50}$ assays for tumours (Ch. 25). These are sampling techniques which should be describable by Poisson statistics. In these cases, the chance of no colony growing will be $e^{-m}$, where m is the mean number of CFU injected. Hence, when 50 per cent of the injected sites show no growth, $e^{-m} = 0.5$, $m = 0.693$, and thus 0.693 viable CFU were injected on average. When $m = 1$, $e^{-m} = 0.37$, and hence the inoculum size resulting in 37 per cent of sites with no colony ($TD_{63}$) gives the plating efficiency directly, i.e. 1 CFU exists within a certain number of cells. In practice, serial dilutions are made and injected into different groups of animals to 'bracket' for example the $TD_{63}$. Sophisticated statistical methods are now commonly available for estimating the most likely value of $TD_{50}$ or $TD_{63}$ with their associated uncertainties. These methods have been developed largely by Finney (1964) and also by others for use in radiobiology (Porter & Berry, 1963; Porter et al, 1973; Gilbert, 1974; Porter, 1980a,b).

Briefly, the probability (F) of lack of growth in an injected site is given by: $F = e^{-m}$ (as above). m is proportional to the inoculum size (z) so that $m = kz$, and: $\ln m = \ln k + \ln z$. As $m = -\ln F$, $\ln(-\ln F) = \ln k + \ln z$. Hence a plot of $\ln(-\ln F)$ against $\ln z$ should give a line with a slope of 1 and an intercept of $\ln k$ on the ordinate, where k is the plating efficiency, i.e. $m/z$.

Also, k is given by the value of 1/z when F = 1/e = 0.37, i.e. at the $TD_{63}$.

Estimates of the parameters with associated error limits can be made using computer programmes with maximum likelihood (Finney, 1964; Porter, 1980a, b) or minimum chi-square techniques (Gilbert, 1969, 1974). We use a modified version of an earlier programme (Gilbert, 1969), as noted elsewhere (Gilbert, 1974), which calculates directly the plating efficiency with the expected standard deviation (sometimes called the standard error) of the mean (Fisher & Hendry, unpublished). Others who are using the above techniques include Clifton & Gould (Ch. 15), Jirtle & Michalopoulos (Ch. 16), Hill (Ch. 25), Rice et al (1980) and Porter (1980a,b). New users are advised to contact one of the current users and either send them the data for fitting if their needs are only occasional or obtain a copy of their computer programme if their respective computers use the same language and format.

Porter et al (1973) also discussed the case where the sigmoid curve is not in accordance with single-cell transplantation kinetics, i.e. when the number of clonogens in an inoculum is not related linearly to inoculum size. In this case other fitting procedures can be used (see Finney, 1964 and Ch. 25) which do not depend on any biological model. However, the effect can be accommodated in the fitting procedure by introducing another parameter 's', where $m = k.z^s$. Values of $s > 1$ could be explained by marked variability between recipients. Values of $s<1$ could result from 2 or more clonogenic cells being required for colony growth, as follows.

If 2 or more clonogenic cells are required, then the probability of there being no growth is the sum of the probabilities of there being (1) no cells, and (2) 1 cell, in the site, i.e. $F = e^{-m} + (m.e^{-m}) = (1 + m)e^{-m}$. Hence $\ln F = [\ln (1 + m)] - m$. An approximate value for 's' can be deduced by approximating $\ln(1 + m)$ by $(m - \frac{m^2}{2})$, when m is much less than 1.

Thus $\ln F = -m' \simeq \frac{-m^2}{2}$,

and $\ln(-\ln F) \simeq \ln 0.5 + 2 \ln m = \ln \frac{k^2}{2} + 2 \ln z$.

Hence, a plot of $\ln(-\ln F)$ against ln z would give a slope of 2 i.e. s = 0.5. However, the approximation is valid only for values of m less than about 0.2, which gives m′ less than 0.02, and which corresponds to less than $TD_2$. For the majority of data, ranging between $TD_5$ (m′ = 0.05) and $TD_{95}$ (m′ = 3), s can be shown by graphical means to approximate to about 0.65. Also, if 3 clonogenic cells or more were required for growth, s would be about 0.5 over the range $TD_5$ to $TD_{95}$.

The above considerations provide a slope that is steeper than expected from single cell transplantation kinetics, as observed with some of the tumour data, but this invokes the peculiar situation where 1 clonogenic cell has zero probability of growth! Thus, either co-operation between 2 cells (or 3 cells etc.) is essential for growth, or single cells are selectively inactivated compared with cells in pairs or in groups. This contrasts with the conventional situation described above, where *every* viable cell has a finite probability of forming a colony or a tumour, but the colony-forming efficiency of the injected cells is less than 1, due for example to the low concentration of viable cells in the inoculum, to a relative lack of helper cells or factors, or to the random cytotoxic action of enzymatic disaggregation procedures and/or the immune system in the grafted host.

With the normal tissue assays, where structures of mammary epithelium and thyroid (Ch. 15) or liver (Ch. 16) are produced in fat pads, values of s = 1 are good evidence for growth from a single cell rather than simply an aggregation of the injected cells. The latter situation would give $s<1$. Aggregation is considered to be a problem when large numbers of cells are injected, but this can be avoided by using multiple injection sites (Ch. 16). The clonal origin of these colonies is discussed in detail in Chapter 16. Values of s = 1 are compatible with the results obtained for all the normal tissues tested so far using this method (mammary CFU — Gould & Clifton, 1977; thyroid CFU — De Mott et al, 1979; liver CFU — Jirtle et al, 1981) and some but not all tumours (Porter et al, 1973). Linearity is also observed for many other CFU, which has been tested directly by the relationship between colony number and cells injected or plated, e.g. spleen colonies (Ch. 2) and other haemopoietic colonies (Ch. 3).

The expressions $F = e^{-m}$ and m = kz are similar to those invoked earlier by Lange and Gilbert (1968) to describe the probability ($\alpha$) of a single neoblast repopulating a planarian. Thus, the probability of a grafted planarian dying (F) = $(1-\alpha)^N$ when N neoblasts had been grafted, and this was approximated to $\exp(-\alpha N)$. Hence, $\alpha$ corresponds to k, and N to z.

### Sampling errors

A characteristic of the Poisson distribution is that the mean M equals the variance which is the square of the standard deviation (SD) i.e. $M = (SD)^2$. The standard deviation of the mean is commonly called the standard error (SE) and $SE = (SD/\sqrt{N})$ where N is the number of estimates of the mean.

Two common procedures in colony experiments are (1) to calculate the number of counts required to reduce the SE to a given percentage of the mean, for example 5 per cent or 10 per cent, (2) to test whether two means are significantly different e.g. for a control and a treated sample. With the first procedure, if the standard error on the mean M is required to be 5 per cent, so that the mean equals M ± 0.05M, then:

$0.05M \times \sqrt{(N \text{ samples})} = SD = \sqrt{M}$ (see above). Hence, $N = 1/(0.0025M)$, and N can be calculated for any value of M. If the mean is 10, as for example in the spleen colony assay, 40 samples (spleens) should be counted, i.e. a total of 400 colonies. If the mean is 100, as in some assays in vitro and in vivo, e.g. crypts per circumference (Ch. 5), only 4 samples need be counted with a total of 400 colonies. If a standard error of 10 per cent of the mean is acceptable, the calculated number of samples can be reduced by a factor of 4. This could mean that only 1 plate in vitro was necessary, although the majority of investigators would in any event use 2 or 3 to cover the possibility of infection or poor growth in a single sample. It is considered that when few samples are counted, the error quoted should be at least the sampling error and not simply the value calculated from the few samples taken (Boag, 1975). This is because the few samples could easily by chance be selected from a narrow range in the distribution and hence appear to have an associated error smaller than was representative. Also, there may well be an additional variance to the sampling error due to pipetting or injection inaccuracies, or variations between recipients, e.g. in mice or in feeder layers in vitro.

For the second procedure, a t-test is commonly used to test whether two means are significantly different. A worked example is given in Mather (1964), and these tests are usually included in computer software packages. The t-test applies when the data are approximately normally distributed. For more than 2 groups, an analysis of variance is used or the non-parametric equivalents (Siegel, 1956). The Poisson distribution deviates from the normal distribution for low values of the mean (much less than 10), and hence the t-test is more appropriate for distributions with higher mean values. If the means are small then the data will be skewed (i.e. non-normal) and either a chi-squared test should be undertaken to compare the observed and the expected frequencies, or the non-parametric equivalent to the unpaired t-test should be used, i.e. the Mann-Whitney U-test. The application of the t-test to colony counts has been discussed by Blackett (1974) and Hazout & Valleron (1977) with special reference to the 'thymidine suicide' technique where small or large differences in the amount of cell kill are used as indicators of the cycling status of the cell population in question (Ch. 2). Relationships are given in Hazout &Valleron (1977) which give the total number of colonies to be counted to give various levels of significance to observed differences between 2 mean values. For example, they calculate that 2000 colonies should be counted for each mean value to make a 10 per cent difference between mean values significant ($P<0.05$) in 90 per cent of cases. 2000 colonies corresponds to 200 spleens with 10 colonies per spleen, or 200 mice with 10 intestinal crypts per circumference, or 20 plates in vitro with 100 colonies per plate. If the difference between the means is 40 per cent, the total number of colonies can be reduced drastically from 2000 to 100.

The fraction of cells surviving is the ratio of the number of colonies in the treated sample to the number in the control sample. If the ratio is (A/B) and the standard errors on A and B are 'a' and 'b' respectively, then the error on the ratio is given by: $\frac{1}{B^2} \cdot \sqrt{(a^2B^2 + b^2A^2)}$.

## CELL SURVIVAL CURVES

### Parameters and models

As energy from ionising radiation is deposited at random in discrete volumes, resulting generally in randomly-distributed biological injury, it is conventional to describe radiation dose-response curves using an exponential (Poisson) distribution. Hence, dose is plotted on a linear scale and cell survival is plotted on a logarithmic scale (see Fig. 1.1). A line on such a plot indicates that equal increments of dose produce equal decrements in log cell survival. If the line extrapolates to the origin, as for example with human fibroblasts (Ch. 19), the relationship between dose and survival can be described by one parameter, the mean lethal (or inactivation) dose $D_{37}$ or $D_o$ (Lea, 1946). If the survival curve is truly exponential, this is the dose which reduces the number of viable cells to 37 per cent of the original number. This value was chosen because according to the Poisson distribution, when all cells have received on average 1 lethal hit at random, $e^{-1}$ (=0.37) should receive no hit i.e. 37 per cent should survive. Because of the exponential nature of the survival curve, this will apply at any level of survival fraction i.e. from 1 to 0.37 or from 0.1 to 0.037.

The term $D_o$ is preferred so as to cover the more general case where, for most mammalian cells, there is a shoulder or initial region demonstrating less sensitivity. Hence, $D_o$ is the mean lethal dose describing the *terminal* exponential region (see left panel, Fig. 1.1). If there is a finite initial slope in the shoulder region this is often described by ${}_1D_o$. The size of the shoulder is described most simply by the extrapolation number 'n' which is the point of extrapolation of the terminal exponential slope on the (log) ordinate. Hence, for many of the survival curves presented in this volume as examples of the use of various assay techniques, comparisons can be made of cell sensitivity ($D_o$) and of shoulder size (n). Alternatively the quasi-threshold dose Dq can be used (Alper et al, 1962), which is the point of extrapolation of the terminal slope on the linear abscissa (Fig. 1.1, left panel), and which equals ($D_o$.ln n). It is not the true threshold dose, and for cells demonstrating a marked initial slope there may be a significant decrease in survival at a dose Dq. Interestingly, with a multi-

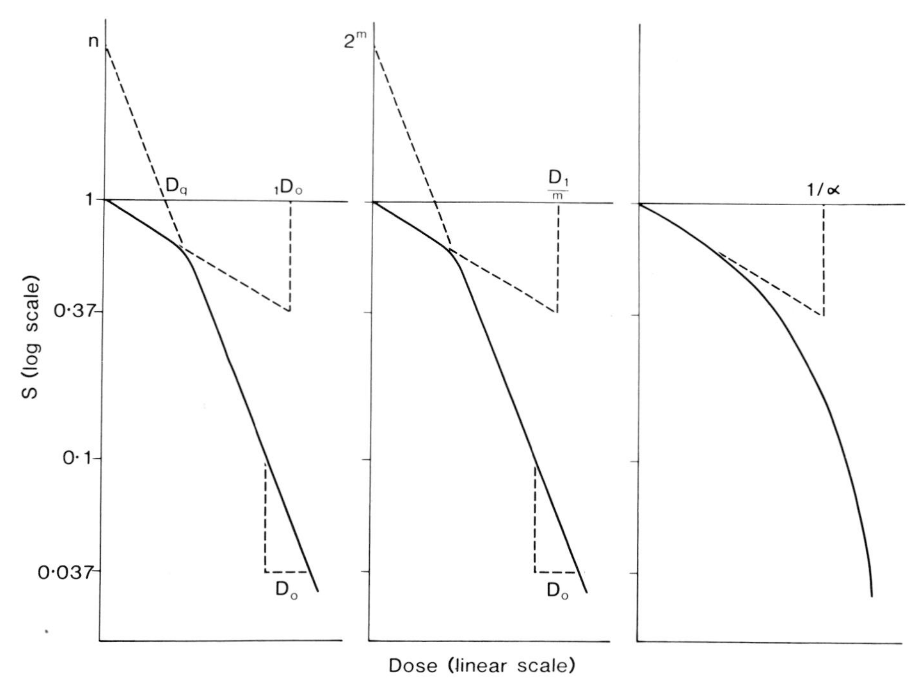


**Fig. 1.1** Shapes of survival curves.
Left panel, conventional multi-target survival curve. n = extrapolation number; $_1D_o$ = mean lethal dose at low doses; Dq = quasi-threshold dose = $D_o$.ln n.
Middle panel, target-pair model. $2^m$ extrapolation number, with m pairs of targets; $\left[\frac{D_1D_2}{m(D_1 + D_2)}\right]$ mean lethal dose at high doses = $D_0$; $[D_{\frac{1}{m}}]$ = mean lethal dose at low doses. For $D_2$ see text.
Right panel, α β curve. Extrapolation number ($2^m$), very high. $1/\alpha$ = mean lethal dose at low doses. Continuous curvature, terminal exponential slope usually never reached.

target curve, Dq is the dose at which the linear decrease in survival per unit dose is maximal (Okumura et al, 1974).

Although assays for mammalian cell survival have been developing over nearly 30 years, the mechanisms responsible for values of n and $D_o$ have not yet been elucidated. However, in the literature there is an increasing use of alternative mathematical models for cell survival curves which are based on various 'plausible'mechanisms. These are often considered to be 'better' models but because of the amount of scatter in most data, the newer models cannot usually be shown to fit the data better when the number of variables is the same as in the old models. As users of colony techniques for dose-response measurements are becoming increasingly confronted with descriptions of survival curve shapes in terms other than $D_o$ and n, for example α and β (Ch. 18), a brief description is given of the meaning and use of such parameters.

All target-type models invoke the Poisson distribution to give the probability of an event not happening, when the mean number of events is specified. A simple case, described above, is where one event kills the cell and where the mean number of events per cell is linear with dose (D) so that survival (S) of the cell is given by:

$$S = \exp(-D/D_0) \qquad \text{equation (1)}$$

If there are 'n' targets in a cell, all of which have to be inactivated to kill that cell, and if the number of inactivating events is linear with increasing dose, then survival of single *targets* could also be described by $[\exp(-D/D_o)]$. The chance of a single target being inactivated is $[1 - \exp(-D/D_o)]$, and the chance of inactivating all n targets is $[1 - \exp(-D/D_o)]^n$. Hence, cell survival (S) is given by:

$$S = 1 - [1 - \exp(-D/D_o)]^n \qquad \text{equation (2)}$$

This is the conventional multi-target equation, which approximates at high doses to $[n.\exp(-D/D_o)]$. A finite slope to the cell survival curve at low doses is often characterised by an additional exponential term, so that:

$$S = \exp(-D/_1D_o) \times (1 - [1 - \exp(-D/D_o)]^n) \qquad \text{equation (3)}$$

(Bender & Gooch, 1962), where $_1D_o$ is the mean lethal dose for the cell from single lethal events, and a different target is implied from the n targets already described. At high doses S becomes $(n. \exp\ [-D(\frac{1}{{}_1D_o} + \frac{1}{D_o})])$. Equation (3) is adequate for empirical description and comparative purposes, but its use for interpretation of the parameters in terms of mechanisms is now not favoured by many people. This is because of the conceptual difficulties in the postulate of a multiplicity of essential targets, e.g. Alper (1979).

A more plausible version of target-type models is based on the assumption that *only one* target of many has to be inactivated to sterilise the cell, e.g. one unrepaired break in a double strand of DNA which leads to lethal chromosomal injury. If such a break can be produced by one radiation event with a probability $\alpha$ and by a combination of two events with a probability $\beta$, then the number of lethal events (L) can be expressed by a quadratic function: $L = \alpha D + \beta D^2$. Hence, the surviving fraction (S) of cells will be given by:

$$S = \exp(-L) = \exp -(\alpha D + \beta D^2) \qquad \text{equation (4)}$$

e.g. Chadwick and Leenhouts (1973). This equation predicts a continuously-bending survival curve, and although it can also accommodate curves which are simple exponentials, in which case $\beta = 0$, it cannot accommodate conventional multi-target-type survival curves which show quite clearly, the presence of a shoulder and a terminal exponential region, e.g. Puck & Marcus (1956).

However, the multi-target curve and the $\alpha$, $\beta$ formulation can be reconciled if the same mathematical principles are applied to derive both. The relationship between them can be demonstrated as follows.

If a potential target-pair is considered, the chance of one strand of the pair not being broken would be given by $\exp(-D/D_2)$, where $D_2$ is the mean number broken per unit dose. The chance of the pair remaining intact would be $(1-[1-\exp(-D/D_2)]^2)$. If there are 'm' potential site-pairs in a cell, the chance of the cell surviving is: $S = (1-[-\exp(-D/D_2)]^2)^m$. If the pair can also be inactivated by *single* events, another exponential factor is added so that:

$$S = [\exp(-D/D_1)]^m \cdot (1-[1-\exp(-D/D_2)]^2)^m \qquad \text{equation (5)}$$

(Gilbert, 1975; Ehrenberg, 1977; Gilbert et al, 1980). At high doses, S approximates to

$$\{2^m \cdot \exp[-mD(\frac{1}{D_1} + \frac{1}{D_2})]\}$$

and hence the survival curve (see middle panel, Fig. 1.1) has an extrapolation number of $2^m$ (Neary, 1965) and a sensitivity $(\frac{1}{D_o})$, of $m(\frac{1}{D_1}+\frac{1}{D_2})$. When $m = 1$, the equation becomes a conventional multi-target equation with $n = 2$. When m is greater than 1, the extrapolation number becomes very large and the terminal sensitivity, which may never be reached in practice because of the enormous shoulder, includes the parameter m. This is because in the original multi-target model, no matter how many targets there were initially in the cell, the sensitivity for lethality is given always by the last surviving target. In the newer version, if there are more potential site-pairs per cell there is a greater chance, in proportion to their number, of hitting one of them. The now widely used expression for survival, $S = \exp-(\alpha D + \beta D^2)$ is an approximation to equation (5), and it applies well either for low doses or for large values of m, but not at high doses when m is small. The approximation applies when D is much less than $D_2$ so that the chance of 2 coincident events on a single site is negligible. In this case S approximates to:

$$S = \exp-(\frac{m}{D_1}.D + \frac{m}{D_2}.D^2) \qquad \text{(Gilbert, 1975).}$$

Hence $\alpha = \frac{m}{D_1}$, and $\beta = \frac{m}{D_2^2}$. Also, $\frac{\alpha}{\beta} = \frac{D_2^2}{D_1}$, and this is the dose at which the contributions of single and double-events to lethality are equal. When m is very large, which is plausible for the number of potential sites where a lethal event may occur, $D_2$ must also be large for values of S to be in the common measurable range. Hence, the approximation will be valid to higher values of dose (D). It is interesting therefore that the $(\alpha,\beta)$ formulation is in effect describing the shape of the huge 'shoulder' of one type of multi-target survival curve. Also, as a quadratic is a good approximation to other equations for survival, e.g. Burch & Chesters (1981), it can be applied with differing limitations in most situations.

The initial slope $\alpha = (m/D_1)$ (see right panel, Fig. 1.1) corresponds to $(1/{}_1D_o)$ in the original multi-target formula (equation 3). As described above, the $(\alpha, \beta)$ formulation can be derived directly from an assumed quadratic relationship for the production of lethal events. However, this does not allow possible interpretation of the values of the parameters in terms of target sizes, nor the accommodation of expected deviations from a quadratic due to considerations of target number and target 'overkill', as already described.

There are many other models, e.g. the 'multihit' and the 'pool' models, which have biological significance but which have not yet been used by many investigators. Readers are referred to radiobiological texts for further information (Elkind & Whitmore, 1967; Proceedings of the 6th L. H. Gray Conference, 1975; Alper, 1979; Proceedings of the 11th L. H. Gray Conference, 1984).

**Cell survival parameters and groups of cells**

If cells are irradiated in vitro after they have all divided once, then the extrapolation number per initial seeded cell should be double the value obtained if they had not divided. After 2 divisions, it should be 4x, and after 3 divisions it should be 8x. However, because of asynchrony in most cell populations, these expectations will not be fulfilled exactly, and there will be a distribution of sizes of cell clusters. Further information on these aspects has been discussed by Elkind &Whitmore (1967).

Many structures in vivo contain presumably similar numbers of clonogenic cells, and hence these could be regarded as multi-target-type structures, where one surviving cell per structure is required to regenerate the structure. Hence data expressed as surviving fractions of structures could be analysed directly using computer methods and multitarget equations (see below).

The response of single cells in these structures can be deduced because if one clonogenic cell or more regenerates a structure, then the fraction (F) of structures ablated will be given by $F = \exp(-m)$ where m is the mean number of clonogenic cells surviving per structure. If m decreases exponentially with increasing dose at high doses, so that $m = [n.\exp(-D/D_o)]$, where n is the total extrapolation number per structure, then:

$$\ln(-\ln F) = \ln n - D/D_o.$$

Hence if semi-logarithmic graph paper is used, and $(-\ln F)$ is plotted on the logarithmic ordinate scale versus dose on the linear abscissa, the line produced would have an intercept of $\ln(n)$ on the ordinate and a slope$(-1/D_o)$ (Withers and Elkind, 1970). Alternatively, $\ln(-\ln F)$ can be plotted on a linear ordinate (Gilbert, 1974). The relationship between these two graphical methods can be found in Chapter 5.

On the other hand, if 2 clonogenic cells are required to regenerate a structure, and it will fail to grow with zero cells [probability = exp(−m)] and with 1 cell [probability = m.exp(−m)] then:

$$F = \exp(-m) + m.\exp(-m) = (1 + m)\exp(-m).$$

$$\text{Hence, } \ln F = \ln(1 + m) - m = -m'.$$

At high doses when m is much less than 1, $\ln(1 + m)$ approximates to $(m - \frac{m^2}{2})$ and hence $m' \simeq \frac{m^2}{2}$.

$$\text{Thus, } m' \simeq \frac{n^2}{2} \cdot \exp(-D.\frac{2}{D_o}) = n'.\exp(\frac{-D}{D_o'})$$

Hence, if m′ is calculated from F, and n′ and $D_o'$ are deduced as described above, the sensitivity will appear greater by a factor of 2 i.e. $D_o' = D_o$ (as noted in Ch. 17), and the extrapolation number will be much greater, i.e. $n' = (\frac{n^2}{2})$.

The change in the threshold dose for the ablation of structures can be calculated, because when say $F = 0.1$, $m = \ln 0.1 = 2.3$. If 2 cells are required to regenerate the structure, then when $F = 0.1$,
$\ln 0.1 = \ln(1 + m_1) - m_1$, and hence $m_1 = 3.89$. As $m = n\exp(-D/D_o)$,

$\ln(\frac{m}{m_1}) = (D_1 - D_2)/D_o$ where $(D_1 - D_2)$ is the reduction in dose between the two models when $F = 0.1$. In the above example, $(D_1 - D_2)/D_o = 0.59$, and hence if $D_o \simeq 1$ Gy, $(D_1 - D_2) \simeq 0.6$ Gy, so that the threshold dose would not change by much.

If the above approximation is used directly, then $D_q' = D_o'.\ln n' = (\frac{D_o}{2}) \cdot \ln(\frac{n^2}{2}) = Dq - 0.35\ D_o$. However, this method is less accurate because the approximation is valid only for values of n less than about 0.2.

This situation, where the survival curve almost 'pivots' about the threshold dose and has a lower $D_o$ and higher n, has indeed been observed in several situations and no satisfactory explanation has yet been found, e.g. the diurnal changes in sensitivity of intestinal crypts (Hendry, 1975).

As already noted, the approximation of $\ln(1 + m)$ by $(\frac{m - m^2}{2})$ is valid only when m is less than about 0.2, i.e. when $F > 0.98$ or when less than 2 per cent of structures survive. In the common measurable range of 95 per cent down to 1 per cent survival of structures, the deduced cell survival curve would appear gently bending with an average sensitivity characterised by $D_o' \simeq \frac{D_o}{1.6}$. Similar reasoning can be applied to the cases where >3, or >4 etc. clonogenic cells are required for regenerating a structure. For x cells, the terminal $D_o'$ would be $(\frac{D_o}{x})$, and the extrapolation number n′would be $(\frac{n^x}{x!})$. However, with increasing values of x, the terminal $D_o'$ would never be reached, and hence the apparent values of $D_o'$ and n′ over the range considered would be greater and lesser respectively. This reasoning is very similar to that applied above in the section on Transplantation Kinetics.

An extension to these ideas is where *many* cells are required to regenerate a larger structure such as a tissue or an animal. The above approach would give a *much* smaller value for $D_o'$, but in this situation a number of surviving cells below a critical level has *zero* probability of rescue. This is clearly unrealistic when the difference between two numbers of surviving cells is very small, and a more logical approach is that proposed by Lange & Gilbert (1968), where *each* cell has a small probability (α) of rescuing the tissue. Hence, the probability of failure is (1− α), and with N surviving cells, the probability of failure (F) would be given by:

$$F = (1-\alpha)^N \simeq \exp(-N\alpha) \text{ when } \alpha \text{ is small.}$$

$$\text{Hence } \ln F \simeq (-N\alpha).$$

If N is related exponentially to dose by $N = No.\exp(-D/D_o)$ where No is the total extrapolation number per tissue, then:

$$\ln(-\ln F) = \ln(\alpha No) - D/D_o.$$

Hence, a plot of $\ln(-\ln F)$ against dose D would give a line with slope $(-1/D_o)$ and an extrapolation number of $\ln(\alpha No)$ on the ordinate. Alternatively, $(-\ln F)$ can be plotted on the logarithmic ordinate of semi-logarithmic graph paper. This approach was developed from ideas concerning tumours presented by Munro & Gilbert (1961), and it has been further applied to the sensitivity of 'target' cells in tumours (Andrews &Mossman, 1976; Wheldon et al, 1977), haemopoietic tissue (Robinson, 1968), intestine (Hendry et al, 1983) and epidermis (Hendry, 1984). The $D_o$ values which can be deduced for the target cells responsible for failure of haemopoietic tissue, intestine and epidermis are very similar to the $D_o$ values measured directly for their respective colony-forming cells. Thus this validates the use of this approach.

As discussed in detail elsewhere (Potten & Hendry, 1983), the necessary killing of all clonogenic cells in a structure in order to ablate it, is analogous to the multi-target theory for *cell* survival where all targets in a cell have to be inactivated to kill it. Hence, the survival (S) of structures will correspond to the survival of (nA) targets, where A target cells per structure each have n subcellular targets, and $S = 1 - [1 - \exp(-D/D_o)]^{nA}$ (see equation 2, p. 6). Thus, computer programmes which fit data to multi-target equations (Gilbert, 1969), can be used directly to fit data for the survival of structure versus dose, and values for the cell $D_o$ and the total number of targets per structure can be calculated. Also, the double-logarithmic transformation described above is a good approximation to the multi-target formula (Watson, 1978; Potten & Hendry, 1983). When (nA) is large and $\exp(-D/D_o)$ is small, the multi-target equation for S approximates to:

$$S \simeq 1 - \exp[-nA.\ \exp(-D/D_o)]$$
$$\text{Hence, } F = 1 - S \simeq \exp[-nA.\ \exp(-D/D_o)]$$
$$\text{and } \ln(-\ln F) \simeq \ln(nA) - D/D_o.$$

nA per structure corresponds to No per tissue (see above). $\alpha$, which is applied above to the ability of a cell to regenerate a structure, is 1 according to the definition of a colony-forming cell.

Thus, the double-logarithmic transformation of F is a good approximation to the multi-target formula not only in the region where cell survival is related exponentially with dose, but also at lower doses in the shoulder region of the cell survival curve, if this is described adequately by a conventional multi-target equation. Specific examples are calculated in Potten & Hendry (1983). The practical use of a linear transform of the multi-target equation has been discussed by Watson (1978).

Finally, the exponential function for cell survival can be replaced by more complex functions and similar analyses of cell survival in multicellular structures can be undertaken. Readers are referred elsewhere for further information on this subject (Yau & Cairnie, 1979; Thames et al, 1981; Potten & Hendry, 1983).

**Curve fitting**

Cell survival data are plotted conventionally on a semi-logarithmic plot of dose versus survival. A line or curve can be drawn by eye through the data and values, for example of $D_o$ and n, can be estimated. This is satisfactory for many purposes but many people employ computer methods so that (1) the statistical weighting of individual datum points can be automatically taken into account, (2) error limits on the fitted parameters can be more easily calculated, (3) curves of specified shape can be fitted to the data.

Mean survival values derived from widely-spread results are clearly not very accurate, and data points are weighted by their inverse variance. If N cells survive out of No at risk, the surviving fraction (P) will be N/No and this fraction could vary between 1 and zero. If No is actually counted (and not estimated from a sample count), binomial statistics apply, and the variance of P is $[\frac{P(1-P)}{No}]$. If No is estimated from a sample count, and No>>1, P<<1, Poisson statistics can be applied, and the variance of P is $[\frac{P(1+P)}{No}]$ when *survivors* are counted, and $[\frac{1}{No}.(1-P).(2-P)]$ when *non-survivors* are counted. Clearly the variance of P is reduced by using (1) large samples, (2) counted numbers of cells at risk rather than numbers estimated from an aliquot, and (3) at low doses (values of P near 1) when non-survivors are counted rather than survivors.

There are several methods of curve fitting. Linear regression can be used for simpler models where a linear transformation of the data can easily be made, for example with an exponential survival curve (Pike & Alper, 1964). In this case, the sums of squares of the differences between the observations and the fitted line are minimised to provide the line of best fit, and sampling errors on the regression constants can be calculated. A worked example can be found in Mather (1964), but computer methods using iterative procedures are now widely available. The main difficulty with the basic technique is in cases where a decision has to be made as to which points to include in the fitting procedure, for example if there is a marked curving shoulder to the survival curve.

Maximum likelihood or minimum chi-square techniques are often preferred. No initial transformation of the data is necessary except if required for subsequent plotting purposes, and the data can be fitted to any equation by iterative procedures. These techniques have been developed notably by Finney since 1947 (updated in Finney, 1978), but also by others specifically for radiobiological applications (Gilbert, 1969; Porter, 1980a,b). Best-fit curves can be obtained, with values for the fitted parameters and their associated error limits.

Several investigators who have used one or other of these techniques can be found in the Proceedings of the 6th L. H. Gray Conference (1975), e.g. Gillespie et al, Bryant & Lansley, Phillips et al, including this author. New users are advised to contact someone who has experience with the use of these techniques and either send them the data for fitting or obtain a copy of their programme. The latter approach nearly always requires the assistance of a computer programmer because of the numerous variations in format between similar computers. Some programmes include options for fitting one or more common parameters to separate survival curves, e.g. fitting a common extrapolation number to extract dose-modifying factors (Pike & Alper, 1964; Gilbert, 1969).

**Effect of colony size**

Many investigators have observed an increase in the number of small colonies after increasing radiation doses. This effect implies that when colonies containing fewer cells are counted the CFC will appear less sensitive. Two examples of this effect are shown in Figure 1.2. Nias & Fox (1968) demonstrated that when progressively smaller colonies were included in the measurement of surviving fraction, the latter increased gradually (left panel, Fig. 1.2). The effect was slightly greater at the higher doses, suggesting that both n and $D_0$ were affected. Also, when compared with the strictest test for survival — back extrapolation of growth curves where numerous cell divisions are needed

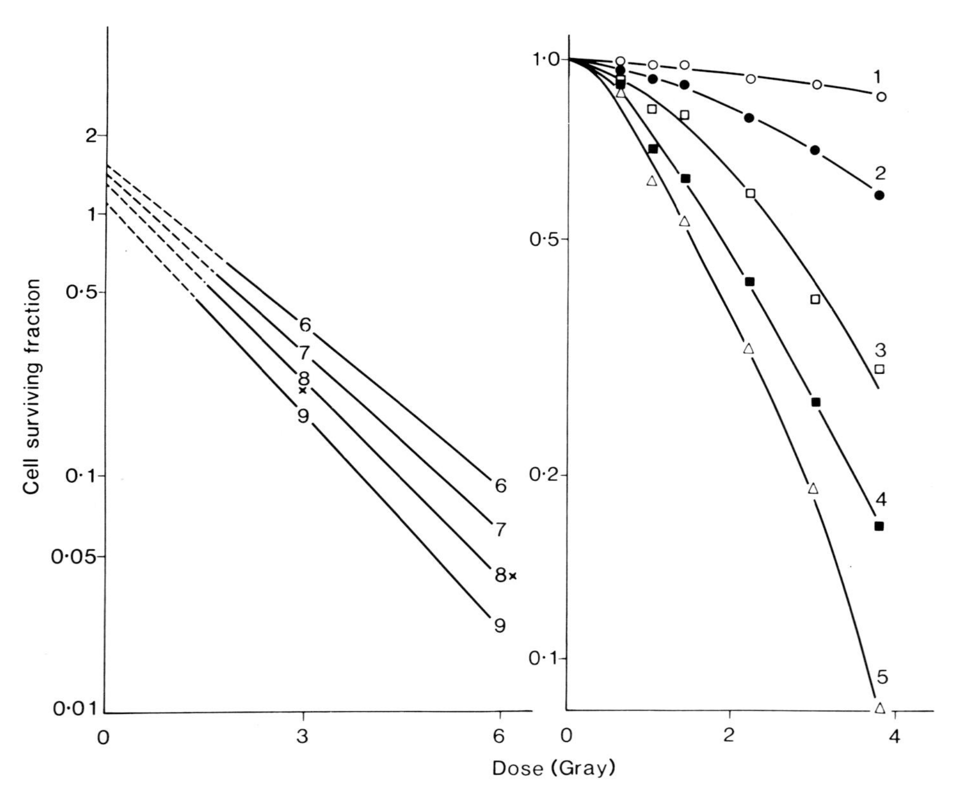


**Fig. 1.2** Effect on a survival curve of different criteria for counting colonies.
Left panel, data obtained using HeLa cells by Nias & Fox (1968). Numbers on lines correspond to minimum number of divisions required for a colony to be counted, i.e. 6 divisions, all colonies with 33 cells or more; 7 divisions, 65 cells or more; 8, 129; 9, 157. Crosses, survival level deduced from a back-extrapolation of growth curves, and corrected for mitotic delay.
Right panel, data obtained using BHK cells, counting maximum number of cells reached per colony up to day $5\frac{1}{2}$ after plating (data courtesy of Dr S. Revell). Numbers on lines correspond to minimum number of divisions required for a colony to be counted. 1 division = 2 cells; 2 divisions = 3–4 cells; 3 divisions = 5–8 cells; 4 divisions = 9–16 cells; 5 divisions = 17–31 cells. Lower limit on number of cells due partly to cells 'floating off' in the culture.

(crosses in left panel, Fig. 1.2) — the levels of survival were similar to those using about 8 divisions. Data at lower doses are shown also in Figure 1.2 (right panel, data courtesy of Dr. S. Revell). These data have been presented in a different form elsewhere (Joshi et al, 1982). Clearly, the effect on survival of different criteria for counting colonies is greatest when changing from 1 to 4 divisions, and thereafter it is smaller but still significant. Further data are given in Chapter 21.

The effect is also seen in vivo (1) with erythropoietin-responsive cells when different degrees of stimulus (amounts of erythropoietin) were given (Lajtha, 1965), (2) when thyroid cell survival is estimated by increase in gland weight after goitrogenic stimulation (inducing about 2 cell divisions on average) compared to the estimation using colony-formation (Malone et al, 1983), (3) with vascular endothelium when different assays are used (Reinhold, 1981, quoted in Hopewell, 1983).

## ACKNOWLEDGEMENT

The work of the author is supported by the Cancer Research Campaign, UK.

## REFERENCES


Alper T, Fowler J F, Morgan R L, Vonberg D D, Ellis F, Oliver R 1962 The characterisation of the type C survival curve. British Journal of Radiology 35: 722

Alper T 1979 Cellular radiobiology. Cambridge University Press, Cambridge, p 320

Andrews J R, Mossman K L 1976 On the human clonogenic cancer cell: the radioresistant fraction and its implications for radiotherapy. International Journal of Radiation Oncology, Biology and Physics 125 Supplement 1: 125

Becker A J, McCulloch E A, Till J E 1963 Cytological demonstration of the clonal nature of spleen colonies derived from transplanted mouse marrow cells. Nature 197: 452

Bender M A, Gooch P C 1962 The kinetics of X-ray survival of mammalian cells in vitro. International Journal of Radiation Biology 5: 133

Blackett N M 1974 Statistical accuracy to be expected from cell colony assays: with special reference to the spleen colony assay. Cell Tissue Kinetics 7: 407

Boag J W 1975 The statistical treatment of survival data. In: Alper T (ed) Cell survival after low doses of radiation: Theoretical and clinical implications. Proceedings of the 6th L H Gray Conference. The Institute of Physics and John Wiley, Chichester, p 40

Botnick L E, Hannon E C, Vigneulle R, Hellmann S 1981 Differential effects of cytotoxic agents on hematopoietic progenitors. Cancer Research 41:2338

Burch P R J, Chesters M S 1981 A 'repair model' of cell survival gives the approximation: $S = \exp - (\alpha D + \beta D^2)$. International Journal of Radiation Biology 39: 221

Chadwick K H, Leenhouts H P 1973 A molecular theory of cell survival. Physics in Medicine and Biology 18: 78

De Mott R K, Mulcahy R T, Clifton K H 1979 The survival of thyroid cells following irradiation: A directly generated single-dose survival curve. Radiation Research 77: 395

Ehrenberg L 1977 A note on the shape of shouldered dose response curves. International Journal of Radiation Biology 31: 503

Elkind M M, Whitmore G F 1967 The radiobiology of cultured mammalian cells. Gordon & Breach, New York, p 615

Finney D J 1947 Probit analysis: A statistical treatment of the sigmoid response curve. Cambridge University Press, Cambridge, p 256

Finney D J 1964 Statistical method in biological assay, 2nd edn. Griffin, London, p 668

Finney D J 1978 Statistical method in biological assay. Griffin, London, p 508

Gilbert C W 1969 Computer programmes for fitting Puck and probit survival curves. International Journal of Radiation Biology 16: 323

Gilbert C W 1974 A double minus log transformation of mortality probabilities. International Journal of Radiation Biology 25: 633

Gilbert C W 1975 Target-type models for survival curves. British Journal of Radiology 48: 1045 (abstract)

Gilbert C W, Hendry J H, Major D 1980 The approximation in the formulation for survival $S = \exp - (\alpha D + \beta D^2)$. International Journal of Radiation Biology 37: 469

Gould M N, Clifton K H 1977 The survival of mammary cells following irradiation in vivo: A directly generated single-dose-survival curve. Radiation Research 72: 343

Hazout S, Valleron A J 1977 Planning the suicide experiments. Cell Tissue Kinetics 10: 569

Hendry J H 1975 Diurnal variations in the radio sensitivity of mouse intestine. British Journal of Radiology 48: 312

Hendry J H 1984 Correlation of the dose-response relationships for epidermal colony-forming units, skin reactions, and healing, in the x-irradiated mouse tail British Journal of Radiology (in press).

Hendry J H, Potten C S and Roberts N P 1983 The gastrointestinal syndrome and mucosal clonogenic cells: relationships between target cell sensitivities, $LD_{50}$ and cell survival, and the effect of antibiotics. Radiation Research 96: 100

Hopewell J W 1983 Radiation effects on vascular tissue. In: Potten C S, Hendry J H (eds) Cytotoxic insult to tissue: effects on cell lineages. Churchill Livingstone, Edinburgh, p 228

Jirtle R L, Michalopoulos G, McLain J R, Crowley J 1981 Transplantation system for determining the clonogenic survival of parenchymal hepatocytes exposed to ionising radiation. Cancer Research 41: 3512

Joshi G P, Nelson W J, Revell S H, Shaw C A 1982 Discrimination of slow growth from non-survival among small colonies of diploid Syrian hamster cells after chromosome damage induced by a range of X-ray doses. International Journal of Radiation Biology 42: 283

Lajtha L G 1965 Response of bone marrow stem cells to ionizing radiation. Current Topics in Radiation Research 1: 139

Lange C S, Gilbert C W 1968 Studies on the cellular basis of radiation lethality III. The measurement of stem cell repopulation probability. International Journal of Radiation Biology 14: 373

Lea D E 1946 Actions of radiations on living cells. Cambridge University Press, Cambridge, p 402

Mackillop W J, Ciampi A, Till J E, Buick R N 1983 A stem cell model of human tumor growth: Implications for tumor cell clonogenic assays. Journal of the National Cancer Institute 70: 1

Malone J F, O'Connor M K, Hendry J H 1983 The cellular basis of radiation effects in thyroid, salivary and adrenal epithelia. In: Potten C S, Hendry J H (eds) Cytotoxic insult to tissue: effects on cell lineages. Churchill Livingston, Edinburgh, p 186

Mather K 1964 Statistical analysis in biology, 5th edn. Methuen, London, p 267

Munro T R, Gilbert C W 1961 The relation between tumour lethal doses and the radiosensitivity of tumour cells. British Journal of Radiology 34: 246

Neary G J 1965 Chromosome aberrations and the theory of RBE. International Journal of Radiation Biology 9: 477

Nias A H W, Fox M 1968 Minimum clone size for estimating normal reproductive capacity of cultured cells. British Journal of Radiology 41: 468

Okumura Y, Uchiyama Y, Morita K 1974 A new significance of quasi-threshold dose and its impact in radiotherapy. Journal of Radiation Research 15: 114

Pike M C, Alper T 1964 A method for determining dose-modification factors. British Journal of Radiology 37: 458

Porter E H, Berry R J 1963 The efficient design of transplantable tumour assays. British Journal of Cancer 17: 583

Porter E H 1980a The statistics of dose/cure relationships for irradiated tumours, Part I. British Journal of Radiology 53: 210

Porter E H 1980b The statistics of dose/cure relationships for irradiated tumours, Part II. British Journal of Radiology 53: 336

Porter E H, Hewitt H B, Blake E R 1973 The transplantation kinetics of tumour cells. British Journal of Cancer 27: 55

Potten C S, Hendry J H 1983 Stem cells in murine small intestine. In: Potten C S (ed) Stem cells: their identification and characterisation. Churchill Livingstone, Edinburgh, p 181

Proceedings of the 6th L H Gray Conference 1975 Alper T (ed) The Institute of Physics and John Wiley, Chichester, p 397

Proceedings of the 11th L H Gray Conference 1984 British Journal of Cancer 49: supplement VI

Puck T T, Marcus P I (1956) Action of X-rays on mammalian cells. Journal of Experimental Medicine 103: 653

Rice L, Urano M, Suit H D 1980 The radiosensitivity of a murine fibrosarcoma as measured by three cell survival assays. British Journal of Cancer 41 (Supplement IV): 240

Robinson C V 1968 Relationship between animal and stem cell dose-survival curves. Radiation Research 35: 318

Schofield R, Lajtha L G 1973 Effect of isopropyl methane sulphonate (IMS) on haemopoietic colony-forming cells. British Journal of Haematology 25: 195

Schofield R, Lord B I, Kyffin S, Gilbert C W 1980 Self-maintenance capacity of CFU-S. Journal of Cellular Physiology 103: 355

Schofield R, Lajtha L G 1983 Determination of the probability of self-renewal in haemopoietic stem cells: A puzzle. Blood Cells 9: 467

Siegel S 1956 Non-parametric statistics for the behavioural sciences. McGraw-Hill, New York, p 312

Steel G G 1968 Cell loss from experimental tumours. Cell and Tissue Kinetics 1: 193

Thames H D, Withers H R, Mason K A, Reid B O 1981 Dose-survival characteristics of mouse jejunal crypt cells. International Journal of Radiation Oncology, Biology and Physics 7: 1591

Till J E, McCulloch E A, Siminovitch L 1964 A stochastic model of stem cell proliferation, based on the growth of spleen colony-forming cells. Proceedings of the National Academy of Science 51: 29

Vogel H, Niewisch H, Matioli G 1968 The self-renewal probability of hemopoietic stem cells. Journal of Cellular Physiology 72: 221

Vogel H, Niewisch H, Matioli G 1969 Stochastic development of stem cells. Journal of Theoretical Biology 22: 249

Watson J V 1978 A linear transform of the multi-target survival curve. British Journal of Radiology 51: 534

Wheldon T E, Abdelaal A S, Nias A H W 1977 Tumour curability, cellular radiosensitivity and clonogenic cell number. British Journal of Radiology 50: 843

Withers H R 1967 The dose-survival relationship for irradiation of epithelial cells of mouse skin. British Journal of Radiology 40: 187

Withers H R, Elkind M M 1970 Micro-colony survival assay for cells of mouse intestinal mucosa exposed to radiation. International Journal of Radiation Biology 17: 261

Xu C X, Hendry J H, Testa N G, Allen T D 1983 Stromal colonies from mouse marrow: characterisation of cell types, optimisation of plating efficiency and its effect on radiosensitivity. Journal of Cell Science 61: 653

Yau H C, Cairnie A B 1979 Cell survival characteristics of intestinal stem cells and crypts of $\gamma$-irradiated mice. Radiation Research 80: 92

*B.I. Lord R. Schofield*

# Haemopoietic spleen colony-forming units

## HISTORICAL BACKGROUND

The functional status of the bone marrow was judged by all haematologists, until comparatively recently, purely by qualitative criteria. The two major criteria were the cellularity and the relative numbers of the various cell types. The existence of a stem cell or stem cells, responsible for the generation of all the recognisable haemopoietic cells, had been invoked by all the major haematologists but its appearance and its potential (in terms of the number of different cell types it could produce) were matters of conjecture and argument. Opinions were paramount, objective evidence was non-existent. The advent of radiation for therapeutic, destructive and creative uses made it necessary to assess the state of the haemopoietic tissues in an objective way since those tissues appeared to be the ones most critically affected by radiation. The observations by Jacobson and his colleagues in Chicago (1949; 1951) that a factor which could circulate could prevent haemopoietic death and accelerate haemopoietic recovery were explained by Ford and his co-workers (1956) on the basis of a repopulating cell. Stroud et al (1955) showed that one could anticipate haemopoietic recovery following radiation exposure (in mice, at least) by the appearance of nodules on the spleens. The probable significance of these nodules was recognised by Till & McCulloch (1961) who showed that the number of them which appeared on the spleens of heavily-irradiated mice was proportional to the number of marrow cells injected in the mice. They suspected that each nodule was formed from a single repopulating unit and that unit might indeed be the putative stem cell. The clonality of this repopulating unit was subsequently confirmed by Becker et al (1963). They studied the chromosome structure of individual colonies resulting from a transplant of marrow which had been damaged by sub-lethal irradiation and found that where a unique marker existed it was carried by virtually all the cells in that colony. These observations, culminating in the development by Till & McCulloch (1961) of the technique for measuring those units which form colonies in the spleens of irradiated mice, form the cornerstone of all subsequent developments in the understanding of haemopoiesis. An illustration of its use for the measurement of radiation dose-survival characteristics for these haemopoietic spleen colony-forming units is shown in Figure 2.1. Many curves similar to this have been published, and they are generally described by a $D_o$ value of 0.9–1.0 Gy and an extrapolation number of 1–3. The particularly-detailed curve shown in Fig. 2.1 could probably be described better by an initial sloping shoulder region ($D_o \simeq 1$ Gy) followed by an increase in sensitivity to a terminal $D_o$ value of at most 0.66 Gy and an extrapolation number of about 10 (Hendry, 1979). Despite attempts to make this technique applicable to other animals, however, the only animal in which reli-

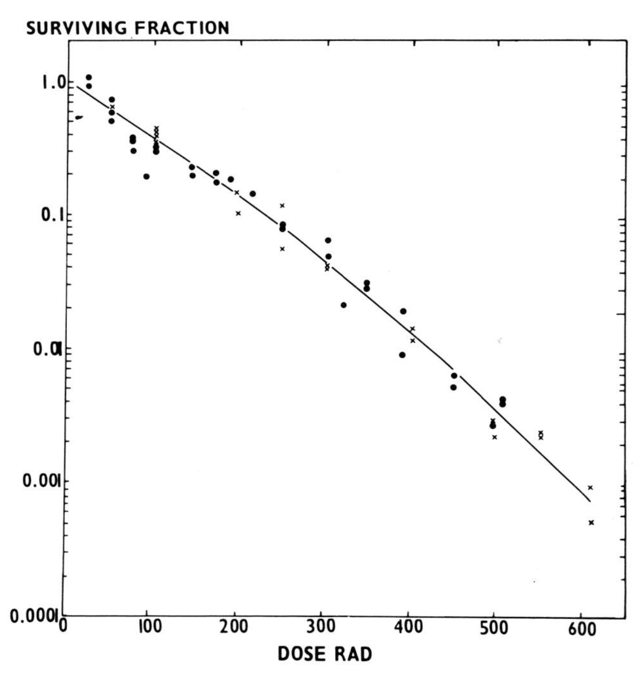


**Fig. 2.1** Survival of femoral CFU-S to $^{137}Cs$ γ-rays. Crosses — irradiation, in vivo; closed circles — irradiation in vitro. Reproduced from Hendry (1979) with permission of Taylor and Francis Ltd. 100 rod = 1Gy.

able and reproducible results can be obtained is the mouse.

## DESCRIPTION OF THE TECHNIQUE

The technique for measuring spleen colony-forming units (CFU-S) requires the use of inbred strains of 'clean' mice so that recipients genetically-identical to the donor of the tissue to be assayed are needed. A radiation source, which can deliver about 10 Gy of radiation within a reasonable period, is required.

The assay for the number of CFU-S in a cell suspension is illustrated in Figure 2.2 and is carried out broadly as follows:

1. Mice are exposed (whole body) to a dose of ionising radiation sufficient to ablate the haemopoietic tissues.
2. An appropriate number of cells in a suspension of haemopoietic tissue is injected within a few hours.
3. The mice are killed 8–10 days later, the spleens are removed and preserved in a suitable fixative.
4. The colonies, which are made more easily visible by the fixation, are counted.

A number of variations on the technique have been employed successfully but certain conditions must be adhered to if results comparable with other published data are to be obtained.

### Mice

The mice, donors and recipients, must be of the same inbred strain or hybrids obtained by crossing two inbred strains, though using donor and recipient of opposite sex does not appear to effect the assay. Injecting a cell suspension from a parent mouse into an irradiated hybrid mouse sometimes results in suppression of colony formation although no host-v-graft reaction occurs (McCulloch & Till, 1963). The suppression is particularly strong when $C_{57}$Bl mice are involved, though it does also occur in other strains.

The mice must also be 'clean', i.e. free from all pathogenic bacteria, viruses and parasites. In order continuously to achieve reproducible and consistent results the mice must be bred and maintained in conditions of high hygenic standards in well ventilated and temperature regulated conditions. It is a common experience that mice kept in conditions which are not very well controlled fail to produce reliable results and many attempts to specify exactly the causes of lack of reproducibility

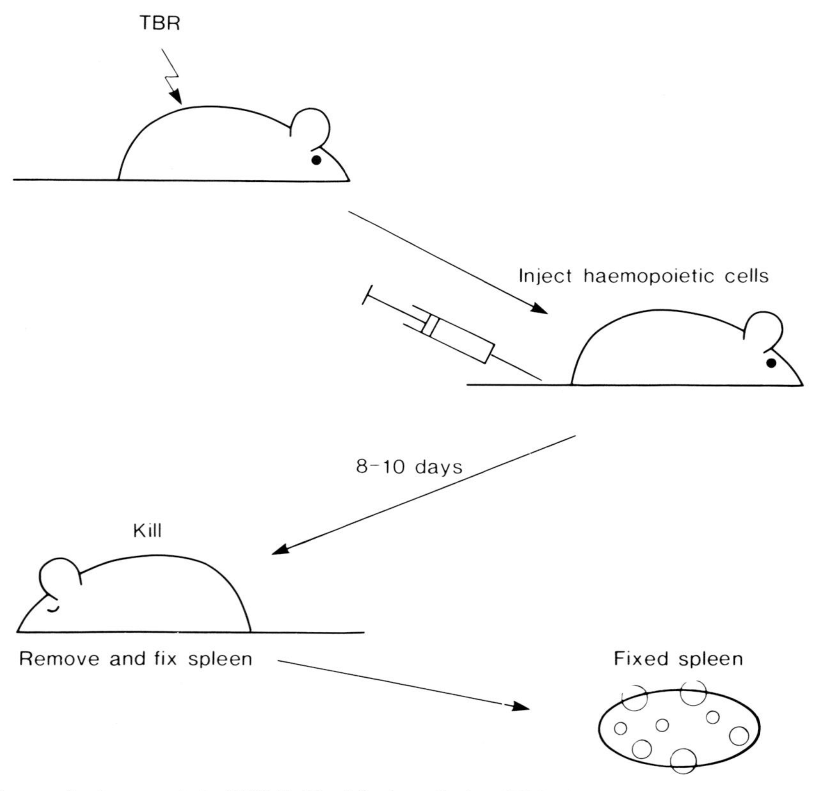


**Fig. 2.2** Method of assay for haemopoietic CFU-S. Total body radiation (TBR) is lethal.

have met with little or no success. Further, the irradiated recipients must be housed in small groups since death of one of the recipients in a group often leads to the death of some or all of the remaining mice in that group. This occurs not only as a result of natural cross-infection but also by cannibalism.

**Irradiation**

The type of radiation used can vary considerably and still yield satisfactory resutls. We have used radiation as diverse as high energy electrons from a linear accelerator in which the radiation is delivered in a few seconds and $^{60}$Cobalt $\gamma$-radiation delivered over a period of 16 hours with equal success. Commonly, X-rays, delivered over a period of less than 1 hour, are used. The major criteria are: (1) the field of radiation must be uniform so that the same dose is absorbed by all parts of the mouse; (2) the haemopoietic tissues must be completely ablated so that no significant numbers of residual CFU-S survive; and (3) the great majority of the mice must survive for at least 9 or 10 days without gross signs of sickness.

When X-rays or electrons are used crowding of the mice can result in inhomogeneity of radiation distribution and it is advisable to irradiate them in compartmentalised boxes such as the one illustrated in Figure 2.3.

**Cell suspension**

The tissues most commonly assayed for CFU-S are femoral bone marrow, spleen and fetal liver. Suspensions are prepared in any medium or balanced salt solution, e.g. Hanks' which is kept, as are the suspensions, until injected, at the temperature of melting ice. A femoral marrow suspension is obtained by cutting off the minimum amount of bone from the ends of the femur, inserting a needle (attached to a 1 ml syringe) which fits the lumen of the femur tightly (e.g. 21 guage), and, by using the femur as an extension of the needle, aspirating the salt solution several times through it by means of the syringe. This produces a suspension consisting of single cells.

Alternatively, marrow from the femur can be obtained efficiently by grinding the bone in a small volume of the suspending fluid using a pestle and mortar. There is much to recommend this method since use of the whole bone eliminates interoperator variation in the amount of material removed from the ends and, in general, it gives a higher cell and CFU-S yield (Carsten & Bond, 1969). To obtain marrow from most other bones, e.g. vertebrae, it is essential to employ this technique.

Spleen suspensions are best made by making a small incision in each end of the spleen and simply expressing the cells into an aliquot of the solution by gently stroking the spleen with a small pair of curved scissors. With care all the cells can be expelled, the capsule of the spleen is left empty and can be discarded. This method is considerably more satisfactory and gives higher cell yields than that of chopping the spleen and removing the debris by sieving (though this is commonly reported). The cells can readily be reduced to a single-cell suspension by aspirating through a small needle (e.g. 25 gauge) several times.

Fetal liver is easily reduced to a single cell suspension by dropping it into an aliquot of the balanced salt solution and then drawing it through a fairly coarse needle (21 gauge) into a syringe and expressing it. After repeating this 2 or 3 times a finer needle is substituted and the process is repeated.

The cell suspensions obtained are counted in the normal way using a haemocytometer or an electronic cell counter. A count of 400 cells will give a standard deviation of about 5% (see Ch. 1).

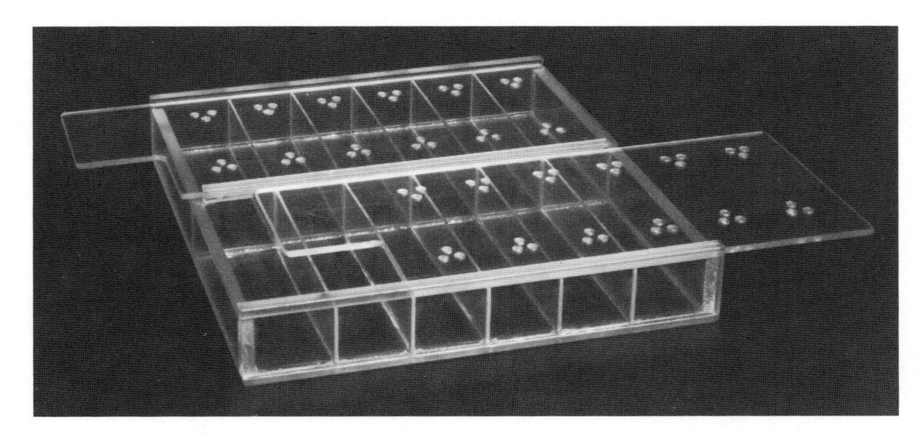

**Fig. 2.3** Perspex box with individual compartments (9.5 cm × 3.0 cm) for 12 mice. The lids, with one end cut to half width, run in grooves on either side of compartment. The narrow section of the lid is slid over each compartment in turn and a mouse inserted under it before completely closing it off.

### Injection

The number of cells injected must be carefully adjusted to obtain a countable number of colonies on the spleens of the recipient (see below). The suspension must be injected intravenously, and in a small volume relative to the total blood volume. In this laboratory we use 0.2 ml as our standard injection dose. Intraperitoneal injections will produce a few colonies but the number is much lower than by the i.v. route and is not so reproducible. The injection is carried out into one of the two lateral tail veins with the mouse held in a suitable restrainer from which the tail projects. The most satisfactory restrainer we have found is illustrated in Figure 2.4 but is not, unfortunately, available commercially. It consists of a heavy base and is adjustable for height (by means of a rod sliding inside a tube and fixed by a knurled screw) and for length of mouse since the restraining compartment consists of a sliding sleeve. The mouse is prevented from backing out by a pivoting door through which the tail projects. The whole is constructed in brass. There are several reasonably satisfactory alternatives on the market. It is a very good idea to warm the mice, before attempting the injection, in order to dilate the veins. We normally stand the boxes containing the mice on an electric warming plate for 15 to 20 minutes. A cheap 'hostess' heater is satisfactory. There is no need to anaesthetise the mice for injection. Apart from the extra time factor, anaesthetic makes the injection more difficult because of the limp tail. It is possible, with practice, to inject well over 100 mice per hour by the method recommended.

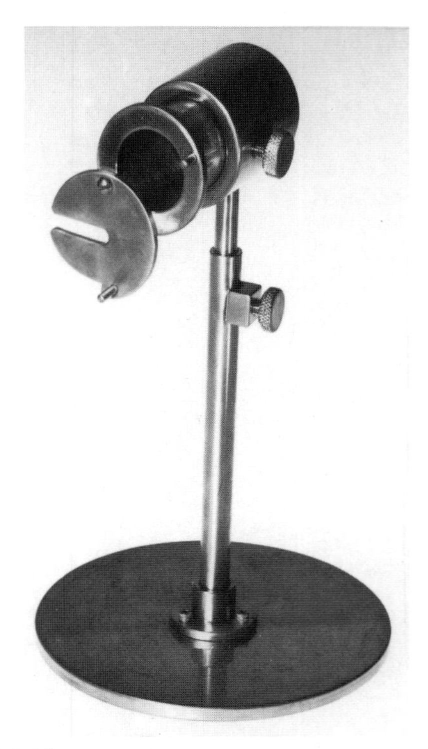

**Fig. 2.4** Mouse restrainer for tail vein injections.

### Fixing the spleens

The mice are killed by cervical dislocation or by exposure to ether and the spleens are removed. They must be kept flat otherwise it becomes very difficult to see all the colonies. This is best achieved by laying them flat in a plastic Petri dish on a filter paper disc wetted with the fixative and then covering them with a second piece of filter paper so that they are kept moist but not allowed to float and twist. The common fixative is Bouin's solution though Telleyesniczky's solution can also be used and the formalin vapour is thus avoided. After a few hours fixation the colonies appear more clearly as pale blobs on a darker ground (Fig. 2.5).

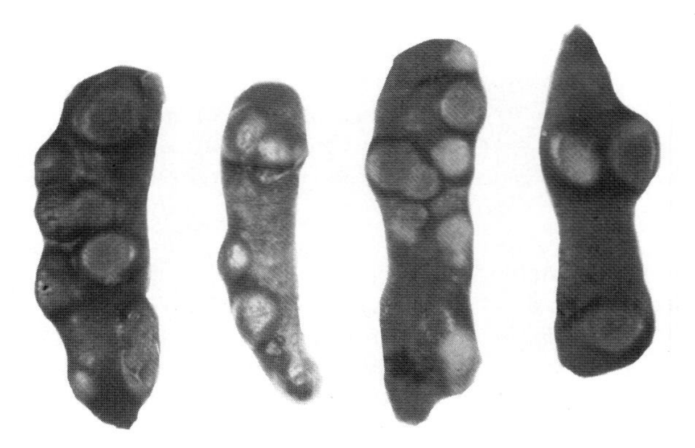

**Fig. 2.5** Mouse spleens containing colonies.

### Counting the colonies

It is possible to count colonies either by eye or, preferably, using low magnification with a dissecting microscope or hand lens. There is a limit to the number of colonies which can be counted on a spleen and in our experience that number is about 15. With increasing numbers a proportion becomes coincident and therefore some are not counted. The literature contains examples of counts (claimed to be precise) of over 50 colonies per spleen. This is simply not possible and, indeed, counts of 20 or more should be regarded with serious suspicion. A mathematical method of correcting colony counts for 'overlap errors' has been published by Till (1972). However, the higher the colony count, the greater the correction and hence, the greater the unreliability of the measurement. Therefore it is extremely important to make the best possible guess at the likely concentration of CFU-S in a suspension to be injected in order to attempt to produce no more than 15 colonies per spleen.

When counting it is also important not to count a colony twice — one colony can protrude from both sides of the spleen! It was reported by McCulloch (1963) and Lewis & Trobaugh (1964) and confirmed by us (Lord & Schofield, 1973) that the number of colonies obtained on the spleens of irradiated mice injected with bone marrow cells is constant from the time the colonies are first detectable (~5 days after injection) up to the time when they begin to become confluent (~13 days). However, as the colonies increase in size throughout this time there is an increasing tendency for some of the colonies to overlap and thus to make their accurate counting difficult. It is most satisfactory to take the spleens for counting on days 8, 9 or 10. The earlier time ensures the best survival whilst the later time makes for the most distinct colonies. Another drawback to attempting to count colonies on day 11 and later is that tiny secondary colonies sometimes begin to appear and confuse the picture.

A recent report by Magli et al (1982) illustrated, however, that some of the colonies seen at 8 days disappear by 10 days. At the same time, some colonies will be visible at 10 days or later which were not seen at 8 days. These authors suggest that the later colonies are more representative of the true pluripotent stem cells and should be scored at 11 days or later. While strictly this is true, the constancy of colony numbers between 5 and 13 days indicates that it is not a necessary restriction for obtaining a reliable index of CFU-S content.

## PRESENTATION OF DATA

Since each recipient mouse is, in effect, an independent measure of the CFU-S content in a given inoculum, the resulting spleen colonies should be quoted as the mean number of colonies in the group of recipient mice together with the standard error of that mean. This standard error should normally be <10 per cent, which is achieved using 10 recipients with about 10 colonies per spleen. The CFU-S concentration in the inoculum (knowing the number of cells injected) is calculated from this mean. CFU-S concentration, however, can be misleading since it does not take account of any changes in cellularity. It is generally more appropriate, therefore, to record the cellularity of the femur, spleen or whatever the source of the CFU-S and convert the CFU-S concentration to total number per femur, spleen, etc.

In view of variations introduced by different radiation techniques and donor treatments it is also advisable to convert CFU-S values to CFC-S to counter possible changes in seeding efficiencies (see below). This is however, very rarely done and, in view of the enormous expenditure of mice and time involved, is probably not over essential unless important intercomparisons are to be made.

## MODIFICATIONS TO THE METHOD

### Bone marrow cellularity

In obtaining marrow from the femur it is usual to cut off both ends and flush out the marrow into medium. How much bone is cut away depends on the operator, so that comparisons of CFU-S per femur between laboratories require some caution. In view of the non-uniform distribution of CFU-S in the marrow (Lord & Hendry, 1972; Lord et al, 1975) it is essential to wash all cells out of the femur shaft very thoroughly. Carsten & Bond (1969) suggested that even then significantly more CFU-S could be obtained by grinding the bone and it is possible that more comparable results could be obtained if all workers adopted a technique of grinding the whole bone including the ends. Such a procedure has not been evaluated but occasional observations in this laboratory have yielded nearly twice the normal number of CFU-S obtained by the normal technique.

### Source of radiation for recipient mice

As stated above, any radiation which reduces background endogenous colonies close to zero while permitting the mouse to live beyond the time required for colony development is satisfactory. Generally, X-rays (~8.0 to 8.5 Gy) or γ-rays (~9.0 to 10.0 Gy) at dose-rates of 0.3 to 6 Gy/min have been found by most people to be appropriate. High energy electrons at dose-rates of 1.7 to 4.0 Gy/sec were first used by Proukakis & Lindop (1967) and subsequently have also been used extensively in these laboratories with success, a dose of 8 to 8.5 Gy being used.

It should be noted that the radiation tolerance of a mouse colony can change with time and periodic checking of the endogenous background colony and survival level should be undertaken. It was found (Lord et al, 1984) that background colonies in our mice following 8.5 Gy electron irradiation had risen to an unacceptable level (more than 0.5 per spleen) and that to eliminate it, it would be necessary to increase the radiation dose to a level which causes premature deaths due to intestinal damage. Very low dose-rate radiation has proved useful in this respect and currently a $^{60}$Cobalt γ-ray unit giving 0.844 Gy/h at 4 metres is used in these laboratories. This allows large numbers of mice to be irradiated in their normal stockboxes overnight, accumulating 13.5 Gy in 16 h. This dose given at the higher rates leads to lethal intestinal damage, the mice dying at about 6 to 7 days. At the low dose-rate there is significant sparing of the gut (Krebs & Leong, 1970) relative to bone marrow (Neal, 1960) so that survival is virtually 100 per cent up to 14 days while endogenous CFU-S are reduced to exceedingly low levels (0.07/spleen). However, the smaller spleen does also result in a lowered spleen seeding efficiency (see below) for injected CFU-S and the appropriate allowances must be made in comparing results with other radiation schedules.

**Recipient-radiation interval**

It is usual to inject haemopoietic cells as soon as possible (usually within 1 h) after completing the recipient irradiation. However, this is not critical. Cells injected up to 48 h after radiation produce colonies as efficiently as those injected within 1 h (Lord & Hendry, 1973). The only criterion is that any delay should not result in the colony development time being extended beyond the animal survival time.

**Maintenance of donor cell suspensions**

The majority of CFU-S assays simply require a cell suspension to be made and injected into recipients without delay. For this purpose, any suspending medium (even basic physiological saline) can be used. Hanks' balanced salt solution is very cheap and convenient. If ex-vivo manipulations are necessary before assaying the cells, such solutions may not be satisfactory and more complete media, possible supplemented with serum, may be necessary. From our own experience, CFU-S decline in number after about 6 h in Hanks' solution at 4°C but survive for at least 24 h in horse serum at 4°C. Different survival rates accompany maintenance in other media or at other temperatures. If variations in technique are required, therefore, it is necessary to check that the handling conditions do not themselves alter CFU-S number and, in the case of suicide experiments, cycling rate.

**Large dose inocula**

When the CFU-S concentration is very low, it is sometimes necessary to inject large numbers of cells to obtain spleen colonies. If the dose required is $>10^7$ cells, the suspension (at $>5 \times 10^7$ cells/ml) is very thick and the cells often tend to clump — particularly if the tissue has a high fat content. Injection of such a suspension in 0.2 ml can lead to convulsions, loss of consciousness and death within about 5 min. This problem can usually be overcome by diluting the suspension and injecting a larger volume, very slowly The total blood volume of a mouse is about 1.7 ml so the injection should be limited to a maximum of 0.5 ml. If this still presents difficulties, 50 units of heparin injected i.p. in 0.2 ml about 30 min prior to the cell injection usually solves the problem and does not alter the spleen seeding-efficiency of the injected CFU-S.

## COMPOSITION OF SPLEEN COLONIES

The cellular composition of spleen colonies can be studied in 2 ways: (1) by histological section and (2) by excision of the colonies from unfixed spleens and squashing them on to a microscope slide.

The first method employs standard histological techniques and gives an excellent picture of the whole tissue. It is, however, an unsatisfactory method both because a large number of serial sections is needed in order to include all the colonies in a spleen and because it is usually difficult, and often impossible, to identify, in section, any but the most obvious cells, e.g. polymorphs, later stages of normoblast and megakaryocytes. Nevertheless, with care, some results can be obtained as was shown by Lewis and his colleagues (1964; 1967; 1968) though their type of work has been used by very few other investigators.

The better way to is to excise the colony under low-power magnification. Provided a reasonably small number of colonies only occurs on the spleen (not more than 10), all the colonies can usually be dissected out. Smears can then be made by adding a drop of Hanks' or similar solution to the colony and producing a suspension simply by mashing it with the end of another slide. A drop of the suspension is then spread by the 'push' technique. Alternatively, the suspension can be squashed between two slides and two 'pull' smears made by sliding the slides apart. We find the latter method gives the more even distribution and preserves the cellular morphology better. After staining by one of the conventional methods e.g. May Grünwald-Giemsa or MacNeal's tetrachrome, the smears are examined microscopically. It has been found (Trentin, 1976) that the vast majority of colonies at the earlier times (7–10 days) after injection are erythroid or predominantly erythroid. However, the proportion of granulocytic cells increases with time and it would be meaningless to compare the composition of colonies in spleens at different times after grafting. Furthermore, the morphological composition of the colony is influenced by microenvironmental determinants within the spleen (Wolf, 1979).

A clear example of the microenvironmental influence can be obtained by comparing the mixture of haemopoietic cells in the spleen and the femoral marrow in the same mouse at the same time after radiation and cell injection. In the marrow there is a much higher proportion of granulocyte development than in the spleen. Nevertheless, if that predominantly granulocytic marrow is injected into a second irradiated mouse the spleen colonies to which it gives rise have a normal erythroid content (Trentin, 1976). Similarly, a pure erythroid spleen colony, retransplanted into a secondary irradiated recipient gives rise to the whole morphological range of differentiated colonies (Lewis & Trobaugh, 1964). Since the cellular composition of haemopoietic tissues after radiation is heavily dependent on both the site and time it would be a mistake to place too great a significance on small variations in the differential composition of the tissue.

In addition to the erythroid and granulocytic cells found in colonies occasionally one finds a colony which contains also (or even exclusively) megakaryocytes. Very commonly a significant proportion of the cells in a colony are undistinguished small or medium-sized round mononuclear cells which would generally be classified by a morphological haematologist as 'lymphocytes'. No doubt this populations contains many haemopoietic progenitor and precursor cells all of which can be detected in spleen colonies. The literature contains examples of studies in which some of these cellular species have been measured and from the measurements a number of deductions has been made (Till et al, 1964; Vogel et al, 1968; Schofield et al, 1980). These are generally esoteric but the method of making the measurements is simple enough. Colonies are excised individually and placed in a small amount (say 0.2 ml) of Hanks' solution in a small Petri dish. A cell suspension is easily obtained by teasing out the cells using a needle attached to a 1 ml plastic syringe. The small residue of connective tissue is discarded and the cell suspension is then drawn into the syringe and diluted to an appropriate volume (0.5–1.0 ml). This suspension is aspirated through the needle four or five times to give a discrete cell suspension. This can now be used in any of the standard clonogenic techniques available for the measurement of haemopoietic cell precursors.

A simplification of this procedure is to measure the mean number of precursor cells per colony. All that is required is to count the numbers of colonies in the fixed spleens from several mice of a given group and also to measure the number of cells of a given precursor cell species in an aliquot of fresh spleen suspension from other mice of the same group. Thus, the total number of CFU-S per spleen can be calculated and since the mean number of colonies per spleen is already known, the mean number of CFU-S/colony is easily found.

For these measurements, it is important that a standard time at which to observe the CFU-S in spleen colonies is used. The number of CFU-S in the spleen approximately doubles every day whilst the number of colonies remains constant. Most of the measurements of this kind have been carried out on day 11 after radiation as a compromise between the highest number of CFU-S and the highest level of mouse survival.

Another factor which must be carefully controlled if the data on CFU-S/colony are to have any meaning is the number of colonies in the spleens. Although the reason is not known, it is clear that the mean CFU-S/colony increases as the number of colonies in the spleen increases, even when different doses of the same cell suspension are injected to produce the colonies (Schofield & Lajtha, 1983).

## MEASUREMENT OF CFU-S PROLIFERATION

Since CFU-S are present in extremely low concentrations and remain morphologically unidentified, the standard autoradiographic techniques cannot be used for measuring their proliferation status. However, the spleen colony technique can be exploited for this purpose. This is done by killing all S-phase CFU-S by exposure to a high dose of an S-phase toxic agent — usually tritium labelled thymidine, and measuring the resulting loss in colony formation. Other agents which are sometimes used are cytosine arabinoside and hydroxyurea though conditions for their use can be slightly different from those for $^{3}$H-thymidine (Byron, 1972). The technique can be carried out in two ways: (1) by in vivo suicide, (Lajtha et al, 1969); (2) by in vitro suicide (Becker et al, 1965).

### Suicide in vivo with $^{3}$H-thymidine

Donor mice, i.e. those in which CFU-S proliferation is to be measured, are injected with 37 MBq (1 mCi) high specific activity (555–840 GBq/mmol) tritiated thymidine ($^{3}$HTdR). A similar control group of mice is injected with saline or an equivalent amount of cold thymidine. The mice are killed one hour later and bone marrow CFU-S assays on the two groups are carried out (Fig. 2.6). The proportion of CFU-S in DNA-synthesis is then given by the formula (C-T)/C where C and T are the respective mean colony counts in the control and thymidine treated groups.

The drawback to this method lies in the fact that two different bone marrow populations, which cannot initially have identical CFU-S concentrations, are compared. The dose of $^{3}$HTdR is not critical. Guzman (unpublished observations) measured the kill using doses of 14.8, 29.6 and 59.2 MBq per mouse and found that 29.6 MBq essentially gave a plateau value.

### Suicide in vitro

This is the more widely used technique for the suicide measurement and has the advantage over the in vivo method that a single bone marrow preparation is used for both the thymidine treated and the control groups. Furthermore, it is economically more practical since it uses only a fraction of the $^{3}$HTdR used in vivo.

The method is illustrated in Figure 2.7. Femora are removed from the donor mice under assay and bone marrow cells suspended in Fischer's medium. The cell concentration is adjusted to $5 \times 10^{6}$ per ml and $2 \times 1$ ml aliquots are dispensed into plastic centrifuge tubes.

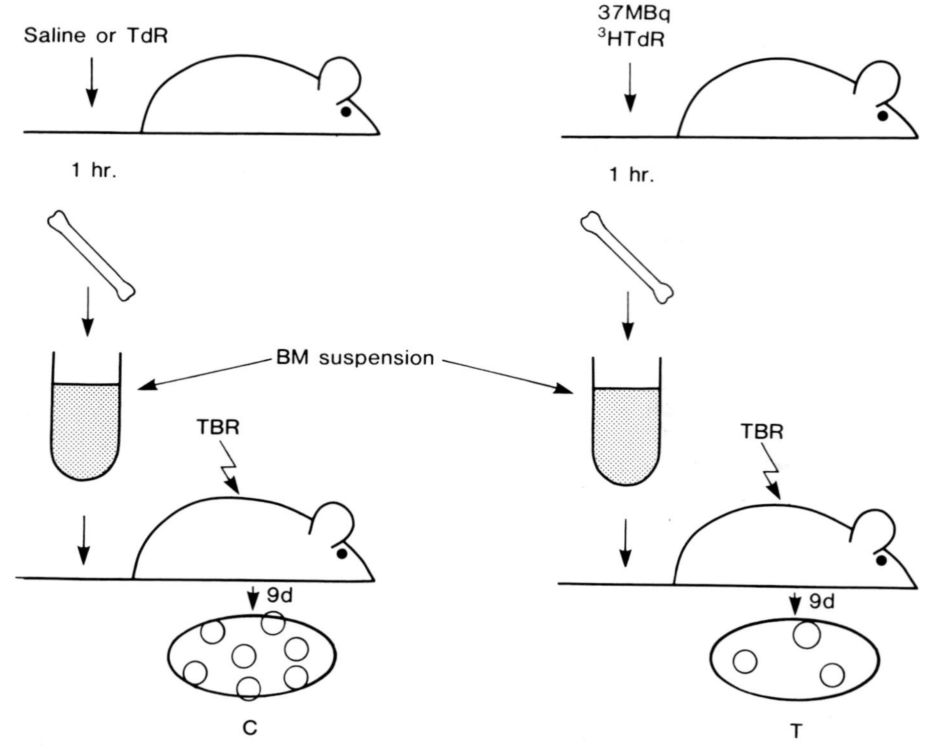


**Fig. 2.6** Suicide technique in vivo for measurement of the proliferation status of CFU-S. TBR = total body radiation, BM = bone marrow, TdR = thymidine, $^{3}$HTdR = tritium labelled thymidine, C = controls, T = $^{3}$HTdR treated sample.

These are heated for 10 mins in a shaking water bath at 37°C for a prewarming period. 7.4 MBq $^{3}$HTdR in 0.2 ml isotonic Fischer's medium are then added to one tube. 0.2 ml of fresh Fischer's medium (with or without cold TdR equivalent to the thymidine in the active tube) is added to the control tube. After a further 30 min incubation, thymidine incorporation is stopped by standing the tubes in ice and the suspension diluted appropriately for CFU-S assay. The proportion of CFU-S in DNA-synthesis is calculated as for the measurement in vivo.

Several practical points should be noted:

1. Fischer's medium is normally used though any good medium should be satisfactory. Simple balanced salt solutions, however, are not adequate for maintenance of CFU-S numbers and proliferation.

2. The type of tube or bottle used for incubation can also be important for maintenance of CFU-S numbers and proliferation. There is no clear indication what factors (e.g., shape, size, material, etc.) are important. Satisfactory types should be chosen by preliminary trial and error. In this laboratory, we routinely use Nunc Tissue Culture Tubes (Gibco Bioculture), size 100 mm × 14 mm with screw cap.

3. Strictly, the correct control for $^{3}$HTdR is an equal amount of cold thymidine. In practice, at the concentration used, it is adequate to use the medium alone.

4. It is not necessary to wash out excess $^{3}$HTdR before injecting cells into recipient mice for CFU-S assay. At the dose used, the dilution is such that the recipient mice each receive less than an autoradiographic labelling dose (<74 kBq per mouse).

5. It is frequently stated that high specific activity $^{3}$HTdR should be used. However, non-S-phase killing can occur if the specific activity is too high and an activity of <740 GBq/mmol is recommended (Lord et al, 1974).

6. A $^{3}$HTdR dose of 7.4 MBq/ml is sufficient to kill all cells in DNA-synthesis (Becker et al, 1965).

## CALCULATION OF STANDARD ERROR ON THE KILL AND SIGNIFICANCE OF THE RESULT

A typical standard error on a spleen colony assay using groups of 10 recipient mice is of the order of 10 per cent or less. In measuring the standard error on a $^{3}$HTdR suicide experiment, however, the kill depends on the difference of two spleen colony counts and the calcu-

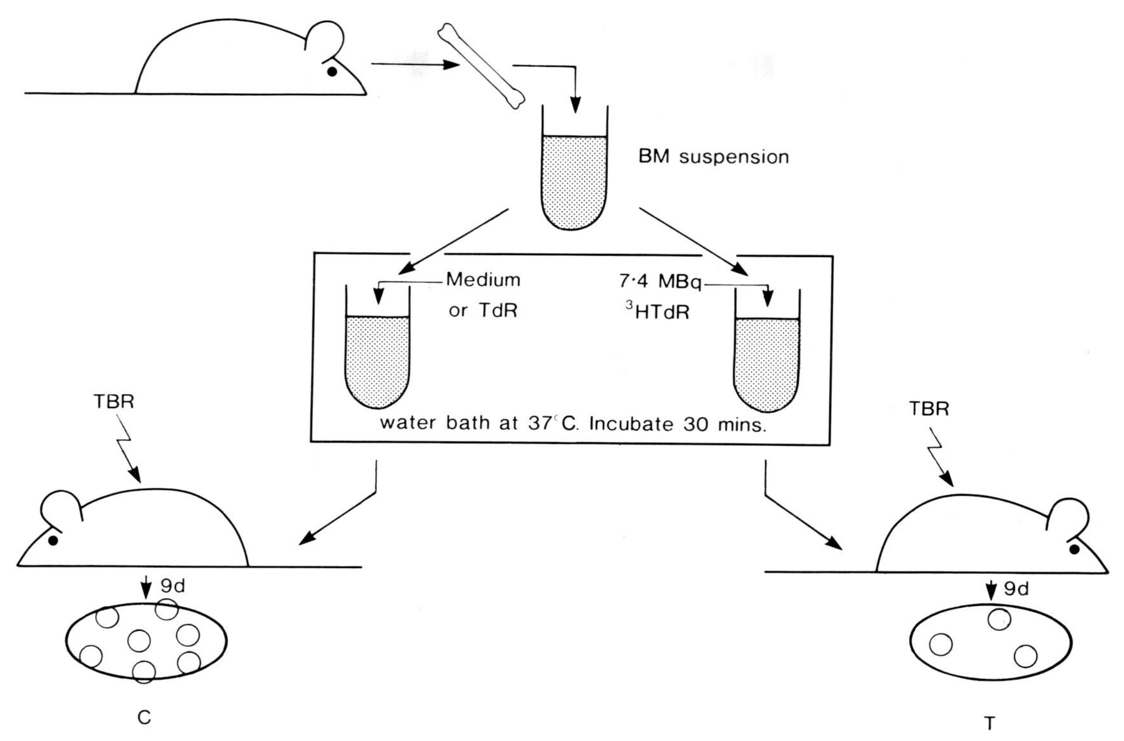


**Fig. 2.7** Suicide technique in vitro for measurement of the proliferation status of CFU–S. Key is the same as Fig. 2.6.

lated error is thus considerably higher. The standard error can be calculated from the formula:

$$\text{s.e.}_K = \frac{1}{C}\sqrt{(\text{s.e.}_T)^2 + \left(\frac{T}{C}\right)^2 \times (\text{s.e.}_C)^2}$$

where C is the total number of control colonies (standard error $\text{s.e.}_C$), T is the total number of colonies after $^3$HTdR treatment (standard error $\text{s.e.}_T$) and $\text{s.e.}_K$ is the standard error on the calculated kill. Since the recipient groups are normally of the same size, the mean number of colonies per spleen (rather than the total colonies per spleen) can be used equally satisfactorily.

The use of this formula indicates limitations on the significance of the kill measurement. If one assumes that $\text{s.e.}_C$ and $\text{s.e.}_T$ are approximately 10 per cent, then the difference in those colony counts must be greater than 12 per cent if the kill is to be significantly different from zero, i.e. if the difference is 12 per cent, the kill is calculated as $0.12 \pm 0.13$; if it is 13 per cent, the kill becomes $0.13 \pm 0.12$. In view of the large errors, it is clear that for the suicide assays, the spleen colony technique is being used at the limit of its resolution. In practice, therefore, in this laboratory kills of $<10$ per cent are generally considered as non-significant.

Only when the kill exceeds 20 per cent is the CFU-S population considered to have a significant proportion of its cells in DNA-synthesis. Furthermore, it is not considered that graded kills can be assessed satisfactorily by this method. To do so would require vast numbers of recipient mice (see Ch. 1 and Hazout & Valleron, 1977): numbers which are not practical in most laboratories.

**Alternative suicide techniques**

Recently, an alternative method for measuring the S-phase fraction of CFU-S has been introduced (Monette & Demers, 1982) which is potentially useful in situations where cell yield is too low for incubations in vitro with $^3$HTdR or when manipulations ex vivo must be avoided. In this method, the cells to be assayed are injected into two groups of recipient mice, one of which is also injected with hydroxyurea (0.9 mg/kg body weight) either simultaneously with, or immediately prior to, the injection of marrow cells. Donor S-phase cells are thus killed in the hydroxyurea-treated recipients and the proportion in S-phase calculated as before. According to this one report, the results obtained for normal and regenerating marrow are compatible with those made using the technique in vivo described above. In this case, hydroxyurea was used rather than $^3$HTdR but no comparison with $^3$HTdR was given.

## SEEDING OF INJECTED COLONY FORMING CELLS IN THE SPLEEN

Only a fraction of the potential spleen colony-forming cells (CFC-S) injected into an irradiated mouse actually settles in the spleen and results in spleen colonies. Such cells are known as spleen colony-forming units (CFU-S) and are related to the number of CFC-S by the simple formula

$$\text{CFU-S} = f \times \text{CFC-S}$$

where f is the seeding factor for colony-forming cells in the spleen.

f is normally measured by a secondary transplantation technique (Siminovitch et al, 1963). A suspension of haemopoietic cells for which f is to be determined is injected into 2 groups of irradiated mice; the first (Group 1, Fig. 2.8) with a number suitable for CFU-S assay (N cells giving n spleen colonies at 9 days), the second (Group 2) at 100 times that inoculum. After allowing time t hours (t = 24, see below) for CFU-S to seed in the spleen, the spleens are removed from this second group and a cell suspension made. The cell concentration is adjusted so that a known fraction (1/p) of a spleen is contained in 0.2 ml and this volume is injected into a further group of irradiated mice (Group 3) for CFU-S assay. The expected 100n colonies in group 2 mice is effectively the number of CFC-S in the spleens which are assayed in group 3 for CFU-S. The seeding factor, therefore, is given by

$$f = \frac{pq}{100\,n}$$

where q is the number of colonies per spleen in the group 3 assay mice. Although the technique is simple, a number of problems make its interpretation far from straightforward.

Two assumptions are essential:

1. That there is no preferential selection of certain types of CFU-S by the spleen (Smith et al, 1968).

Figure 2.9 illustrates the principle showing that differential seeding of A and B type CFU-S would result in a different experimental f value, measured by the second transplant than the true one indicated by the first transplant. The growing appreciation of the heterogeneity of the CFU-S population (Schofield, 1978; Rosendaal et al, 1979) makes this an important consideration. Fred & Smith (1968), however, measured a third transplant fraction and found comparable values with the second transplant fraction thus giving some confidence that the first and second transplant fractions are also very similar.

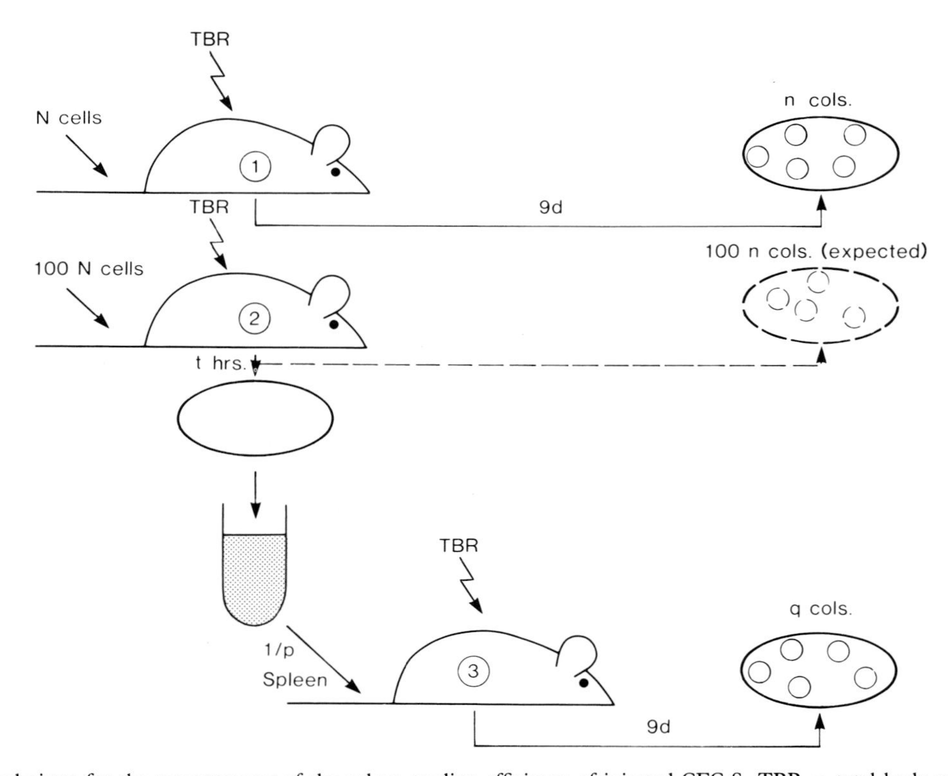


**Fig. 2.8** Technique for the measurement of the spleen seeding efficiency of injected CFC-S. TBR = total body radiation.

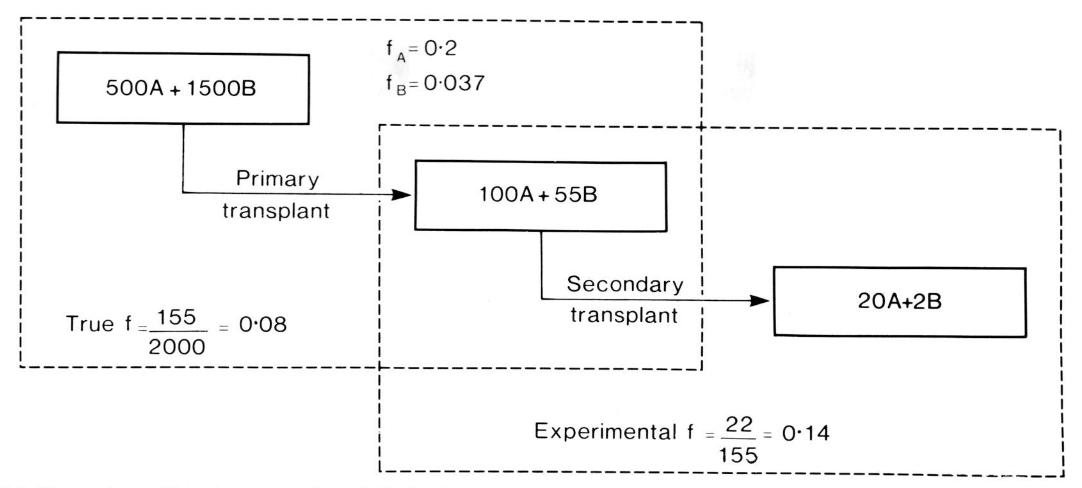


**Fig. 2.9** Illustration of how heterogeneity of CFC-S in the primary transplanted inoculum could lead to an erroneous experimental measurement of the spleen seeding efficiency. The figures used are purely illustrative and are not based on any experimental data.

Hendry (1971) described an alternative approach which, although appropriate only for normal spleen CFU-S, is nevertheless an important independent check on the standard method. Hendry found that the slopes to the CFU-S radiation dose/survival curves were the same whether he measured the 'fraction surviving' by transplantation into lethally irradiated mice (the standard CFU-S assay) or the number of endogenously derived colonies growing in sublethally irradiated mice. Hence, he was able to equate the number of endogenous spleen colonies at a given radiation dose with the surviving fraction at that dose and extrapolate back to 100 per cent survival to determine the equivalent number of endogenous colony formers (splenic CFC-S) theoretically present at zero radiation dose. Knowing the number of CFU-S obtainable from the normal spleen (the zero radiation dose/100 per cent survival point) it was then possible to calculate a first-transplant f value of $0.03 \pm 0.02$ which is comparable with those of 0.06 (Lahiri & van Putten, 1969) and 0.05 (Lord, unpublished data) determined by the usual method.

2. That the number of spleen colonies is linearly related to the number of CFU-S injected.

This relationship holds true over the range for which colonies are readily counted, i.e. up to about 15 colonies per spleen (Till & McCulloch, 1961). Although not countable, it must be assumed that if N cells give n colonies then 100 N cells would give 100 n colonies if left to develop.

**Spleen mixing time, t**
Originally, Siminovitch et al (1963) waited only 2–3 hours between injecting bone marrow and assaying the spleens. This is the point at which the maximum number of CFU-S have accumulated in the spleen and which gives a value of f for normal bone marrow of ~ 0.17. Playfair & Cole (1965) waited 24 h and found a value of 0.08. Lord (1971) showed that this difference was directly related to loss of CFU-S due to the post-irradiation collapse of the spleen (which was complete by about 24 h) after initially seeding there. In $W/W^v$ mice which are genetically deficient in CFU-S, spleen colonies will develop without the need for prior irradiation. The post irradiation collapse of the spleen cellularity as seen in normal mice is, therefore, avoided. As a result, the f value, measured in $W/W^v$ mice, is the same at both 2 and 24 h after transplantation (Lord, 1971). If irradiated, however, the $W/W^v$ mice behave like normal mice; the post irradiation collapse occurs and the measured f value falls between 2 and 24 h (Lord & Hendry, 1973). On the other hand, boosting the spleen size by injections of foreign protein results in an increased seeding efficiency. Thus, colonies appear to develop from the reduced number of CFU-S left in the spleen after its collapse rather than the higher number found after 2–3 hours (see Table 2.1) and a 24 hour interval is considered to be more realistic for the determination of f. Lahiri et al (1970) achieved the same effect by irradiating the primary recipients (Groups 1 & 2, Fig. 2.7) 24–48 h before injecting the cells and then allowing a mixing interval of 2–3 h. In this way, the primary inoculum sees only the collapsed spleen and consequently fewer CFC-S initially enter the spleen.

**Induced changes in spleen seeding**
Fred & Smith (1968) further pointed out that various pre-treatments of the marrow could result in a change in the fraction of CFU-S seeding in the spleen, for ex-

**Table 2.1** Measured seeedings factors for CFC-S from different sources

| Source of CFC-S | Measured seeding factor f | Mixing interval t (h) | Reference |
|---|---|---|---|
| NBM | 0.17 | 2–3 | Siminovitch et al, 1963 |
| NBM | 0.20 | 3 | Fred & Smith, 1968 |
| BM — 1d/post endotoxin | 0.12 | 3 | Fred & Smith, 1968 |
| BM — 2d/post endotoxin | 0.08 | 3 | Fred & Smith, 1968 |
| BM — 1–2d/post 2 Gy | 0.11 | 3 | Fred & Smith, 1968 |
| NBM | 0.25 | 2 | Lahiri & van Putten, 1969 |
| N spleen | 0.11 | 2 | Lahiri & van Putten, 1969 |
| NBM | 0.085 | 24 | Lahiri & van Putten, 1969 |
| N spleen | 0.06 | 24 | Lahiri & van Putten, 1969 |
| NBM | 0.33 | 2 | Lord, 1971 |
| NBM | 0.11 | 24 | Lord, 1971 |
| NBM or quiescent CFU-S | 0.13 | 2 | Monette & DeMello, 1979 |
| NBM | 0.04 | 24 | Monette & DeMello, 1979 |
| RBM or proliferating CFU-S | 0.06 | 2 | Monette & DeMello, 1979 |
| RBM | 0.02 | 24 | Monette & DeMello, 1979 |
| NBM | 0.095 | 24 | Lord & Hendry, 1973 |
| N spleen | 0.05 | 24 | Lord (unpublished) |
| N fetal liver | 0.03 | 24 | Lord (unpublished) |

NBM = normal bone marrow
RBM = regenerating bone marrow
N = normal

ample, endotoxin, vinblastine or irradiation (see Table 2.1). Monette & DeMello (1979) observed different seeding factors for quiescent and proliferating CFU-S. In addition, CFU-S from different haemopoietic tissues seed with different efficiencies (Lahiri & Van Putten, 1969; Lord, unpublished — see Table 2.1). This is not simply a result of the different cellular constitutions in bone marrow, spleen and fetal liver because, for example, the colony-forming capacity of bone marrow CFC-S is not altered by the addition of large numbers of irradiated spleen cells.

A change in the radiation conditions can also affect the seeding. For example, a low dose-rate $^{60}$Cobalt irradiation unit is now in regular use in these laboratories for the preparation of recipient mice. It has been found that the extremely low dose-rate (0.844 Gy/h) from this unit results in small spleens with subsequently lowered seeding (Table 2.2) and lowered colony counts from the same size bone marrow inoculum when compared with the very high dose-rate (1.4 Gy/sec) obtained using electrons from a linear accelerator.

**Table 2.2** Effect of recipient radiation quality on the seeding of CFC-S in the spleen

| Radiation | Spleen seeding factor f |
|---|---|
| 12MeV linear accelerator electrons (dose-rate 1.4 Gy/sec) | 0.095 |
| $^{60}$Cobalt γ-rays (dose rate 0.844 Gy/h) | 0.064 |

It is clear, therefore, that no absolute value can be quoted for the spleen seeding factor of CFC-S. Strictly speaking, it is necessary always to measure f when different experimental conditions or treatments are being used and assessed.

## ASSAY OF LYMPHOMA

Bruce & van der Gaag reported in 1963 a modification of the spleen colony assay described by Till &McCulloch (1961) which could be used for the measurement of lymphoma colony-forming cells. The major difference is that the recipient mice do not necessarily have to be irradiated in order for lymphoma cell colonies to develop, although they can be if required. The technique was applied by them to spontaneously-arising lymphomas in AKR/J, $C_3H$ and ($C_3H$ × AKR)$F_1$ mice and to mouse lymphomas induced by the Gross virus. It has also been used for the assay of transplanted lymphomas, e.g. in AKR/J (Bruce & Meeker 1965) and L1210 (Wodinsky et al, 1967). Preparation of cell suspensions is similar to that described earlier except that lymphomatous thymus requires to be chopped and suspended in appropriate medium by use of a glass homogeniser and/or fine wire mesh screening. Colonies become confluent at an earlier time than do bone marrow colonies and therefore the mice should be killed and the spleens fixed (in Bouin's solution) on the 8th day after the lymphoma cell injection. Further, to eliminate the danger of confluence the number of colonies in the spleens must be kept low, i.e.

to 4 or 5. About ($8 \times 10^2$) to $10^3$ cells only are needed from a lymphomatous thymus to produce this number of colonies but the number of bone marrow or spleen cells injected has to be adjusted appropriately.

In order to assay the number of normal CFU-S and lymphoma CFU-S in the same sample of, for example, bone marrow, it is necessary to carry out parallel assays in irradiated and unirradiated mice. Normal CFU-S numbers are assessed as the difference between the number of colonies on the spleen of irradiated mice (in which both normal and lymphoma colonies develop) and the spleens of unirradiated mice (in which only lymphoma colonies grow).

The method by which the cycling state of lymphoma CFU-S has been measured is that of in vivo tritiated thymidine 'suicide' (Bruce & Meeker, 1965). They injected a maximum of 222 MBq $^3$HTdR per mouse in aliquots equally spaced throughout 24 hours and showed that there was a much greater mortality amongst lymphoma CFU (up to 99.9 per cent) than in the normal CFU-S (up to 80 per cent). However, there seems no good reason why the $^3$HTdR suicide of lymphoma CFU-S should not be measured by exposure in vitro to $^3$HTdR in a similar way to that which is used for bone marrow CFU-S.

## REFERENCES


Becker A J, McCulloch E A, Siminovitch L, Till J E 1965 The effect of differing demands for blood cell production on DNA synthesis by hemopoietic colony-forming cells of mice. Blood 26:296

Becker A J, McCulloch E A, Till J E 1963 Cytological demonstration of the clonal nature of spleen colonies derived from mouse marrow cells. Nature 197:452

Bruce W R, van der Gaag H 1963 A quantitative assay for the number of murine lymphoma cells capable of proliferation in vivo. Nature 199:79

Bruce W R, Meeker B E 1965 Comparison of the sensitivity of normal hematopoietic and transplanted lymphoma colony-forming cells to tritiated thymidine. Journal of the National Cancer Institute 34:849

Byron J W 1972 Comparison of the action of $^3$H-thymidine and hydroxyurea on testosterone-treated hemopoietic stem cells. Blood 40:198

Carsten A L, Bond V P 1969 Colony-forming units in the bone marrow of partial-body irradiated mice. Recent Results in Cancer Research 17:21

Ford C E, Hamerton J L, Barnes D W H, Loutit J F 1956 Cytological identification of radiation chimeras. Nature 177:452

Fred S S, Smith W W 1968 Induced changes in transplantability of haemopoietic colony-forming cells, Proceedings of the Society for Experimental Biology and Medicine 128:364

Hazout S, Valleron A J 1977 Planning the suicide experiments. Cell and Tissue Kinetics 10:569

Hendry J H 1971 The f number of primary transplanted splenic colony-forming cells. Cell and Tissue Kinetics 4:217

Hendry J H 1979 The dose-dependence of the split-dose response of marrow colony-forming units (CFU-S): similarity to other tissues. International Journal of Radiation Biology 36:631

Jacobson L O, Marks E K, Gaston E O, Robson M J, Zirkle R E 1949 Role of the spleen in radiation injury. Proceedings of the Society for Experimental Biology and Medicine 70:7440

Jacobson L O, Simmons E L, Marks E K, Gaston E O, Robson M J, Eldridge J H 1951 Further studies on recovery from radiation injury. Journal of Laboratory and Clinical Medicine 37:683

Krebs J S, Leong G F 1970 Effect of exposure rate on the LD50/5 of mice exposed to $^{60}$Co $\gamma$-rays or 250 kV X-rays. Radiation Research 42:601

Lahiri S K, Keizer H J, van Putten L M 1970 The efficiency of the assay for haemopoietic colony-forming cells. Cell and Tissue Kinetics 3:355

Lahiri S K, van Putten L M 1969 Distribution and multiplication of colony-forming units from bone marrow and spleen after injection in irradiated mice. Cell and Tissue Kinetics 2:21

Lajtha L G, Pozzi L V, Schofield R, Fox M 1969 Kinetic properties of haemopoietic stem cells. Cell and Tissue Kinetics 2:39

Lewis J P, Trobaugh F E Jr. 1964 Haematopoietic stem cells. Nature 204:589

Lewis J P, O'Grady L F, Trobaugh F E Jr. 1967 Surface colonies as a measure of total number of hemopoietic colonies in murine spleens. Experimental Hematology 12:33

Lewis J P, Passovoy M, Freeman M, Trobaugh F E Jr. 1968 The repopulating potential and differentiation capacity of hematopoietic stem cells from the blood and bone marrow of normal mice. Journal of Cellular Physiology 71:121

Lord B I 1971 The relationship between spleen colony production and spleen cellularity. Cell and Tissue Kinetics 4:211

Lord B I, Hendry J H 1972 The distribution of haemopoietic colony-forming units in the mouse femur and its modification by X-rays. British Journal of Radiology 45:110

Lord B I, Hendry J H 1973 Observations on the settling and recoverability of transplanted hemopoietic colony-forming units in the mouse spleen. Blood 41:409

Lord B I, Schofield R 1973 The influence of thymus cells in hemopoiesis: stimulation of hemopoietic stem cells in a syngeneic, in vivo, situation. Blood 42:395

Lord B I, Lajtha L G, Gidali J 1974 Measurement of the kinetic status of bone marrow precursor cells: three cautionary tales. Cell and Tissue Kinetics 7:507

Lord B I, Testa N G, Hendry J H 1975 The relative spatial distribution of CFU-S and CFU-C in the normal mouse femur. Blood 46:65

Lord B I, Hendry J H, Keere J P, Hodgson B W, Xu CX, Rezvani M, Jordon T J 1984 A comparison of low and high dose-rate radiation for recipient mice in spleen colony studies. Cell and Tissue Kinetics 17:323

Magli M C, Iscove N N, Odartchenko N 1982 Transient nature of early haemopoietic spleen colonies. Nature 295:527

McCulloch E A 1963 Les clones de cellules hématopoiétique

in vivo. Revues Française Etudes Cliniques et Biologie 8:15

McCulloch E A, Till J E 1963 Repression of colony-forming ability of $C_{57}$B1 hematopoietic cells transplanted into non-isologous hosts. Journal of Cellular and Comparative Physiology 61:301

Monette F C, DeMello J B 1979 The relationship between stem cell seeding efficiency and position in cell cycle. Cell and Tissue Kinetics 12:161

Monette F C, Demers M L 1982 An alternative method for determining the proliferation status of transplantable murine stem cells. Experimental Hematology 10:307

Neal F E 1960 Variation of acute mortality with dose-rate in mice exposed to single large doses of whole-body X-irradiation. International Journal of Radiation Biology 2:295

Playfair J H L, Cole L J 1965 Quantitative studies on colony forming units in isogenic radiation chimeras. Journal of Cellular and Comparative Physiology 65:7

Proukakis C, Lindop P J 1967 Age dependence of radiation sensitivity of haemopoietic cells in the mouse. Nature 215:655

Rosendaal M, Hodgson G S, Bradley T R 1979 Organization of haemopoietic stem cells: the generation-age hypothesis. Cell and Tissue Kinetics 12:17

Schofield R 1978 The relationship between the spleen colony-forming cell and the haemopoietic stem cell: a hypothesis. Blood Cells 4:7

Schofield R, Lajtha L G 1983 Determination of the probability of self-renewal in haemopoietic stem cells: a puzzle. Blood Cells 9:467

Schofield R, Lord B I, Kyffin S, Gilbert C W 1980 Self-maintenance capacity of CFU-S. Journal of Cellular Physiology 103:355

Siminovitch L, McCulloch E A, Till J E 1963 Distribution of colony-forming cells among spleen colonies. Journal of Cellular and Comparative Physiology 62:327

Smith W W, Wilson S M, Fred S S 1968 Kinetics of stem cell depletion and proliferation: effects of vinblastine and vincristine in irradiated mice. Journal of the National Cancer Institute 40:847

Stroud A N, Chatterley D M, Summers M M, Brues A M 1955 Regeneration and recovery of the spleen and thymus after single doses of X-irradiation. Report of the Argonne National Laboratory 5456:15

Till J E 1972 Overlap error in counts of splenic colonies. Series Haematologica 5:5

Till J E, McCulloch E A 1961 A direct measurement of the radiation sensitivity of normal mouse bone marrow cells. Radiation Research 14:213

Till J E, McCulloch E A, Siminovitch L 1964 A stochastic model of stem cell proliferation based on the growth of spleen colony-forming cells. Proceedings of the National Academy of Sciences USA 51:29

Trentin J J 1976 Hemopoietic inductive microenvironments. In: Cairnie A B, Lala P K, Osmond D G (eds) Stem cells of renewing cell populations. Academic Press, New York, p 255

Vogel H, Niewisch H, Matioli G 1968 The self-renewal probability of hemopoietic stem cells. Journal of Cellular Physiology 72:221

Wodinsky I, Swiniarski J, Kensler C J 1967 Spleen colony studies of leukaemia L1210. I. Growth kinetics of lymphocytic L1210 cells in vivo as determined by spleen colony assay. Cancer Chemotherapy Reports 51:415

Wolf N S 1979 The haemopoietic microenvironment. In: Lajtha L G (ed) Cellular dynamics of haemopoiesis. Clinics in haematology, vol 8. Saunders, Philadelphia, p 469

# Clonal assays for haemopoietic and lymphoid cells in vitro

## GLOSSARY OF ABBREVIATIONS

BFU-E: burst-forming unit, erythroid.
BL-CFC: colony-forming cell which originates a colony of B lymphocytes.
BPA: burst-promoting activity.
BSA: bovine serum albumin.
CFC: colony-forming cell. A prefix denotes the line of differentiation, Ba = basophilic, Eo = eosinophilic, GM = granulocyte-macrophage, Meg = megakaryocytic, Mix-CFC = see below.
CFU: colony-forming unit. Differs from CFC to denote that only a proportion of the cells with the potential to form a colony may form one. Historically, it is derived from the CFU-S assay, in which only a proportion of the cells with such potential will lodge in the spleen and form a colony. It is also used when it is not known if the colony originates from only one cell.
CFU-E: colony-forming unit, erythroid.
CFU-F: colony-forming unit, fibroblastic.
CFU-S: colony-forming unit, spleen.
CM: conditioned medium, i.e. medium in which cells have been grown and into which they may have released growth factors.
CSF: colony stimulating factor. A prefix denotes the cell lineage stimulated. GM = granulocyte-macrophage, M = macrophage.
LCM: medium conditioned by leukocytes.
Mix-CFC: colony-forming cell which originates a colony containing cells of several lineages.
PHA-LCM: medium conditioned by leukocytes stimulated by phytohaemagglutinin.
TL-CFU: colony-forming unit which originates a colony of T lymphocytes.
FCA: fetal calf serum
Epo: erythropoietin
SCCM: spleen cell conditioned medium
LPS: lipopolysacharide

## HISTORICAL BACKGROUND

The development of in vitro clonal assays for haemopoietic cells has produced a dramatic increase in our knowledge and understanding of the structure and physiology of the haemopoietic tissue. Using these techniques, distinct populations of primitive cells have been defined by their capacity to give rise to recognisable progeny. Following the first reports (Bradley & Metcalf, 1966; Pluznick & Sachs, 1965) of the growth of colonies of neutrophilic granulocytes and macrophages in soft agar, techniques were developed for growing other classes of progenitor cells in vitro: erythroid (Stephenson et al, 1971; Axelrad et al, 1973), megakaryocytic (Metcalf et al, 1975), eosinophilic (Iscove et al, 1971; Metcalf et al, 1974) and basophilic (Nakahata et al, 1982a) cell assays have been described. More recently, the development of a technique for mixed colonies, containing erythroid, granulocytic, macrophagic and megakaryocytic cells from human as well as from murine tissues, has been achieved (Metcalf et al, 1979; Fauser & Messner 1978; 1979). As there is formal proof that the colonies in vitro are derived from single cells (Moore et al, 1972; Prchal et al, 1976; Cormack, 1976; Fauser & Messner, 1978; Metcalf et al, 1980) these clonal assays allow quantitative studies to be made of a whole range of haemopoietic cells, from multipotential cells not yet restricted to one lineage of differentiation, which give rise to colonies containing several thousands of recognisable mature cells, to progenitor cells restricted to one line of differentiation, and which may have very limited proliferation capacity. The former are close to (some perhaps overlapping with) stem cells, and the latter, to recognisable blast cells. In addition, 'stem cell' colonies have been described (Nakahata & Ogawa, 1982a). These colonies contain only 40–1000 cells which do not show signs of differentiation, but which on replating give rise to large numbers of secondary colonies which could be of the 'stem cell' type, mixed, erythroid or granulocytic. A general scheme of the haemopoietic system is shown in Figure 3.1. Although most of the work in this field has been carried out with murine or human cells, haemo-

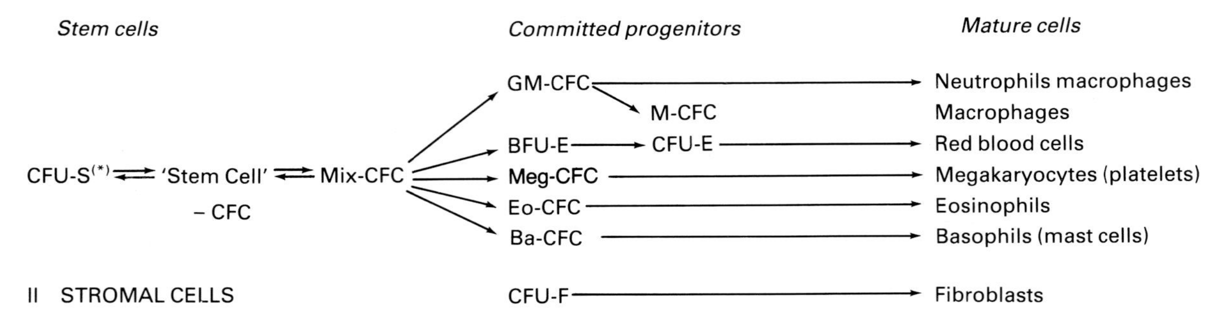


**Fig. 3.1** Scheme of clonal assays in haemopoiesis.
CFU = colony-forming unit; S = spleen; E = erythroid, F = fibroblastic. CFC = colony-forming cell. The prefix denotes the line of differentiation. BFU-E = burst-forming unit, erythroid. *Note*: CFU was used instead of CFC when the single cell-origin of the colonies had not yet been demonstrated, but continues to be used. * = in vivo assay (see Ch. 2).

poietic colony formation has been described in several species of mammals: dog (Adamson & Brown, 1978), sheep (Darbre et al, 1979), rabbit (Ohno & Fischer, 1977), rat (Bradley & Siemienovicz, 1968), cat (Testa et al, 1983) and monkey (Moore et al, 1972).

As the proliferation of colony forming cells in vitro is dependent on the continuous presence in the culture of stimulatory molecules which may be specific regulators, these colony assays have also been used to study the colony-forming cells with regard to sensitivity and specificity of response to such stimulators. In addition, colony assays have been used as test systems for the definition and purification of those factors, as well as for the investigations of regulatory networks (Guilbert & Stanley, 1980; Metcalf, 1981; Iscove et al, 1982; Bazill et al, 1983).

The role of stromal cells in the bone marrow in the maintenance of normal haemopoiesis is at present the subject of extensive studies. Although a functional test of a complex arrangement of different types of stromal cells can be performed in long-term bone marrow cultures (Dexter, 1979) only one clonal assay is so far available to study one type of stromal cell, the precursor of fibroblastic cells in the marrow (Friedenstein et al, 1976). This chapter describes the general requirements for the clonal cultures of haemopoietic and stromal cells, the specific requirements of some of the assays, and the problems most likely to be encountered when working with these techniques. In addition, the assays for cells which give rise to colonies of B and T lymphocytes (Metcalf et al, 1975), will be described briefly.

## METHODOLOGY

### Laboratory facilities

The general principles which apply to the culture of mammalian cells are of relevance for primary cultures of haemopoietic cells. Most laboratories use laminar flow hoods to do cell culture work. The hoods have to comply with safety regulations which are related to the kind of biological materials handled in them. The hoods which deliver filtered air to the working surface are designed to protect the procedure underway from air-carried contamination, but afford no protection to the experimental worker, and are known as sterile cabinets. These are used to set cultures with biological material which is considered safe for the operator (for example, bone marrow from experimental mice). Cultures from such materials may also be set up in clean rooms, with UV lights and/or filtered air, with only little risk of contamination. The handling of biological material which may carry risk to the operator should comply, in the UK at least, with the Safety at Work etc. Act of 1974. Laboratories have a list of rules which comply with the requirements of the act, and investigators starting in this field should familiarise themselves with the current regulations at their place of work. For the handling of human or other potentially dangerous (e.g. monkey) tissues, vertical flow cabinets or hoods which have an air flow high enough to retain aerosols within the cabinet are needed to protect the operator. Manufacturers now comply with these safety requirements in the designing of cabinets.

### Tissue culture materials

1. Glassware is used for the preparation and storage of culture components. A common problem encountered is incorrect washing after use: minute traces of detergent are enough to inhibit cell growth completely. Depending on the shape of the containers, from 5 to 10 rinses may be necessary after washing: in our laboratory culture tubes are rinsed in tap water 10 times. This is followed by soaking in distilled water for 1 hour, plus another

hour in double distilled water. Storage bottles with a wider neck require six rinses, the last two always in distilled and double distilled water. During washing, glassware should be completely filled with the washing solution. Trapped air bubbles would result in incomplete washing.

2. Plastic petri dishes are generally used for the plating of haemopoietic cells; for a final volume of 1 ml per dish, dishes of 3 cm diameter are adequate. When larger volumes per dish are desired, 3 ml can be plated in dishes of 5 cm diameter. Smaller volumes per culture (0.25–0.50 ml) may be cultured in wells. When small volumes are plated in relatively large dishes, evaporation will result in an increased molar concentration of the constituents of the medium, cell growth may be limited or stopped.

Bacterial or tissue-culture-grade petri dishes are adequate for the cultures in which the cells are grown in suspension in soft agar. For stromal fibroblastic cells, which grow attached to the surface of the culture vessel, tissue-culture-grade flasks are more convenient. When disposing of used materials which contain potentially dangerous biological material, current safety regulations have to be observed.

**Cloning techniques in general**

The cloning of cells in vitro involves plating a suspension of single cells from haemopoietic tissues, with adequate nutrients, in a balanced pH, together with the stimulatory factors needed for cell proliferation and/or differentiation in a suitable matrix (usually soft agar or methylcellulose). Following a period of incubation, the progeny of single cells (colony-forming cells) are observed as colonies in which mature haemopoietic cells can be identified. In this section, the technical aspects of each step of these clonal cultures will be discussed, starting with the general principles which are valid for all the different assays, and following in the next section with specific points related to particular assays.

*Preparation of cell suspensions*

The techniques for the preparation of cell suspensions from mouse bone marrow, spleen or fetal liver are described in detail elsewhere in Chapter 2, and are similar to those used for other experimental animals. It is of critical importance that single cell suspensions are plated for all clonal assays. In some circumstances (high cell concentration, numerous red cells), cell aggregates tend to form quickly. Repeated gentle pipetting immediately before plating will usually disaggregate cell clumps.

Aspirated samples from patients' bone marrow are collected in tubes which contain 10 ml of culture medium, (see below), plus 200 Units of preservative-free heparin. The resulting suspension is centrifuged for 10 minutes at 800 G, the supernatant is removed, and the cells resuspended in culture medium. For some assays (for example, CFU-E) a thorough separation of red cells may be required. This can be achieved by layering the bone marrow cell suspension on a 2 per cent solution of methylcellulose, and allowing the red cells to sediment through it for 30 minutes to 2 hours. The nucleated cells remain on top, and can be recovered, washed and resuspended for culturing. Similar results can be achieved with a 6 per cent solution of fibrinogen, or with commercially available Fycol or Percoll preparations. Alternatively, osmotic shock can be used to lyse the red cells without deleterious effect on the colony-forming cells.

*Storage of cell suspensions*

It is desirable to keep the time between preparation of cell suspensions and plating as short as possible. In most determinations that time can be kept below 30 minutes. However, some experimental designs may require a longer time; while some colony-forming cells (GM-CFC, BFU-E) survive at least 4–5 hours suspended in buffered culture medium containing 2–5 per cent serum, kept at 4°C on ice, others, like CFU-F, may show a rapid decrease with time (Xu et al, 1983). This decrease is not dependent on cell adherence to the container, and may be as high as 30 per cent after 1 hour.

**Culture medium**

Several synthetic culture media support adequate colony growth of haemopoietic cells. They are usually supplemented with serum, although, as will be seen below, work on serum-free or low-serum cultures is also being carried out. The media most commonly used for each culture assay are shown in Table 3.1. For an excellent review on the components of synthetic media, and the rationale for choosing the best conditions for a particular cell type, the reader is referred to Ham (1981). As important as the medium selected, is to ensure that the medium is prepared correctly, is adequately buffered and is reasonably fresh. The first point is especially important when concentrated medium is purchased and prepared as single strength medium in the laboratory. Even when it is prepared according to manufacturers instructions, it is necessary to check the osmolarity, which may vary as much as 50 mosm/l between batches.

The osmolarity of the medium has to be adjusted as necessary for the assay intended. The range which allows optimal colony formation may be different for different assays and for different species. Some examples are given in Table 3.2 The data refer to the osmolarity of the synthetic culture medium, not to the final plating mixture (which also contains horse or fetal calf serum). Buffering of the culture media is usually achieved by a bicarbonate-$CO_2$ system, in which the $CO_2$ is provided

**Table 3.1** Culture media most commonly used

| Assay | Species | Medium | Illustrative references |
|---|---|---|---|
| GM-CFC | Mouse | Dulbecco's | Metcalf, 1977 |
| | | Fischer's | Dexter & Testa, 1976 |
| | Human | McCoy's | |
| | | Alpha | Aye et al, 1975 |
| Eo-CFC | Mouse | Dulbecco's | Johnson & Metcalf, 1980 |
| Ba-CFC | Mouse | Alpha | Nakahata et al, 1982a |
| Meg-CFC | Mouse | McCoy's | Williams et al, 1981 |
| | Human | Iscove's | Messner et al, 1982 |
| | | Alpha | Vainchenker et al, 1979 |
| Mix-CFC | Mouse | Dulbecco's | Metcalf et al, 1979 |
| | | Alpha | Nakahata et al, 1982a |
| | Human | Alpha | Fauser & Messner, 1979 |
| | | Iscove's | Ash et al, 1981 |
| BFU-E, CFU-E | Mouse | Iscove's | Iscove et al, 1982 |
| | | NCTC 109 | Stephenson et al, 1971 |
| | | Alpha | Testa & Dexter, 1978 |
| | Human | Alpha | Gregory & Eaves, 1978 |
| | | Alpha | Nakahata et al, 1982a |
| CFU-F | Mouse | Fischer's | Friedenstein et al, 1982 |
| | | Alpha | Xu et al, 1983 |
| | Human | RPMI 1640 | Nagao et al, 1981 |
| | | Alpha | Howarth et al, 1982 |
| BL-CFC | Mouse | McCoy's | Metcalf et al, 1975 |
| | | | Kincade et al, 1976 |
| | Human | RPMI 1640 | Shredni et al, 1981 |
| TL-CFC | Mouse | Alpha | Ching et al, 1981 |
| | Human | Alpha | Price et al, 1977 |

**Table 3.2** Osmolarity of the culture medium

| Cell type | Species | Range (mosm/1) | Optimum | Reference |
|---|---|---|---|---|
| BFU-E, CFU-E GM-CFC | Mouse | — | 280 | Iscove et al, 1980 |
| BFU-E | Mouse | 280–310 | 300 | Testa, unpublished |
| | Human | 280–330 | — | |
| | Cat | 280–310 | — | |
| CFU-F | Mouse | 280–312 | — | Xu et al, 1983 |

by the gas mixture used in the incubator. In crowded cultures, enough $CO_2$ is produced by cellular metabolism to result in an acidified medium. This has not been found to be detrimental for colony growth (Metcalf, 1977). When a bicarbonate-$CO_2$ buffer system is used, detrimental changes to alkaline pH may occur when cultures are taken out of the incubator for repeated examination (this is not advisable unless there is a very good reason because of the risks of contamination, change in pH, and dehydration). One alternative is to add a synthetic organic buffer, like HEPES, while still conserving the bicarbonate-$CO_2$ system. The osmolarity should be adjusted if HEPES is added.

The storage times which are considered safe by manufacturers may be based on the capacity of culture medium to support the growth of cell lines which are adapted to conditions in vitro, and which may have less stringent requirements for growth than haemopoietic and bone marrow stromal cells in primary cultures. We aim to use medium within a week of preparation although one-month-old medium will still support growth of GM-CFC and BFU-E. However, the plating efficiency in the CFU-F assay for mouse cells is markedly decreased if 10-day-old medium is used (Xu et al, 1983). Although the plating efficiency may be partially restored by addition of extra vitamins and amino acids, this introduces an unwelcome source of variation into the assay.

**Serum**

Although some colony assays may be performed in supplemented synthetic medium without serum (see below), most assays are still performed in cultures which contain serum. Unfortunately, different batches of serum differ widely in their capacity to support colony growth, and this is one of the recurrent problems encountered in this work. Pre-testing of serum batches, using the system for which it is intended, is always necessary, as a good performance in one assay system does not ensure a similar result in other colony assays (or in the same assay in a different species) (Dexter & Testa, 1976). Furthermore, the capacity of a batch to support the growth of established cell lines is not a reliable indication of its usefulness in the systems described here. Historically, fetal calf serum has been the first choice for cell culture. However, we have found that in our laboratory, only the occasional batch supported the growth of GM-CFC from mouse, and that horse serum (which is cheaper) was consistently better (Dexter & Testa, 1976). Fortunately, conventional suppliers are becoming aware of the increasing use of bone marrow cultures in research, and it is now possible to purchase serum which has been pre-tested for a particular assay.

Another aspect which should be considered in selecting a batch of serum is the level of stimulatory (or inhibitory) material which may influence colony growth: for example, fetal calf serum may contain enough erythropoietin to stimulate optimum erythroid growth in mixed colonies from mouse fetal liver (Metcalf et al, 1979). It is also possible to select batches which have low levels of burst-promoting-activity (BPA) (see Table 3.3) so that an exogenous source of BPA can be titrated in a more sensitive system.

Another approach is to modify culture conditions in such a way that little or no serum is required for colony growth.

**Serum-free cultures**

The concentration of serum can be drastically reduced, and for some assays completely eliminated, by some additions to the culture medium (Guilbert & Iscove, 1976; Iscove & Melchers, 1978; Iscove et al, 1980). CFU-E and BL-CFC can be grown in the absence of serum in culture media supplemented with 10 mg/ml of bovine serum albumin, 3 mg/ml transferrin, 160 μg/ml cholesterol and 160 μg/ml soybean lipid. Similar conditions will result in satisfactory BFU-E growth in only 4 per cent and of GM-CFC in 2.5 per cent fetal calf serum (Iscove et al, 1982).

**Stimulation of colony growth**

Haemopoietic colony-forming cells proliferate in culture

**Table 3.3** Stimulating factors for haemopoietic colony-forming cells

| Assay | Expressed potential for differentiation | Progeny | Number per $10^5$ bone marrow cells | Incubation time (days) | Stimulation |
|---|---|---|---|---|---|
| Mix-CFC | multipotential | G, M, Meg, E | 5–20 (mouse) | 7 | factor(s) in CM (BPA?)* |
| | | | 1–15 (human) | 14–20 | factor(s) in CM (BPA?)* |
| early BFU-E | bi or multipotential | E, Meg | 10–30 (mouse) | 8–14 | factor(s) in CM (BPA?)* |
| | | E, Eo | 0.5–1 (human) | 14–20 | factor(s) in CM (BPA?)* |
| late BFU-E | unipotential | E | 50–100 (mouse) | 3–8 | BPA, Epo |
| CFU-E | unipotential | E | 100–500 (mouse) | 2 | Epo |
| | | | 100–200 (human) | 7 | |
| GM-CFC | bipotential | G,M | 100–200 (mouse) | 7 | GM-CSF |
| | | | 30–120 (human) | 7–10 | |
| M-CFC | unipotential | M | 50–100 (mouse) | 7–20 | M-CSF |
| Eo-CFC | unipotential | Eo | 5–10 (mouse) | 7–14 | Eo-CSF |
| | | | 15–60 (human) | 14–21 | |
| Bas-CFC | unipotential | Bas | 10–20 (mouse) | 16–20 | factors in CM* |
| Meg-CFC | unipotential | Meg | 5–40 (mouse) | 6–7 | Meg-CSF |
| | | | 10–50 (human) | 7–24 | |

* CM produced by stimulated spleen cells or by WEHI-3B cells (seel text)

G — granulocytes; M — macrophages; Meg — megakaryocytes; E — erythroid; Eo — eosinophils; Ba — basophils. 'Stem cell' colonies have not been included here, since they express their miltipotential capacity for differentiation only after subculturing (Nakahata et al, 1982a). CSF = colony stimulating factor; BPA = burst promoting activity; Epo = erythropoietin in response to stimulatory factors which are essential for colony growth, and which are believed to be specific regulators for haemopoiesis (reviewed by Metcalf, 1981). A list of factors, and their target cells, are shown in Table 3.3. While those acting on relatively late colony-forming cells are lineage-restricted (erythropoietin, macrophage colony stimulating factor) and can be purified without other contaminating moieties (Miyake et al, 1977; Guilbert & Stanley, 1980), those which act on the earlier colony-forming cells have not been separated in all the purification procedures carried out to date (Iscove et al, 1982; Bazill et al, 1983). This suggests that there may be one molecule (or a family of closely related molecules), which acts on early colony-forming cells of several lineages (erythroid, megakaryocytic, granulocytic) and also on the multipotential cells which originate mixed colonies. The biology and characteristic of those factors have been reviewed recently (Graber & Krantz, 1978; Metcalf, 1981; Iscove et al, 1982).

One of the most widely used sources of factor acting on early cells (essential for the growth of mixed colonies) is medium conditioned (CM) by cells from the mouse spleen stimulated by lectin. Either concavalin A (1 μg/ml) or pokeweed mitogen (1:3000–1000 final concentration) is added to a spleen cell suspension ($2 \times 10^6$ cells per ml, in culture medium with 5 per cent heat-inactivated human serum or fetal calf serum). The conditioned medium is harvested after 7 days, filter sterilised and kept frozen until needed (Metcalf & Johnson, 1978). The cells can also be cultured without serum, adding 800 μg/ml bovine serum albumin, 3 mg/ml transferrin, 100 μg/ml soybean lipid and 25 μg/ml cholesterol (Iscove et al, 1982). Another good source of CM for mouse cells is provided by WEHI-3B cells, grown with 0.1 per cent to 10 per cent fetal calf serum (Iscove et al, 1982; Dexter et al, 1980).

Granulocyte-macrophage colony-stimulating factor (GM-CSF) may be obtained from medium conditioned by a variety of mouse tissues: lung, muscle, uterus, bone shafts etc. (reviewed by Metcalf, 1977). CM from WEHI-3B cells is also a good source.

For human mixed, erythroid, and granulocyte-macrophage colonies, the most widely used source is medium conditioned by peripheral blood leukocytes stimulated by 1 per cent phytohaemagglutinin cultured for 7 days in synthetic medium and 10 per cent fetal calf serum (PHA-LCM) (Aye et al, 1975). Another source of GM-CSF is medium conditioned by human placenta (Metcalf, 1981). All the CM described vary in activity from batch to batch, and should be tested at several concentrations before use.

Erythropoietin (Epo) is necessary for the later stages of erythroid growth, both in pure erythroid and in mixed colonies. Although Mix-CFC and BFU-E are not themselves sensitive to Epo their progeny requires it for full maturation leading to recognition of the colonies (Iscove et al, 1982). In contrast, CFU-E require Epo not only to proliferate, but also to survive in culture.

The usual sources of partially purified Epo which are commercially available are plasma from anaemic sheep, or urine from patients with aplastic anaemia. One useful source of Epo is serum from mice made anaemic by phenylhydrazine (60 mg/kg i.p.) followed 24 hours later by 6 Gy whole-body irradiation. Blood is collected 8 to 10 days later. The serum may contain up to 40 Units of Epo per ml (Tambourin et al, 1973). Most Epo preparations contain inhibitors of colony growth, like endotoxin; they may also contain other stimulators of colony formation, like CSF or BPA,. This makes it necessary to titrate each batch before use. Optimum CFU-E growth is achieved at 0.1–0.2Units per ml (Gregory & Eaves, 1978). The requirements for the BFU-E assay may vary: some workers find it necessary to use up to 5–10 Units per ml. However, cultures which contain an optimum amount of BPA support good growth of BFU-E with 1 Unit of Epo per ml (Iscove et al, 1980).

Stimulation of murine B-lymphocyte colonies may be obtained either by adding to the culture endotoxin, its lipopolysacharide, or peritoneal macrophages, all of which stimulate B-cell proliferation, or just by using as a gelling agent unpurified Difco or Bacto agar which contain mitogens (Kincade et al, 1976; and see below). Human B-lymphocyte colonies are stimulated by irradiated T cells, and by PHA or pokeweed mitogen, as will be detailed below. Colonies of T cells develop in response to T cell growth factor (also called Interleukin II in the literature) which may be obtained commercially, and which is produced by mitogen-stimulated T cells. However, the culture systems used to date have relied on stimulation by PHA or Concavalin-A added directly to the cultures, or by medium conditioned by peripheral blood cells stimulated by PHA as described above for mixed colonies (Aye et al, 1975 Ching et al, 1981). The production of cloned cell lines of different populations of T lymphocytes in response to T cell growth factor is proving very useful and is widely used at present. However, as this is outside the scope of this book, the reader is referred to a recent review by Bach et al (1981).

Many substances which may be added to cultures will modify colony growth: they may be nutrients which supplement sub-optimal culture media (in cultures with or without serum), hormones that enhance colony growth, carrier molecules, etc. Inhibitors may be present in the serum, or be produced by accessory cell populations. For extensive reviews on these subjects see Metcalf (1977), Moore (1979), Testa (1979, 1982), and Kincade (1981).

### Matrix for colony growth

*Agar*

In our laboratory, Bacto agar (Difco) is prepared as a stock solution at 3 per cent by mixing agar powder with double distilled water and heating the mixture in a water bath at about 90°C for one hour, stirring occasionally. The solution is dispensed in aliquots of convenient volume (20–30 ml), autoclaved, and stored at 4°C until used. When needed for plating, it is melted in a bath of boiling water, and cooled to about 45°C before being added to the plating mixture to obtain a final concentration of 0.3 per cent.

In several laboratories, a fresh 0.6 per cent solution of agar is prepared when needed and boiled for 2 minutes. For plating it is mixed with an equal volume of double strength medium (held at 37°C) to reach a final concentration of 0.3 per cent (Metcalf, 1977). In this case, the agar solution also has to be held at 37°C before mixing, to avoid a decrease in plating efficiency due to the effect of heat on the colony-forming cells. Gelling will take place a few minutes after mixing. If either the culture medium or the agar solution is colder than 37° when mixed, uneven gelling may occur, which results in a liquid phase containing lumps of hard agar. If the laboratory is too hot, gelling may not occur. In that case, the culture dishes may be placed in a refrigerator for a few minutes before moving them to an incubator. It is always necessary to check that the gel has formed before they are placed in the incubator as gelling will not take place at 37°C.

It should be noted that agar solutions used for B-cell colonies should not be autoclaved, as this results in inactivation of mitogens which induce proliferation of B-cells.

*Methylcellulose*

Carboxymethylcellulose powder with a viscosity of 4000 cps is prepared as a 2.2 per cent stock solution. For 1 litre, a flask with 0.5 litre of double distilled water heated to just below boiling point is placed on a magnetic stirrer and 22 g of methylcellulose powder is dispensed on the surface. This mixture is then boiled for one minute and left to cool to 37°C before adding 0.5 litre of double strength culture medium. The solution is again placed on a magnetic stirrer and stirred overnight at 4°C. It can then be divided into aliquots and kept frozen for up to about 2 months. Some practice may be required before a good methylcellulose batch is prepared: an unclear or lumpy preparation will not allow colony growth. Each batch should be tested before being used for experiments: the optimal final concentration may vary between 0.8 per cent–0.9 per cent. It is advisable to thaw it for about 48 hours before needed. Other components of the plating mixture (for example, L-glutamine, L-asparagine, bovine serum albumin, sodium bicarbonate) are added 24 hours before the cultures are started. Any remaining solution may be used within a week.

*Plasma clot and collagen gel*

The use of a plasma clot for the growth of erythroid colonies was first reported by Stephenson et al (1971). The plating mixture (consisting of medium, serum, stimulatory factors, plus cells) contains 10 per cent of bovine embryo extract, and is cooled on ice. To this is added citrated bovine plasma, at a concentration of 10 per cent of the final volume. Aliquots are pipetted into wells immediately, since a clot forms quickly at room temperature. Thrombin (0.1 Units/ml) may be substituted by bovine embryo extract, which has on occasion been found to reduce the plating efficiency of colony-forming cells (Testa, unpublished data).

Gels of native collagen fibres have been used as a substrate for the growth of a variety of cell types (Schor, 1980), including haemopoietic colony-forming cells (Lanotte et al, 1981). Gels of Type I collagen, extracted from rat tail tendons are prepared by mixing 5.0 ml of an aqueous solution, 0.5 ml of 4.4 per cent sodium bicarbonate, 0.1 ml of 200 mM L-glutamine and 2.5 ml of horse serum, plus 4 ml of 10x concentrated tissue culture medium and the stimulatory factors desired (for example, CSF, BPA) in 1.5 ml of medium. All the materials are kept at 4°C. The cells to be cultured are added and mixed rapidly, and adequate aliquots plated in petri dishes. A convenient volume is 1 ml per dish of 3 cm diameter. Gelling occurs within 5 minutes at room temperature (Lanotte et al, 1981).

*Selection of a matrix*

The selection of the matrix for colony growth depends on the aims of the experiment: for practical purposes, agar is easy to prepare, store and manipulate, and achieves better immobilisation of cells in the gels than any of the other substrata discussed above. The last point is important to ascertain the single-cell origin of the colonies. On the other hand, it contains sulphated polysaccharides that stimulate colony formation from B lymphocytes of the mouse (Kincade et al, 1976). These colonies are indistinguishable by gross appearance from granulocyte-macrophage colonies. Methylcellulose allows easy retrieval of cultured cells for examination, but also allows a fair degree of cell mobility, which may result in colonies which are not clonal. As the proportion of those may be as high as 60 per cent (Singer et al, 1979), this may be an important problem when comparatively large cell inocula are used, as is the case for Mix-CFC in some techniques (see p. 35).

Plasma clots and collagen gels may be considered to be more physiological substrata than agar or methylcellulose, but both allow cell migration along the fibres of the matrix which may result in cell aggregation, for example, of stromal cells in collagen gel. While this may be desirable in some cases (Lanotte et al, 1981), it has to be taken into consideration when deciding on the matrix to be used: they will not be the best choice to study the effect of regulatory factors, because of possible interference with physiological responses. Another problem found with the plasma clot or the collagen gel is that either may contract and float on a liquid phase if the handling of the culture dish is not very gentle. Furthermore, when incubated for longer than 7–10 days, cultured cells (macrophages and granulocytes) may produce enzymes which will partially digest the matrix.

**Incubation**

Incubation is carried out at 37° in a fully humidified atmosphere. Tissue culture incubators gassed with 5 or 10 per cent $CO_2$ in air with a flushing system of pure $CO_2$ after the incubator door has been opened are generally used. The main problems encountered are drying of cultures and temperature gradients. If drying happens, the surface of the agar or methylcellulose becomes uneven, and dried rings may be seen around the edge of the dishes. Humidification may be improved by placing strips of filter paper along the walls, with one end immersed in the tray with distilled water at the bottom of the incubator. An excessive gas flow may contribute to a drier atmosphere, as can frequent opening of the incubator.

A way of avoiding evaporation is to cover the surface of each dish with a thin layer of paraffin oil, which does not stop gas exchange and is not toxic for GM-CFC and BFU-E growth (Potter & Capellini, 1983). In addition, this slows down changes in pH when cultures are removed from the incubator for repeated examination.

An increase of 0.1–0.3°C above 37°C may impair growth. Temperatures below 37°C, however, are better tolerated. BFU-E and CFU-E grow apparently normally at 33°C (Testa & Dexter, 1978). As an alternative to conventional incubators, airtight boxes, with 2 valves for gas inflow and outflow may be used. A few petri dishes filled with distilled water are enough to humidify a small box. Gassing for a few minutes with a gas mixture with 5 per cent $CO_2$ is enough to maintain an adequate pH for at least 2 weeks in a sealed box. This system has the advantage that if one box per experiment is used, the cultures remain undisturbed during the whole incubation period.

**Gassing**

Most workers use a mixture of 5 or 10 per cent $CO_2$ in air for the gas phase in incubators. However, decreasing the $O_2$ concentration to 5 per cent usually results in an increased plating efficiency by about a factor of 2–4 for GM-CFC (Bradley et al, 1978), BFU-E, CFU-E (Rich &Kubanek, 1982), and CFU-F (Xu et al, 1983).

**Scoring**

An Olympus dissecting microscope with a zoom lens is adequate for the scoring of most types of colony at 30x or 40x magnification. One exception is the CFU-E-derived colonies which may be composed of as few as 8 cells. They can be scored using an inverted microscope at 80–100 magnification.

**Cytological examination**

Individual colonies may be sampled by picking them out of the cultures using a microhaematocrit tube or a Pasteur pipette. The colonies are deposited on a glass slide and, after drying may be stained with 0.6 per cent orcein in 60 per cent acetic acid run under a coverslip (Metcalf, 1977). These wet preparations should be examined within 1–3 hours. Individual colonies may be squashed and stained permanently with May-Grunwald Giemsa as described by Testa & Lord (1970). Alternatively, whole agar gels may be slid off the petri dish, fixed on a slide and stained. Several stains have been tried; α-napthyl acetate-esterase, napthol AS-D chloroacetate-esterase, and Luxol Fast blue to identify monocytic, neutrophilic and eosinophilic colonies (Phillips et al, 1983).

## COLONY ASSAYS

Table 3.3 provides a summary of the different assays, together with some of the characteristics of the haemopoietic colony-forming cells.

**Mix-CFC**

The colony-forming cells able to generate large colonies containing several lineages of differentiated cells are called Mix-CFC, or GEMM-(granulocyte-erythroid-macrophage-megakaryocytic) CFC. Variations of the prefix denote the different cell lineages found in the colonies: for example, GMM (granulocyte-macrophage-megakaryocytic) (Metcalf et al, 1979; Fauser & Messner, 1979; Nakahata & Ogawa, 1982a). Growth of mixed colonies from murine cells depends on stimulatory factors present in medium conditioned by lectin-stimulated spleen cells or by WEHI-3B cells as described in the previous section.

Human Mix-CFC are stimulated by medium conditioned by PHA stimulated leukocytes. Bone marrow or peripheral blood cells may be separated using Percol, Fycol or bovine serum albumin to obtain a fraction of cells of density less than 1.077 g $cm^{-1}$ which contains the Mix-CFC. Cells are washed 2x, counted and plated at

$2 \times 10^5$ cells per ml in 0.9 per cent methylcellulose, 5 per cent PHA-LCM (pre-tested) $5 \times 10^{-5}$M β-mercaptoethanol, 1–2.5 Units Epo (pre-tested) and 30 per cent FCS in alpha or Iscove's medium. Triplicate 1 ml aliquots are cultured in 3 cm petri dishes for 14 days. An example of a mixed colony is shown in Figure 3.2A. Agar may be used instead of methylcellulose to support colony growth (Metcalf, personal communication, Howarth & Testa, unpublished).

Mix-CFC from murine haemopoietic tissues are cultured for 7 days in 0.3 per cent agar in Dulbecco's medium modified by the addition of L-asparagine (20 μg/ml and DEAE Dextran (75 μg/ml) as described by Metcalf et al (1979), with 12.5 per cent heat-inactivated human plasma (pre-tested) or 20 per cent FCS (pre-tested) and 20 per cent CM produced by lectin-stimulated spleen cells (SCCM). Cells from the CBA strain of mice produce mixed colonies with a recognisable erythroid component in cultures without exogenous erythropoietin. However, cells from other strains of mice require about 2 Units per ml of Epo (Nakahata & Ogawa, 1982a). Mixed colonies without any recognisable erythroid component (GMM-CFC) or small 'stem-cell' colonies, composed only of 50 to a few hundred cells have been found in similar cultures using methylcellulose instead of agar (Nakahata & Ogawa, 1982a; 1982b). This latter type of colony does not contain mature cells, but is composed of immature blast-type cells, and is characterised by having a high efficiency in originating new daughter colonies, including large mixed colonies, upon replating.

## BFU-E

The burst-forming units, erythroid, give rise to colonies which contain from several hundred to about $10^4$ cells after 8–14 days of incubation for murine cells and 14–20 for human, in cultures similar to those described above for mixed colonies. The BFU-E require a factor, burst-promoting activity or BPA, to proliferate in cultures in the earlier stages of colony development; the latter stages require erythropoietin to achieve full maturation of the erythroid cells. BPA is found in the same preparations used to stimulate mixed colonies; lectin-stimulated SCCM, WEHI-3B cells CM, PHA-stimulated leukocyte CM (PHA-LCM), etc. (see above). Although most workers use one of those exogenous sources of BPA, enough BPA may be provided by FCS or by bone marrow cells to ensure optimum burst formation (Wagemaker, 1978).

The BFU-E population encompasses a wide spectrum of cells, the earliest perhaps overlapping with the Mix-CFC: certainly, a large proportion of bursts, up to 40 per cent, contain megakaryocytes (McLeod et al, 1980). These large bursts are detected after the long incubation times stated above, and are probably better referred to as mixed colonies. More mature BFU-E which form pure erythroid colonies are detected at shorter incubation times, and are usually identified by a prefix which denotes the length of incubation, i.e. Day 3 — BFU-E, Day 8 — BFU-E (Gregory & Eaves, 1978). Figure 3.2C shows an example of an erythroid burst.

## CFU-E

These colony-forming cells are erythroid precursors more mature than the BFU-E, which give rise to small erythroid colonies of 8–60 cells after only 2 or 7 days of incubation for murine or human cells respectively (Stephenson et al, 1971; Iscove et al, 1974). They are grown in a plasma clot or in methylcellulose with smaller doses of erythropoietin than those used for CFU-E. The optimum dose varies from 0.01 Units per ml for mouse fetal liver, to about 0.2 Units for human bone marrow. As conditioned medium which contains BPA, necessary for the growth of BFU-E, does not inhibit BFU-E growth, both assays can be performed on the same cultures, scoring the relevant colonies at the appropriate times. Impure preparations of Epo may, however, inhibit CFU-E when used at the doses necessary for BFU-E growth. Colonies are recognised as tight clusters of haemoglobin-containing cells. Mouse erythroid cells are also much smaller than granulocytes or macrophages.

## GM-CFC

These cells, which give rise to colonies of granulocytes and macrophages, were the first haemopoietic cells to be grown in vitro in a clonal assay (Bradley & Metcalf, 1966). This assay is probably the most widely used and is carried out using a variety of culture conditions: murine GM-CFC are assayed in agar, methylcellulose or collagen gel, using synthetic media supplemented with 15–30 per cent of fetal calf or horse serum, and one of several sources of colony stimulating factor (CSF). These colonies may contain both granulocytes and macrophages or only one of those cell types. Impure CSF will give different proportions of GM, M or G colonies depending not only on the source, but also on the concentration of CSF (reviewed by Burgess & Metcalf, 1980). CSF that stimulates only M-CFC has been purified (Guilbert & Stanley, 1980). Preparations with highly purified GM-CSF, however, also stimulate mixed colonies and primitive bursts (Burgess & Metcalf, 1980). Human GM-CFC are assayed in agar or methylcellulose, usually with human placenta CM (Burgess & Metcalf, 1980) or PHA-LCM (Aye et al, 1975).

Murine colonies are scored after 7 days of culture, but human GM-CFC cells may take longer. Usually, the minimum size of a clone to be scored as a colony is 40–50 cells. Smaller clones are called clusters; they outnumber colonies, and are likely to be derived from a more mature population than GM-CFC (Metcalf, 1977).

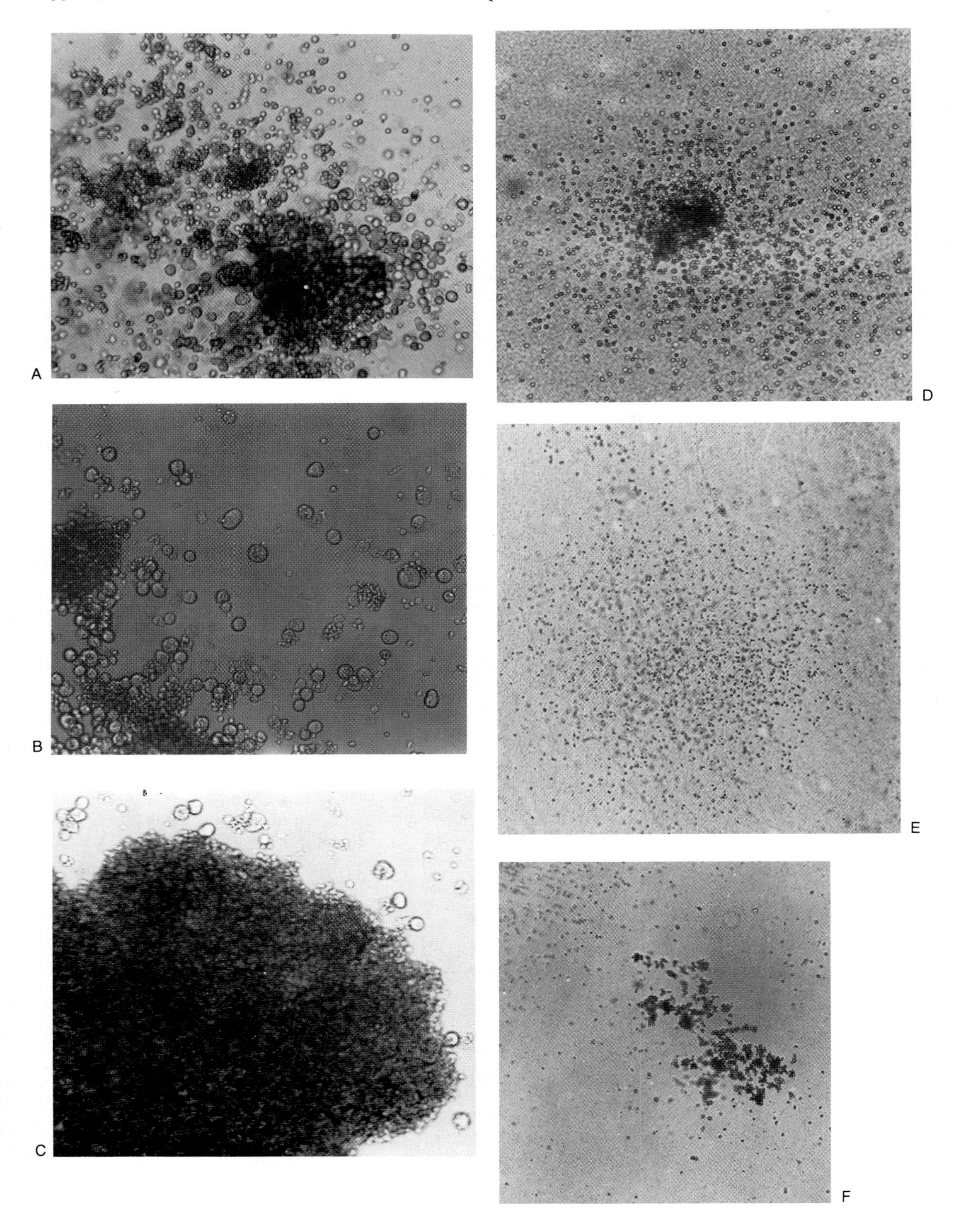
A
B
C
D
E
F

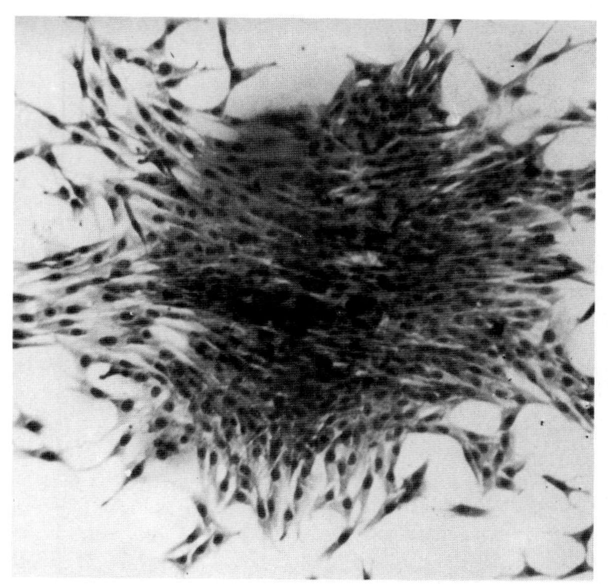


**Fig. 3.2A** An 11-days-old mixed colony derived from mouse bone marrow. Tight dark clusters of small haemoglobin-containing erythroid cells, large megakaryocytes and light granulocytes and macrophages can be seen.
**B** Part of a mixed colony with a large megakaryocytic component.
**C** Part of a large erythroid colony derived from human bone marrow. BFU-E derived colonies may also be composed of several smaller tight clusters of erythroid cells. Megakaryocytes are seen on the periphery of the colony.
**D** A 7-days-old granulocyte-macrophage colony derived from mouse bone marrow.
**F** Diffuse 7-days-old eosinophilic colony derived from mouse bone marrow.
**F** Multicentric 10-days-old eosinophilic colony derived from mouse bone marrow.
**G** A 7-days-old fibroblastic colony derived from cat bone marrow, stained with May Grunwald-Giemsa.

Granulocytic colonies disappear at longer incubation times, as eosinophilic or mast cell colonies emerge (see below). This may lead to a fairly constant total colony count, but one must be aware that the colony-forming cells originating from them belong to different subpopulations. For an extensive review, the reader is referred to Metcalf (1977). An example of a granulocyte-macrophage colony is seen in Figure 3.2D.

## Eo-CFC

Eosinophilic colonies have been described in cultures of murine and human bone marrow (Metcalf et al, 1974; Chervenick & Boggs, 1971). The colonies are usually either multicentric, composed of tight clusters or very difuse ones (Fig. 3.2E and F). They are grown in similar conditions to those for GM-CFC. However, Eo-CSF for murine cells appear to be different from GM-CSF: the usual sources are CM produced by lectin-stimulated spleen or lymph node cells, or WEHI-3B-CM (Metcalf et al, 1974). In optimal bone marrow cultures, Eo-CFC are about 5–10 per cent of the total number of CFC, and are 41–60 per cent of the total number found in human blood cultured for 18–20 days (Chervenick & Boggs, 1971).

On cytological examination, the colony cells contain the characteristic eosinophilic granules; both specific staining with Luxol Fast blue and electron microscopy have confirmed their nature (Johnson & Metcalf, 1980; Zucker-Franklin & Grusky, 1974).

## Ba-CFC

A clonal assay for mouse basophils (probably mast cells) has recently been described (Nakahata et al, 1982a). They grow in methylcellulose cultures similar to those for GM-CFC containing 10 per cent of medium conditioned by mitogen-stimulated spleen cells. The colonies take 16–20 days to reach a plateau in numbers, and are characterised in situ by the homogenous appearance of refractile cells with smooth contours. It is advisable, however, to confirm a presumptive diagnosis by cytological examination of the colony cells.

## Meg-CFC

Megakaryocytic colonies may be grown in 0.25 per cent agar in McCoy's medium supplemented with 25 per cent FCS, 0.4 per cent each of MEM essential and non-essential aminoacids, 2 mM L-glutamine, 16 $\mu$g/ml asparagine, 8 $\mu$g/ml L-serine, 1 mM sodium piruvate, $10^{-4}$M $\beta$-mercaptoethanol and $3 \times 10^{-7}$ prostaglandin $E_2$, plus a source of Meg-CSF (Williams et al, 1981). Several sources of CSF contain moieties which stimulate megakaryocytic colonies (Metcalf et al, 1975; Nakeff & Daniels-McQueen, 1976), including WEHI-3B-CM (Williams et al, 1981). It is important to notice that colony formation may be suboptimal at low cell inocula (less than $3 \times 10^5$ bone marrow cells) which results in loss of linearity between the numbers of cells plated and the colonies produced. A linear relationship may be restored by the addition of irradiated (9 Gy) bone marrow cells (Williams et al, 1981). Colonies usually contain less than 50 cells, and are readily identified by the large size of the individual cells. They may be picked out and stained with acethylcholinesterase (Nakeff & Daniels-McQueen, 1976). Megakaryocytic colonies from human bone marrow cells may be grown in a plasma clot as for murine colonies (Vainchenker et al, 1979; McLeod et al, 1980) substituting AB human serum for fetal calf serum, or in methylcellulose in Iscove's medium supplemented with 30 per cent fresh human plasma and 5 per cent of PHA-LCM (Messner et al, 1982). Colonies composed solely of megakaryocytes may be recognised after 7–10 days of culture, and continue to grow for a further 10 days, when they may contain 20 to 500 cells. There is no accepted criterion for the minimum size of a megakaryocytic colony. Some workers think that, because of the endomitosis in megakaryocytes, as few as 3 cells qualify to be counted as a colony.

## CFU-F

This clonal assay for a population of cells which give rise to colonies of fibroblastic cells was first described for guinea pig marrow (Friedenstein et al, 1970) and later for murine and human bone marrow (Friedenstein et al, 1976; Castro-Malaspina et al, 1980).

CFU-F from mouse bone marrow are grown in 5–10 ml of alpha medium plus 5–15 per cent pre-tested FCS, at a cell inoculum of (2–10) $\times 10^6$ cells. The cells are dispensed into flasks, gassed with 5 per cent $CO_2$ and 5 per cent $O_2$ in nitrogen, and are incubated for 7–10 days. The medium is replaced by fresh medium on day 7.

The fibroblastic colonies, which contain 50–1000 cells, grow attached to the surface of the flask. They are usually scored after fixing and staining. Murine CFU-F seem to be more sensitive than haemopoietic colony-forming cells to manipulation, changes in pH, aged culture medium and short-term storage before plating (Xu et al, 1983).

CFU-F from human bone marrow are cultured in medium (RPMI or alpha) plus 20 per cent FCS (Castro-Malaspina et al, 1980; Nagao et al, 1981) and are incubated for 8 days with a change of medium at day 4. There is no general agreement on the minimum number of cells that constitute a colony: some authors score a minimum of 6 cells (Nagao et al, 1981) which is perhaps too low, while other fix a more standard threshold of 50 cells (Xu et al, 1983). In colonies derived from murine bone marrow, macrophages and endothelioid cells may be found in association with fibroblasts. The endothelioid cells, however, are observed at later times (10 days or longer) and are rarely more than 10 per colony (Xu et al, 1983).

## BL-CFC

Colonies of murine B-cells are usually grown in 0.3 per cent soft agar in McCoys medium enriched with several additions (Kincade, 1983). Maximum plating efficiency is obtained using pre-tested fetal calf serum, usually added to obtain 15 per cent final concentration, plus sodium pyruvate, L-glutamine, MEM essential and non-essential aminoacids and vitamins. The growth of B-lymphocytes is dependent on the presence of mercaptoethanol, optimally at $5 \times 10^{-5}$M (Metcalf et al, 1975). In addition, mitogens should be present in the culture. If the gelling agar is Difco or Bacto agar, mitogens present in this unpurified agar will stimulate B-lymphocyte proliferation (Kincade et al, 1981). However, under these conditions colony formation may not be linearly related to the number of cells plated. Linearity may be restored by the addition of one or more agents: pre-tested endotoxin (10–20 $\mu$g/ml), its lipopolysacharide (LPS), or sheep red blood cells obtained commercially and used within one month of purchase at 0.05 ml of a 5 per cent suspension of red cells per ml of culture (Metcalf, 1977; Kincade, 1981).

B-cell colony formation has been reported in serum-free cultures, in medium enriched with serum albumin, transferrin and soybean lipid, or when stimulated by LPS and using methylcellulose instead of agar (Iscove & Melchers, 1978). This has the advantage that the action of putative regulators can be studied in cultures where the stimulation exerted by serum and agar has been eliminated.

B-cell colonies are scored after 6–7 days of incubation because they degenerate if incubated longer. If red blood cells have been added, they are lysed before scoring by the addition of a few drops of 3 per cent glacial acetic acid in water. If the B-cell colonies are derived from bone marrow, adequate controls should be run to ensure that GM-CFC derived colonies are not growing under the conditions used, as they have the same gross appearance and may be scored wrongly. Human B-cell colonies may be grown using a double layer technique: stimulating factors provided by medium conditioned by human peripheral blood cells stimulated by PHA or pokeweed mitogen are added to an underlayer of an agar concentration of 0.5 per cent. Within 24 hours, the test cells (BL-CFC) are plated on top in 0.3 per cent agar. BL-CFC are obtained from the mononuclear fraction of peripheral blood cells, which has been depleted of a large proportion of T-cells and has been cultured in suspension for 3 days. The growth medium for this step also contains mitogens at similar concentrations to those used in the underlayer (Shredni et al, 1981). It should be noted that complete depletion of T cells in the mononuclear cell population results in lack of colony growth. This may be corrected by the addition of lethally-irradiated T cells.

Some characteristics of the BL-CFC assay are shown in Table 3.4.

## TL-CFU

Most of the early assays necessitated a two-step culture system with a pre-incubation in a liquid phase in the presence of mitogens, before the cells were plated for colony formation. This has several disadvantages, in addition to the practical ones: there is cell death and cell replication, which are difficult to quantitate, and possible cell interactions which may alter the results of the assay. Therefore, only more recent one-step cultures will be described here.

Mouse bone marrow or spleen cells are plated on 0.8 per cent methylcellulose in alpha medium supplemented

**Table 3.4** Lymphoid colony-forming cells

| Assay | Progeny cells[a] | Incidence per $10^5$ cells | Source of cells | Incubation time (days) | Stimulation |
|---|---|---|---|---|---|
| BL-CFU | Igm+, IgG+ | 500–1000 | Mouse spleen or lymph nodes | 6–7 | Endotoxin, LPS, macrophages, RBC, mitogens present in agar |
| | | 50–100 | mouse bone marrow | | |
| | Igm+ | 20–30 | human mononuclear blood cells | 4–6 | PHA-M, PWM, conditioned medium |
| TL-CFU | Thyl+, Lytl+ | 60 | mouse bone marrow | 6 | PHA-LCM |
| | Thyl+, Lytl+,Lyt2+ | 25 | mouse spleen | 6 | PHA-LCM |
| | E-rosette + | 70–80 | human mononuclear blood cells | 5–7 | PHA-LCM or PHA |

[a] Markers expressed by a large proportion (30–90%) of colony cells. References given in the text. PWH = pokeweed mitogen

with 20 per cent pre-tested fetal calf serum, $5 \times 10^{-5}$M mercaptoethanol and 0.29 mg/ml of L-glutamine, plus 40 per cent of pre-tested conditioned medium from PHA-stimulated human leukocytes (PHA-LCM) and incubated for 6 days. At that time, lymphoid colonies appear as tight spheres, floating in the methylcellulose and containing 1000–5000 cells. About 10 per cent of the colonies derived from spleen and 50 per cent from bone marrow are derived from GM-CFC, and can be detected by their dispersed nature. This can be confirmed by staining for peroxidase, as the lymphoid colonies are negative (Ching et al, 1981). Some characteristics of TL-CFU are shown in Table 3.4. Human mononuclear cells, separated by a Ficoll-Hypaque density gradient, give rise to T cell colonies in a similar culture. The addition of 0.5 per cent PHA may replace the use of PHA-LCM (Price et al, 1977).

### General comments

As should be obvious from the above, there rarely is one 'best' recipe to grow any type of colony or one best time to score it. Good results depend on following the general principles stated in the methods section, rather than on any specific 'magic' step. Pre-testing of sera, BPA, CSF, Epo, etc., to select conditions that support best growth, and to ensure the optimum concentration of stimulators, are necessary steps in standardising assays. The selection of synthetic medium is in general less critical than to ensure its freshness and correct preparation. Furthermore, it is likely that many of the additions used in serum-containing cultures are not essential for colony growth, and will only restore colony formation to the levels achieved with freshly prepared medium, supplemented by a good batch of serum. However, serum-free cultures are preferable for some experimental designs, for example, when investigating the sensitivity of colony-forming cells to regulators.

## STRUCTURE & REGULATION OF THE HAEMOPOIETIC SYSTEM

An orderly chain of processes involving cell proliferation, differentiation and maturation leads to the production of functionally mature blood cells and platelets from the haemopoietic stem cells. The earliest haemopoietic cell type which can be studied experimentally is the murine CFU-S (see Ch. 2). CFU-S may fulfill the definition of multipotential stem cells: they have extensive capacity for self-reproduction and for giving rise to differentiated progeny. The closest cell types which can be studied by in vitro colony assays are the Mix-CFC and the cells that give rise to 'stem cell' colonies (Metcalf et al, 1979; Nakahata & Ogawa, 1982a). Both express self-reproduction capacity and multipotentiality, although the latter will only give rise to recognisable differentiated cells after subculturing. As CFU-S have been found in these colonies, there is no doubt that a certain degree of overlap exists amoung CFU-S, Mix-CFC and 'stem cell' CFC. The extent of the overlap is difficult to establish: it is known that CFU-S are heterogenous, and that not all of them fulfil the definition of stem cells (Schofield, 1978; Magli et al, 1982), some having matured and largely lost the capacity for self-renewal. The extent of the capacity to self-renew for Mix-CFC and 'stem cell'-CFC has been studied less extensively; generally only one and never more than two sequential replating experiments have been done to asses it, while for CFU-S, serial transplantations in vivo have been performed. However, available data make it feasible to propose a tentative hierarchy: CFU-S $\leftrightarrows$ 'stem cell'-CFC $\rightarrow$ Mix-CFC, for progression towards maturation (Fig. 3.1). In this context, it should be pointed out that 'stem cell'-CFC is probably not an adequate name for that cell type, unless it is established that it can originate the most primitive type of CFU-S, i.e. with high self-renewal capacity. If the latter is indeed the case, 'stem-cell'-CFC may precede CFU-S in the maturation sequence. The heterogeneity of Mix-CFC, and the observation that some may be restricted to a few lineages (for example GMM) places Mix-CFC after both the CFU-S and the 'stem-cell'-CFC.

More mature cells lose their multipotentiality, and become restricted to fewer, eventually to one, line of differentiation. During this progression, various combinations of differentiated progeny are recognised in single colonies: granulocyte (neutrophilic and/or eosinophilic), erythroid-macrophage-megakaryocytic (GEMM; Fauser & Messner, 1979), erythroid-eosinophilic (EEo; Nakahata et al, 1982b), granulocyte-macrophage-megakarypcytic (GMM; Nakahata & Ogawa, 1982b), erythroid-megakaryocytic (E-Meg; McLeod et al, 1980). The expression of multipotentiality, as assessed by the presence of recognisable cells in the colonies, and of self-reproduction, assessed by the capacity to generate daughter colonies when replating, has to be taken as a minimum estimate, as not all the potential is necessarily expressed in vitro. One example is the 'stem cell' colony, which is composed only of blast-type cells, but originates not only new similar colonies, but large Mix-CFC and CFU-S among others (Nakahata & Ogawa, 1982a).

As cells progress along the differentiation-maturation pathway, more mature progenitor cells, usually considered committed to one or two lines of differentiation can be investigated: BFU-E, GM-CFC (Fig. 3.1). Up to this point, the colony-forming cells respond to a factor (or a family of growth factors) which appear to act on multipotential as well as on early progenitors, in a lineage-independent fashion (Iscove et al, 1982). As cells mature they gradually become responsive to lineage-specific factors: Eo-CSF, M-CSF, erythropoietin, etc. This dual-regulation model is best illustrated by the erythroid lineage; early BFU-E are sensitive to BPA (one of the names for the lineage-indifferent factor (Iscove et al, 1982). As the BFU-E mature, they progressively acquire sensitivity to Epo, when the CFU-E stage is reached, sensitivity to Epo is at its peak, and sensitivity to BPA is lost (Iscove et al, 1982). A similar sequence may be postulated for the granulocyte-macrophage series, but is less well defined at present. For comprehensive reviews on CSF, see Burgess & Metcalf (1980) and Metcalf (1981).

The late progenitors (for example, CFU-E) have a more limited proliferation capacity and are restricted to one line of differentiation (Table 3.3). These are the most mature colony-forming cells assayed in vitro, and are placed close to the recognisable proerythroblasts in the maturation sequence.

In addition to the factors already mentioned, complex cell interactions are important in the regulation of haemopoiesis: macrophages, lymphocytes, mature granulocytes, exert complex influences (direct or indirect) on colony formation (for reviews, see Moore, 1979; Broxmeyer, 1982). Stromal cells in the bone marrow (fibroblasts, adipocytes, endothelial cells) are essential for haematopoiesis in long-term cultures (Dexter, 1979), although their roles are not known. The CFU-F assay is the only clonal assay available to study stromal cells. However, non-clonal cells also may be associated with these colonies (Xu et al, 1983). Work on the characteristics of CFU-F and their alterations in disease is being carried out in experimental systems and in human bone marrow cultures (Nagao et al, 1981; Friedenstein et al, 1982; Xu et al, 1983).

## LYMPHOID COLONY-FORMING CELLS

The processes involved in cell differentiation and maturation in the lymphoid tissue are not yet as well characterised as those in the haemopoietic system. Cloning assays for lymphoid cells tend to detect relatively mature cell populations. This has directed their application towards analysing the immune response, and the subpopulations of immunocompetent cells. The BL-CFC show characteristics of mature B cells: they express Ia antigen, appear late in embryonic development in mice, and are able to recirculate (reviewed by Metcalf, 1977). It is known that BL-CFC are heterogeneous: when two mitogens are used together in the assay, the number of colonies obtained may exceed the number stimulated by each mitogen alone, and may approximate the sum of those obtained with each agent used separately (Kincade, 1981). This suggest that different subpopulations may be stimulated by different agents.

The clonal origin of B cell colonies has been established by analysis of membrane immunoglobulin and the characteristics of antigen binding of the colony cells. T cell colonies grown from murine spleen and bone marrow appear to be produced from a $Thyl^-$ precursor cell (Ching et al, 1981), and contain $Thyl^+$ cells, which may show helper, cytotoxic (of more than one specificity) or suppressor functions. More than one cell type may coexist in one colony (Ching & Miller, 1980). In other culture systems, colonies are derived from $Thyl^+$ cells; they may be, as discussed by Ching et al (1981), derived from more mature cells.

Work with enzyme markers has shown that T-cell colonies may be of clonal origin (Gerassi & Sachs, 1976). However, data obtained with marker chromosomes and mixtures of cells with different $H_2$ types show that a TL-CFU may be composed of 1–3 cells (Ching & Miller, 1980; Ching et al, 1981). The heterogeneity of the TL-CFU population is also indicated by the different cell progenies found in the colonies.

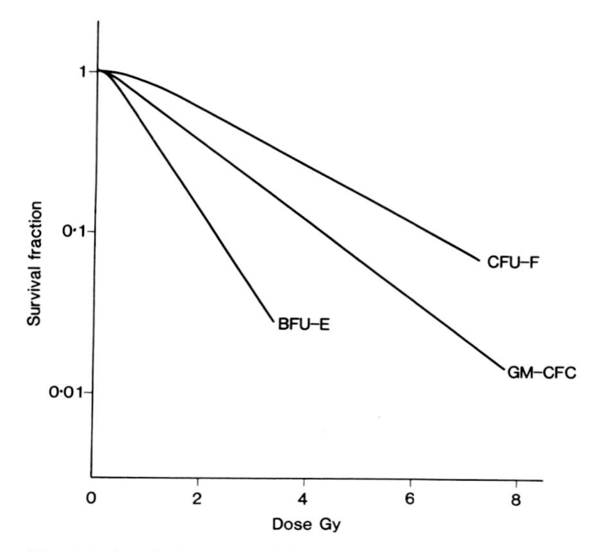


**Fig. 3.3** Survival curves of femoral colony-forming cells from mice after irradiation in vitro with γ-rays. Curves for GM-CFC and CFU-F adapted from Hendry & Lord (1983), and these represent the mean of published curves. The curve for BFU-E survival is that reported by Wagemaker et al (1979).

The $D_o$ of murine BL-CFC is 0.6 Gy (Metcalf et al, 1976). The radiosensitivity of TL-CFU varies according to the source of the cells: the $D_o$ is 0.98, 1.25 and 1.42 Gy for CFU derived from spleen depleted of Thyl+ cells, normal spleen or normal bone marrow respectively. There are also variations in the extrapolation number, which has a value of 1 in the first two cases and 1.5 in the last example (Ching et al, 1981).

Examples of survival curves after acute irradiation are shown in Figure 3.3. The data available on survival of GM-CFC and CFU-F from several species have been reviewed recently by Hendry & Lord (1983). For other haemopoietic colony-forming cells in vitro, little or no data have yet been reported.

The sensitivity of some populations may vary widely between species; for example, GM-CFC from dog bone-marrow have a $D_o$ of 0.61–0.70 Gy, showing greater sensitivity than mouse or human GM-CFC ($D_o$ = 1.37–1.60). Also, different culture conditions may select different subpopulations of colony-forming cells or result in suboptimal colony formation. Both factors may be reflected in the survival curves. A lower plating efficiency results in a lower sensitivity of murine CFU-F (Xu et al, 1983) and an increased sensitivity of human GM-CFC (Broxmeyer et al, 1976).

## CONCLUSION

A vast amount of knowledge has been accumulated since the first clonal assays for haemopoietic cells were described. The study of the colony-forming cells and of the regulators that influence them has helped our understanding of the normal haemopoietic system and its alteration in disease. This chapter has dealt with the problems encountered most frequently. The importance of technical points and the standardisation of assays to obtain a good plating efficiency of colony-forming cells cannot be emphasised too much. Some of the conflicting data in the literature would not have seen the light of day if those points had been taken into account.

## REFERENCES


Adamson J W, Brown J E 1978 Aspects of erythroid differentiation and proliferation. In: Papaconstantinou J, Rutter W J (eds) Molecular control of proliferation and differentiation. Academic Press, London, p 161

Ash R C, Detrick R A Zanjani E D 1981 Studies of human pluripotential hemopoietic stem cells (CFU-GEMM) in vitro. Blood 58:309

Axelrad A A, McLeod D L, Shreeve M M, Heath D S 1973 Properties of cells that produce erythrocyte colonies in vitro. In: Robinson W A (ed) Hemopoiesis in culture. DHWE Publication (NIH), p 226

Aye M T, Till J E, McCulloch E A 1975 Interacting populations affecting proliferation of leukemic cells in culture. Blood 45:485

Bach F H, Alter B J, Widmer M B, Segall M, Dunlap B 1981 Cloned cytotoxic and not cytotoxic lymphocytes in mouse and man: their reactivities and a large cell surface membrane protein (LMP) differentiation marker system. Immunological Reviews 54:5

Bazill G W, Haynes M, Garland J, Dexter T M 1983 Characterization and partial purification of a haemopoietic cell growth factor in WEHI-3 cell conditioned medium. Biochemical Journal 210:747

Bradley T R, Metcalf D 1966 The growth of mouse bone marrow cells in vitro. Australian Journal of Experimental Biology and Medical Sciences 44:287

Bradley T R, Siemienowicz R 1968 Colony growth of rat bone marrow cells in vitro. Australian Journal of Experimental Biology and Medical Sciences 46:595

Bradley T R, Hodgson G S, Rosendaal M 1978 The effect of oxygen tension on haemopoietic and fibroblast cell proliferation in vitro. Journal of Cellular Physiology 97:517

Broxmeyer H E 1982 Granulopoiesis. In: Trubowitz S, Davis S (eds) The human bone marrow. CRC Press, Florida, p 145

Broxmeyer H E, Galbraith P R, Baker F L 1976 Relationship of colony stimulating activity to apparent kill of human colony forming cells by irradiation and by hydroxyurea. Blood 47:403

Burgess A W, Metcalf D 1980 The nature and action of granulocyte-macrophage colony stimulating factors. Blood 56:947

Castro-Malaspina H, Gay R E, Resnick G, Kapoor N, Meyers P, Chiarieri D, McKenzie S, Broxmeyer H E, Moore M A S 1980 Characterization of human bone marrow fibroblast colony forming cells (CFU-F) and their progeny. Blood 56:289

Chervenick P A, Boggs D R 1971 In vitro growth of granulocytic and mononuclear cell colonies from blood of normal individuals, Blood 37:131.

Ching L M, Miller R G 1980 Characterisation of in vitro T lymphocyte colonies from normal mouse spleen cells: Colonies containing cytotoxic lymphocyte precursors. Journal of Immunology 124:696

Ching L M, Muraoka S, Miller R G 1981 Differentiation of T cells from immature precursors in murine T cell colonies. Journal of Immunology 127:2156

Cormack D 1976 Time-lapse characterization of erythrocytic colony forming cells in plasma cultures. Experimental Hematology 4:319

Darbre P D, Adamson J W, Wood W G, Weatherall D J, Robinson J S 1979 Patterns of globin chain synthesis in erythroid colonies grown from sheep marrow of different developmental stages. British Journal of Haematology 41:459

Dexter T M 1979 Cell interactions in vitro. Clinics in Haematology 8:453

Dexter T M, Testa N G 1876 Differentiation and proliferation of haemopoietic cells in culture. In: Prescott D (ed) Methods in cell biology. Academic Press, New York, p 387

Dexter T M, Garland J M, Scott D, Scolnick E, Metcalf D 1980 Growth of factor-dependent hemopoietic precursor cell lines. Journal of Experimental Medicine 152:1036

Fauser A A, Messner H A 1978 Granuloerythropoietic colonies in human bone marrow, peripheral blood and cord blood. Blood 52:1243

Fauser A A, Messner H A 1979 Identification of megakaryocytes, macrophages and eosinophils in colonies of human bone marrow containing neutrophilic granulocytes and erythroblasts. Blood 53:1023

Friedenstein A J, Chailakhjan R K, Lalykina K S 1970 The development of fibroblast colonies in monolayer cultures of guinea pig bone marrow and spleen cells. Cell & Tissue Kinetics 3:393

Friedenstein A J, Gorskaja U F, Kulagina N N 1976 Fibroblast precursors in normal and irradiated mouse haemopoietic organs. Experimental Hematology 4:267

Friedenstein A J, Latsinik N W, Grosheva A G, Gorskaya U F 1982 Marrow microenvironment transfer by heterotopic transplantation of freshly isolated and cultured cells in porous sponges. Experimental Hematology 10:217

Gerassi E, Sachs L 1976 Regulation of the induction of colonies in vitro by normal human lymphocytes. Proceeding of the National Academy of Sciences, USA 73:4546

Graber S E, Krantz S B 1978 Erythropoietin and the control of red cell production. Annual Revue of Medicine 29:51

Gregory C, Eaves A C 1978 Three stages of erythropoietic progenitor cell differentiation distinguished by a number of physical and biologic properties. Blood 51:527

Guilbert L J, Iscove N N 1976 Partial replacement of serum by selenite, transferrin, albumin and lecithin in haemopoietic cultures. Nature 263:594

Guilbert L J, Stanley R 1980 Specific interaction of murine colony stimulating factor with mononuclear phagocytic cells. Journal of Cell Biology 85:153

Ham R G 1981 Survival and growth requirements of non-transformed cells. In: Baserga R (ed) Tissue growth factors. Springer-Verlag, New York, p 13

Hendry J H, Lord B I 1983 The analysis of the early and late response to cytotoxic insults in the haemopoietic cell hierarchy. In: Potten C S, Hendry J H (eds) Cytotxic insult to tissue. Churchill Livingstone, Edinburgh, p 1

Howarth C, Morris-Jones P, Testa N G 1982 Long-term bone marrow damage in children treated for ALL. Evidence from in vitro colony assays (GM-CFC and CFU-F). British Journal of Cancer 46:918

Iscove N N, Melchers F 1978 Complete replacement of serum by albumin, transferrin and soybean lipid in cultures of lipopolysaccharide-reactive B-lymphocytes. Journal of Experimental Medicine 147:923

Iscove N N, Senn J S, Till J E, McCulloch E A 1971 Colony formation by normal and leukaemic human marrow cells in culture: effect of conditioned medium from human leukocytes. Blood 37:1

Iscove N N, Sieber F, Winterhalter K H 1974 Erythroid colony formation in cultures of mouse and human bone marrow; analysis of the requirement for erythropoiethin by gel filtration and affinity chromatography on agarose-concavalin. Journal of Cellular Physiology 83:309

Iscove N N, Guilbert L S, Wayman C 1980 Complete replacement of serum in primary cultures of erythropoietin dependent red cell precursors (CFU-E) by albumin, transferrin, iron, unsaturated fatty acid, lecithin and cholesterol. Experimental Cell Research 126:121

Iscove N N, Roitsch C A, Williams N, Guilbert L J 1982 Molecules stimulating early red cell, granulocyte, macrophage and megakaryocyte precursors in culture: similarity in size, hydrophobicity and charge. Journal of Cellular Physiology, Supplement 1:65

Johnson G R, Metcalf D 1980 Detection of a new type of mouse eosinophil colony by luxol-fast blue staining. Experimental Hematology 8:549

Kincade P W 1981 Practical aspects of murine B lymphocyte cloning. In: Adler W H, Nordin A A (eds) Immunological techniques applied to aging research. CRC, Boca Raton, p 85

Kincade P W, Ralph P, Moore M A S 1976 Growth of B-lymphocyte clones in semisolid culture is mitogen dependent. Journal of Experimental Medicine 143:1265

Lanotte M, Schor S, Dexter T M 1981 Collagen gel as a matrix for haemopoiesis. Journal of Cellular Physiology 106:269

Magli M C, Iscove N N, Odartchenko N 1982 Transient nature of early haemopoietic spleen colonies. Nature 295:527

McLeod D L, Shreeve M M, Axelrad A A 1980 Chromosome evidence for the bipotentiality of BFU-E. Blood 56:318

Messner H A, Jamal N, Izaguirre C 1982 The growth of large megakaryocyte colonies from human bone marrow. Journal of Cellular Physiology, Supplement 1:45

Metcalf D 1977 Hemopoietic colonies. Springer-Verlag, New York p 227

Metcalf D 1981 Hemopoietic colony stimulating factors. In: Baserga R (ed) Tissue growth factors. Springer Verlag, New York, p 343

Metcalf D, Johnson G R 1978 Production by spleen and lymph node cells of conditioned medium with erythroid and other haemopoietic colony stimulating activity. Journal of Cellular Physiology 96:31

Metcalf D, Johnson G R, Burgess A W 1980 Direct stimulation by purified GM-CSF of the proliferation of multipotential and erythroid precursor cells. Blood 55:138

Metcalf D, Johnson G R, Mandel T E 1979 Colony formation in agar by multipotential hemopoietic cells. Journal of Cellular Physiology 98:401

Metcalf D, MacDonald H R, Odartchenko N, Sordat B 1975 Growth of mouse megakaryocytic colonies in vitro. Proceedings of the National Academy of Sciences, USA 72:1744

Metcalf D, Parker J, Chester H M, Kinkade P W 1974 Formation of eosinophilic-like granulocytic colonies by mouse bone marrow cells in vitro. Journal of Cellular Physiology 84:275

Metcalf D, Wilson J W, Shortman K, Miller J F A P, Stoker J 1976 The nature of the cells generating B-lymphocyte colonies in vitro. Journal of Cellular Physiology 88:107

Metcalf D, Nossal G J V, Warner N L, Miller J F A P, Mandel T E, Layton J E, Gutman G A 1975 Growth of B-lymphocyte colonies in vitro. Journal of Experimental Medicine 142:1534

Miyake T, Yung C K, Goldwasser E 1977 Purification of human erythropoietin. Journal of Biological Chemistry 252:5558

Moore M A S 1979 Humoral regulation of granulopoiesis. Clinics in Haematology 8:287

Moore M A S, Williams N, Metcalf D 1972 Purification and characterization of the in vitro colony-forming cell in monkey haemopoietic tissue. Journal of Cellular Physiology 79:283

Nagao T, Komatsuda M, Yamauchi K, Arimori S 1981 Fibroblast colonies in monolayer cultures of human bone marrow. Journal of Cellular Physiology 108:155

Nakahata T, Ogawa M 1982a Identification in culture of a class of haemopoietic colony-forming units with extensive capacity to self-renew and generate multipotential hemopoietic colonies. Proceedings of the National Academy of Sciences, USA 79:3843

Nakahata T, Ogawa M 1982b Clonal origin of murine hematopoietic colonies with apparent restriction to granulocyte-macrophage-megakaryocyte (GMM) differentiation. Journal of Cellular Physiology 111:239

Nakahata T, Spincer S S, Canter J R, Ogawa M 1982a Clonal assays of mouse mast cell colonies in methylcellulose culture. Blood 60:352

Nakahata T, Spincer S S, Ogawa M 1982b Clonal origin of human erythro-eosinophilic colonies in culture. Blood 59:857

Nakeff A, Daniels-McQueen S 1976 In vitro colony assay for a new class of megakaryocyte precursor: Colony-forming unit megakaryocyte (CFU-M). Proceedings of the Society of Experimental Biology and Medicine 151:587

Ohno Y, Fischer J W 1977 Inhibition of bone marrow erythroid colony-forming cells (CFU-E) by serum-free chronic uremic rabbits. Proceedings of the Society of Experimental Biology and Medicine 156:56

Phillips P G, Chikkappa G, Brinson P S 1983 A triple stain technique to evaluate monocyte, neutrophil and eosinophil proliferation in soft agar cultures. Experimental Hematology 11:10

Pluznik D, Sachs L 1965 The cloning of normal 'mast' cells in tissue culture. Journal of Cellular Physiology 66:319

Potter C G, Cappellini M D 1983 Improved culture of BFU-E and CFC-GM by the use of an oil seal. British Journal of Haematology 54: 153–154

Prchal J F, Adamson J W, Steinman L, Fialkow P J 1976 Human erythroid colony formation in vitro: evidence for clonal origin, Journal of Cellular Physiology 89:489

Price G B, Teh H S, Minder M, McCulloch E A 1977 Induction of human T-lymphoid cytotoxic lymphocytes by components of leukocyte conditioned medium. Experimental Hematology Supplement 2:54

Rich I N, Kubanek B 1982 The effect of reduced oxygen tension on colony formation of erythropoietic cells in vitro. British Journal of Haematology 52:579

Schofield R 1978 The relationship between the spleen colony-forming cell and the haemopoietic stem cell. Blood Cells 4:7

Schor S 1980 Cell proliferation and migration on collagen substrate in vitro. Journal of Cell Science 41:159

Singer J W, Fialkow P J, Dow L W, Ernst C, Steinman L 1979 Unicellular or multicellular origin of human granulocyte-macrophage colonies in vitro. Blood 54:1395

Stephenson J R, Axelrad A A, McLeod D L, Shreeve M M 1971 Induction of colonies of hemoglobin-synthesizing cells by erythropoietin in vitro. Proceedings of the National Academy of Sciences, USA 68:1542

Sredni B, Sieckman D G, Kumagi S, House S, Green I, Paul W E 1981 Long-term culture and cloning of non-transformed human B lymphocytes. Journal of Experimental Medicine 154:1500

Tambourin P E, Wendling F, Gallien-Lartique O, Huaulme D 1973 Production of high plasma levels of erythropoietin in mice. Biomedicine 19:112

Testa N G 1979 Erythroid progenitor cells: their relevance for the study of haematological disease. Clinics in Haematology 8:311

Testa N G 1982 The erythroid system at the level of progenitor cells: structure and regulation. In: Trubowitz S, Davis S (eds) The human bone marrow. CRC, Florida, p 125

Testa N G, Lord B I 1970 A technique for the morphological examination of haemopoietic cells grown in agar. Blood 36:586

Testa N G, Dexter T M 1978 Production of erythroid precursor cells (BFU) in vitro. In: Murphy M J (ed) In vitro aspects of erythropoiesis. Springer-Verlag, New York, p 72

Testa N G, Onions D, Jarrett O, Frassoni F, Eliason J F 1983 Haemopoietic colony formation (BFU-E, GM-CFC) during the development of pure red cell hypoplasia induced in the cat by Feline Leukaemia Virus. Leukaemia Research 7:103

# 4

S. Hornsey

# The macro-colony assay in small intestine

## HISTORICAL BACKGROUND

In 1967, Withers developed an in situ colony technique which could be used to establish cell survival curves directly for epithelial cells in skin. He noticed that following ulcerative doses of X-rays given to the skin in radiotherapy patients 'islands' of regeneration could be seen in the area of ulceration. Applying this observation to experimental mice he showed that cell survival could be measured by counting the colonies arising after graded doses to a small field in an otherwise ulcerated area (Withers, 1967). The colonies could be counted with a hand lens. He later extended this macro-colony method to the small intestine of mice (Withers & Elkind, 1968, 1969).

The macro-colony technique developed by Withers for the small intestine (Withers & Elkind, 1968, 1969) as with that developed for skin, involved the growth of colonies in an otherwise ulcerated area of tissue. To ulcerate the surrounding tissue required a high radiation dose (16–30 Gy) and the macro-colonies took about 11–13 days to grow to sufficient size to be visible. As the doses of 16–30 Gy used in the macro-colony assay, if given to the whole animal or to the abdomen, would cause the animal to die in 4–5 days from the acute intestinal syndrome (Quastler, 1956), the irradiation could only be applied to an exteriorised portion of intestine whilst the rest of the animal was protected. Withers quickly realised that by microscopic examination, the damage could be assessed much earlier, indeed it could be assessed before death from the acute intestinal syndrome following whole-body irradiation. This meant that the technical complications of exteriorising the intestine could be avoided, enabling experimental workers to use larger numbers of animals and improve the statistical reliability of their data. It also enabled workers to investigate a lower dose response range which may have seemed more relevant to the dose ranges used in radiotherapy. For these reasons the macro-colony technique in the intestine has been rarely used. This is perhaps regrettable, because, if used in combination with the micro-colony assay it would have considerably increased the dose range which could have been investigated and would have given greater reliability to the assessment of the various parameters of the stem cell survival curve.

## TECHNIQUES AND METHODS

The macro-colony technique in the intestine involves the scoring of nodules of regenerating epithelial cells growing within a sterilised area at a time when they have grown sufficiently large to be detected by observation with minimal magnification (x 6 to x 15, using a dissecting microscope) and before they have grown sufficiently large to fuse into confluent colonies. It has been used only for the jejunum which can be exteriorised with greater ease than the ileum or duodenum. These latter lie closer to the backbone of the animal, i.e. dorsal to the jejunum, and more closely attached to the dorsal peritoneal wall.

### Irradiation techniques

A loop of jejunum approximately 1.3 cm in radius is exteriorised from the anaesthetised mouse, and placed on a supporting platform whilst the mouse is protected by lead from irradiation. A 'test' area in the centre of the loop with its supporting blood supply is protected by a lead strip whilst the areas lateral to this (the 'moat'

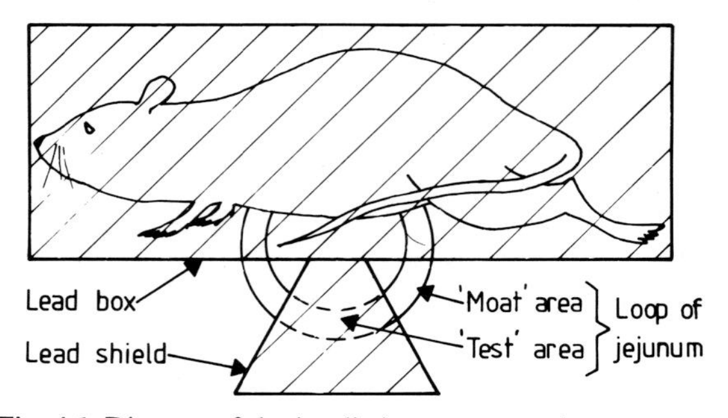


**Fig. 4.1** Diagram of the irradiation system used on an exteriorised loop of intestine in the macro-colony assay technique.

areas) are irradiated (Fig. 4.1). The lead strip is removed and the whole loop is X-irradiated with the 'test' dose. The doses are arranged so that the total dose to the 'moat' is 30 Gy which includes the 'test' dose. The lead strip protecting the central 'test' area of the loop of jejunum should not rest on the intestine or mesentery so that there is no impairment of the blood supply to the intestine. It should be supported clear of the intestine but as close as possible to give sharp edges of the radiation beam to the 'test' area. The thickness of the lead strip will depend on the radiation quality used for irradiating the 'moat' areas. A balance is needed between the thickness of the lead to give adequate protection to the 'test' area and thinness needed to reduce shadows at the edges of the lead strip which will give a fuzzy edge to the 'test' area if the lead is too thick. For 200 KVp X rays Withers found 1.7 mm lead adequate, in which case the dose to the protected area was less than one per cent.

The dose-rate should be checked in situ under the normal irradiation conditions. Tubes or discs of lithium fluoride or a similar substance for thermoluminescent dosimetry can be inserted into the loop of intestine so that any dose absorption or scattering by overlying saline soaked material, or back-scattering from the supporting platform can be taken into account.

During the irradiation, precautions must be taken to ensure that the exteriorised loop is kept moist, either by gassing both animal and loop, enclosed in a containing box, with a moisturised, warm gas mixture (Withers & Elkind, 1969) or by moisturizing the intestinal loop with saline during irradiation (Hornsey, 1970a). Precaution should always be taken to see that the anaesthetised animal in its protective box is adequately supplied with oxygen. A gas supply, to the box containing the animal, which flows out over the exteriorised loop, facilitates any study of the effects of oxygen concentration on radiosensitivity of the intestine (Hornsey, 1970b). In all the studies using the macro-colony technique on the mouse intestine, gas mixtures without five per cent $CO_2$ have been chosen because it has been shown that a better blood flow is maintained to the rectum (and possibly therefore to the rest of the intestine) by oxygen alone rather than oxygen with $CO_2$ (Kruuv et al, 1967).

The additional dose of radiation given to the 'moat' areas to top up the 'test' dose to the sterilising dose of 30 Gy can be given either before or after the 'test' dose. The order of 'test' and 'moat' doses does not affect sensitivity (Fig. 4.2). Different qualities of radiation, or different dose rates, may be used for 'test' and 'moat' doses providing the topping-up 'moat' dose is sufficient to sterilise the lateral edges of the irradiated area. Six to 30 animals per dose point were used, increasing the number of animals with dose, to obtain the cures shown in Figure 4.2.

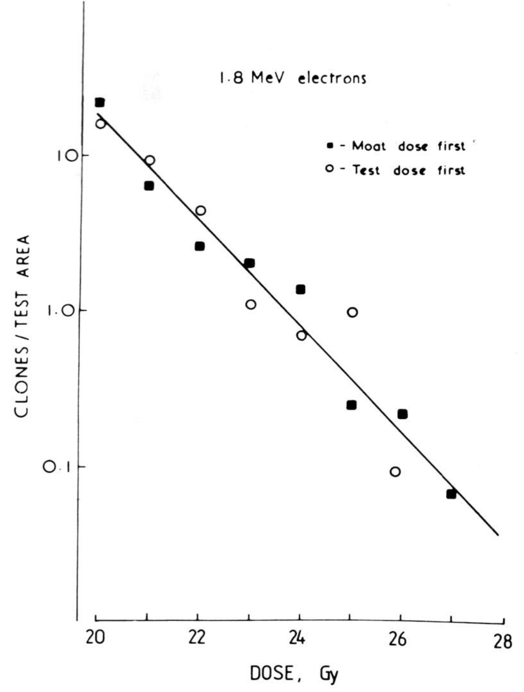


**Fig. 4.2** The mean number of clones produced in the central 'test' area after irradiation when the lateral 'moat' area has been irradiated immediately before [■] or after [○] the 'test' dose (from Hornsey, 1970a).

After irradiation, the loop of the intestine is eased back into the peritoneal cavity and the peritoneal wall and skin are sutured.

**Preparation of specimen**

After killing the animal, the irradiated loop of intestine is removed, opened longitudinally and gently washed. The mucilaginous film overyling the intestinal surface with the dying epithelial surface in the irradiated area, can be removed with vigorous washing after immersion for about 10 min in a solution of buffered pancreatin (five per cent pancreatin). When the mucilaginous cover and dying epithelium has sloughed off, the regenerating clones are clearly visible against the exposed subcutaneous surface. The specimen can be gently stretched and pinned on cardboard with the epithelial surface uppermost, and fixed. Fixation in Bouin's solution maintains the shade and texture difference between colony and sterilised or ulcerated surface so that the colony assay can be carried out without any further differential staining (Fig. 4.3). If necessary the specimens can be stored in 70 per cent alcohol after fixation.

**Colony assay, timing and dose range**

The time post-irradiation at which macro-colonies can best be counted is dependent on a balance between the

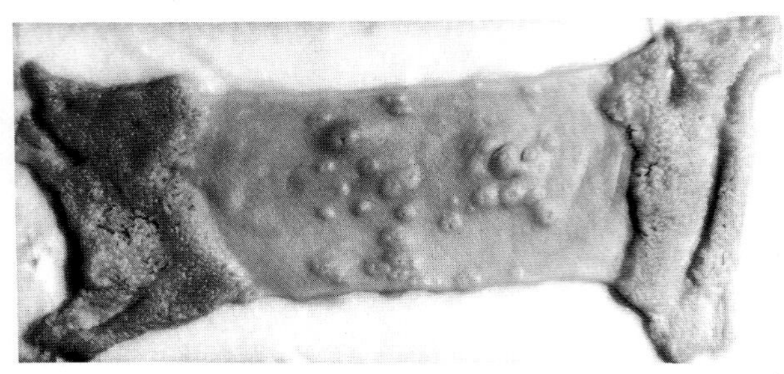

**Fig. 4.3** Macro-colonies at 13 days post-irradiation in a loop of jejunum (from Withers & Elkind, 1969).

developing ulceration of the irradiated area and the growth of the colonies from the surviving crypt cells into colonies large enough to be visible with minimal magnification under a dissection microscope. If the killing of the animal is delayed an intestinal obstruction may occur which may kill the animal; the intestinal muscles in the irradiated area go into spasm, causing the obstruction, when the dying epithelial cover is lost.

The irradiated loop of intestine is just detectable on opening the abdomen at 9–10 days post-irradiation by a reddening of the irradiated area which is slightly flaccid. At 11–13 days the irradiated loop is immediately obvious because of a tight contraction of the musculature both longitudinal and circular. This results in a shrinkage of the surface of the ulcerated area to between half to one-third of the total irradiated length.

The top end of the dose range which can be assayed is limited by the 30 Gy X-rays or fast electrons which will completely sterilise the loop of gut. At doses approaching this dose there will be so few colonies that large numbers of animals would be needed to give statistically reliable results. The original rationale for using the 'moat' was to prevent any longitudinal migration of epithelial cells into the test area from the edges of the unirradiated intestine adjacent to the irradiated loop. The 'moat' area free of any colonies is a very useful delineation of the 'test' area to be assayed. Regardless of whether there is any stretching or shrinking of the irradiated area during the development of damage and fixation of the specimen, the 'moat' defines a standard 'test' area which has been irradiated.

The lower end of the dose range which may be used will be largely determined by the time post-irradiation at which colonies can most easily be counted. The earlier the time, the smaller the colonies and the greater resolution of one colony from those adjacent to it so that more colonies per unit area may be counted.

The timing and the dose range which may be assayed may be strain-dependent. Withers & Elkind (1969) used hybrid CDF female mice with a post-irradiation fixation time of 13 days and assayed over a dose range of 15.5 to 22.5 Gy 200kVp X-rays. Hornsey (1970a & b) using $C_3H$ female mice, although assaying at 12 days post-irradiation, could nevertheless only assay over a dose range from 20 to 28 Gy fast electrons. If an RBE factor of 0.8 for the effectiveness of fast electrons relative to 200 KVp X-rays is accepted, then the dose range used in these latter experiments is equivalent to 16–22.5 Gy X-rays. Although the test area used by Withers & Elkind (1969) was 2 cm in length, twice the length used in the Hornsey study, they must have counted smaller colonies for they assayed up to 100 colonies/test area compared with 20/test area in the Hornsey studies. Colonies were visible under the dissection microscope at 11 days post-irradiation in the mice used in the Hornsey studies and if 11 days post-irradiation had been used as the assay time more colonies/test area could have been resolved and therefore lower doses could have been assayed also. In a recent study, Hume & Marigold (personal communication), using female HC/CFLP mice, found that after irradiation with 250 KVp X-rays to a test area 0.8 cm in length, colonies could be assayed at 11 days over a dose range of 15–24 Gy. By 12 days post-irradiation, in these mice, the colonies were confluent at the lower doses. It is clear that the colonies appear earlier in the HC/CFLP mice used by Hume & Marigold and the $C_3H$ mice used by Hornsey (1976) than in the CDF mice used by Withers & Elkind (1969).

The interpretation of the colony data from single and split dose experiments in the original papers of Withers & Elkind (1968, 1969) describing the macro-colony technique, are based on the assumption that the macro-colony arises from a single clonogenic stem cell in the intestinal crypt. It was this assumption and the realisation that the crypts could be assayed histologically at much earlier times and within a lower dose range that led to the development of the micro-colony assay (Withers & Elkind, 1970; see also Ch. 5).

**Analysis of results**

Histologically the macro-colonies are areas of actively proliferating mucosal epithelial cells. By giving graded doses of radiation to the 'test' area a relationship between the number of nodules and the dose can be established. Because the 'test' dose used in the macro-colony technique is large it is probable that the nodules or colonies arise from single clonogenic stem cells in the crypts of the small intestine. Withers & Elkind (1969) estimated that a 2 cm section of jejunum contained about $10^7$ crypt stem cells. The maximum number of macro-colonies they counted was about 100, a surviving fraction of 1 in $10^5$. With survival levels of this order, the probability of more than one surviving stem cell being responsible for the origins of the macro-colony is negligible. Therefore, if it is assumed that the crypt stem cells survive independently, the dose/effect curve for macro-colony counts can be regarded as a cell survival curve for the crypt stem cells. With the micro-colony technique which covers a much lower dose range it cannot be assumed that each surviving crypt arises from one stem cell only, and a correction must be made using Poisson statistics (Withers et al, 1970).

**Comparison of macro- and micro-colony survival assay**

In a paper introducing the micro-colony technique for estimating cell survival in the small intestine, Withers & Elkind (1970) compared the dose effect curves obtained with the micro- and macro-colony techniques. These curves are shown in Figure 4.4. The micro-colony assay has had a Poisson correction applied to the colonies/circumference to give survivors/circumference and this was then multiplied by the number of crypts/2 cm length of jejunum to give the values for comparison with the macro-colony assay. The $D_0$ values using both the macro-colony and micro-colony assay are closely similar. For both techniques, irradiations in $O_2$ are more effective than in air. The fact that there was such close agreement in the values for $D_0$ obtained from the two techniques in this early study (Table 4.1) is remarkable considering the differences in irradiation geometry and dose rate (1.3 Gy/min for the micro-colony assay and 6.85 Gy/min for the macro-colony assay), and in anaesthesia, and in the blood flow which might be expected to be disturbed for the macro- and undisturbed for the micro-colony technique. It has been shown that slight changes in oxygenation or dose-rate can alter survival levels when respiration is reduced by anaesthesia (Hornsey, 1970b). In a study with neutrons (16 MeV d on Be) where small differences in oxygenation might not be significant, a single exponential curve fitted the survival data obtained from both micro-colony and macro-colony

**Table 4.1** Comparison of $D_0$ and n for mouse jejunum irradiated with 200 kVp X-rays derived from micro- or macro-colony assay or from the two methods combined. The $D_0$ values with standard errors are taken from Withers and Elkind (1970), and the values without errors are derived graphically from Fig. 4.4

| Gas | Assay | $D_0$ (Gy) | n |
|---|---|---|---|
| Air | Micro-colonies | 1.09 (0.99–1.22) | 70 |
| | Macro-colonies | 1.07 (0.98–1.12) | 30 |
| | Micro-& Macro- | 1.20 | 20 |
| Oxygen | Micro-colonies | 1.03 (0.92–1.15) | 80 |
| | Macro-colonies | 0.93 (0.86–1.02) | 170 |
| | Micro- & Macro- | 1.01 | 100 |

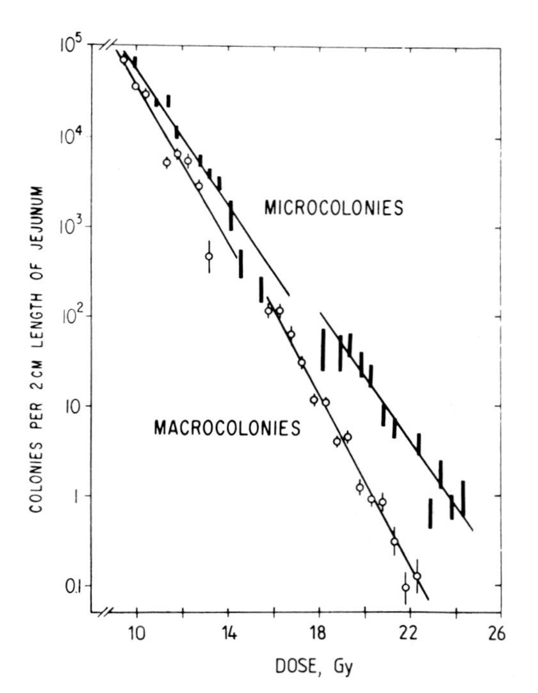


**Fig. 4.4** The dose-related number of micro-colonies and macro-colonies arising in mouse jejunum after irradiation in air or oxygen (from Withers & Elkind, 1970). Values of $D_0$ and n are given in Table 4.1.

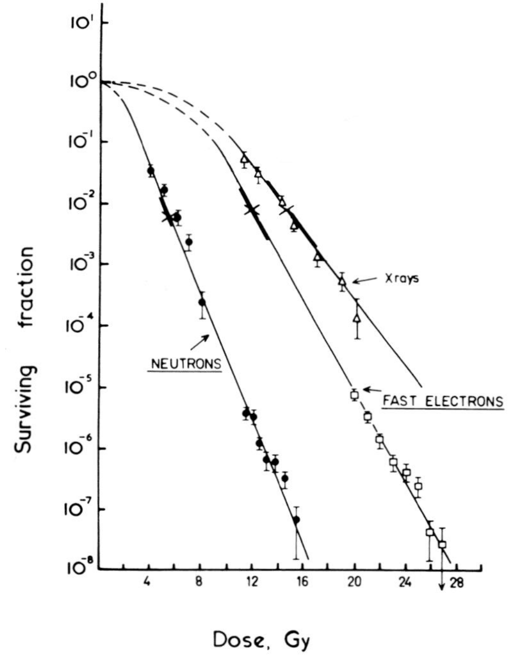


**Fig. 4.5** Jejunal stem cell survival curves following irradiation with neutrons (16 MeV d on Be), 1.8 MeV electrons, and 8 MeV X-rays. The neutron survival curve [$D_0$ = 0.82 (0.73–0.93) Gy; n = 6 (1.0–33)] was estimated from micro-colony and macro-colony counts, the electron survival curve [$D_0$ = 1.25 (1.05–1.30) Gy; n = 64 (4–1121)] from macro-colony counts only, and the X-ray survival curve [$D_0$ = 1.61 (1.49–1.77) Gy; n = 70 (30–164)] from micro-colony counts only (from Hornsey, 1973).

assay over the survival range of $5 \times 10^{-2}$ to $10^{-7}$ (Fig. 4.5) (Hornsey, 1973). When the data from the micro- and macro-colony techniques are combined, accurate average values of $D_o$, N, and $D_q$ can be calculated.

Masuda et al (1977) by analysis of micro-colony data after fractionated doses, have suggested that the stem cell survival curves following X-rays for jejunum, colon and gastric mucosa are gently but continuously bending even at high doses. The fractionation data on jejunum used in this analysis were for fractions separated by 3 hours (Withers et al, 1975) and the results may be subject to errors or artifacts introduced by the fractionation regimen. For example, the blood supply in the venules and arterioles carrying blood to the intestinal loop is reduced or interrupted for four hours post-irradiation following a dose of 9 Gy (Falk, personal communication). This disruption of the blood supply may be dose-related and affect survival levels for such short fractionation regimes. However, there is a tendency for micro-colony counts following single doses to be lower than the values on the fitted line at high doses (Withers et al, 1975; see Fig. 5.3). It is not clear whether this is real or associated with the fixed assay time and delays in repopulation at these high doses, as survival curves obtained by the macro-colony assay all appear to be exponential. Recently Moore & Maunda (1983) have shown that the crypts over Peyer's patches are more resistant to radiation than other crypts. This may lead to the survival at high doses in the macro-colony assay being dominated by survival of these crypts.

In converting colony counts from both micro- and macro-colony techniques into cell survival values, a value for the number of clonogenic or stem cells in the crypts has to be assumed. There is still some uncertainty about the size of the crypt stem cell population, values quoted varying from 12 (Zywietz et al, 1979) to 200 (Lesher & Baumann, 1968). A recent mean value of 80 estimated from published values has been quoted by Potten & Hendry (1983). In estimating the number of surviving cells from the colony counts a correction must be made at the higher survival levels in the micro-colony technique to allow for colonies arising from more than one cell.

The higher the survival, the greater the adjustment which is needed for the survival value, which steepens the survival curve. The correction can be made from Poisson statistics (Withers et al, 1970). Because of the low survival levels reached with the macro-colony technique no correction is needed in the estimate of the number of surviving stem cells. The $D_o$ value obtained directly from the macro-colony dose/effect curve will therefore be the same as that for the cell survival curve. The other parameters of the cell survival curve N and $D_q$ derived from the macro-colony assay however will be affected by the number of crypt stem cells which has been assumed, or which is deduced by other means (see Ch. 5). The micro-colony assay is now being used in studies on hyperthermia and radiation on the jejunum using exteriorised loops of intestine (Merino et al, 1978; Hume & Marigold, 1981).

For future experimental work on the sensitivity of the intestine to various toxic agents it seems likely that the micro-colony assay technique will continue to be the preferred technique. It is technically simpler, less time-consuming and needs fewer animals than the macro-colony technique. However, the macro-colony technique may still be useful in helping to set limits to stem cell numbers and cell survival parameters which have not yet been definitively determined. It also provides a more definitive test of the ability of cells to undergo repeated division, and produce a nodule of differentiating epithelium.

## ACKNOWLEDGEMENT

The author wishes to thank Dr H R Withers for the photograph shown in Figure 4.3.

## REFERENCES

Hornsey S 1970a Differences in survival of jejunal crypt cells after radiation delivered at different dose rates. British Journal of Radiology 43:802

Hornsey S 1970b The effect of hypoxia on the sensitivity of the epithelial cells of the jejunum. International Journal of Radiation Biology 18:539

Hornsey S 1973 The effectiveness of fast neutrons compared with low LET radiation on cell survival measured in the mouse jejunum. Radiation Research 55:58

Hume S P, Marigold J C L 1981 The response of mouse intestine to combined hyperthermia and irradiation: the contribution of direct thermal damage in assessment of the thermal enhancement ratio. International Journal of Radiation Biology 39:347

Kruuv J A, Inch W R, McCredie J A 1967 Blood flow and oxygenation of breathing gases containing carbon dioxide at atmospheric pressure. Cancer 20:51

Lesher S, Baumann J 1968 Recovery of reproductive activity and the maintenance of structural integrity in the intestinal epithelium of the mouse after single-dose wholebody $^{60}Co$ gamma-ray exposures. In: Effects of radiation on cellular proliferation and differentiate. IAEA, Vienna, p 507

Masuda K, Withers H R, Mason K A, Chen K Y 1977

Single dose response-curves of murine gastrointestinal crypt stem cells. Radiation Research 69:65

Merino, O R, Peters, L J, Mason, K A and Withers, H R 1978. Effect of hyperthermia on the radiation response of the mouse jejunum. International Journal of Radiation Oncology, Biology and Physics 4:407

Moore, J V, and Maunda, K K Y 1983. Topographic variations in the clonogenic response of intestinal crypts to cytotoxic treatments. British Journal of Radiology 56:193

Potten C S, Hendry J H 1983 Stem cells in murine small intestine. In: Potten C S (ed) Stem cells, their identification and clearacterisation. Churchill Livingstone, Edinburgh, p 155

Quastler H 1956 The nature of intestinal radiation death. Radiation Research 4:303

Withers H R 1967 The dose-survival relationship for irradiation of epithelial cells of mouse skin. British Journal of Radiology 40:187

Withers H R and Elkind M M 1968 Dose-survival characteristics of epithelial cells of mouse intestinal mucosa. Radiology 91:998

Withers H R and Elkind M M 1969 Radiosensitivity and fractionation response of crypt cells of mouse jejunum. Radiation Research 38:598

Withers H R and Elkind M M 1970 Micro-colony survival assay for cells of mouse intestinal mucosa exposed to radiation. International Journal of Radiation Biology 17:261

Withers H R, Brennan J T and Elkind M M 1970 The response of stem cells of intestinal mucosa to irradiation with 14 MeV neutrons. British Journal of Radiology 43:796

Withers H R, Chu A M, Reid B O and Hussey D H 1975. Response of mouse jejunum to multifraction radiation. International Journal of Radiation Oncology Biology and Physics 1:41

Zywietz F, Jung H, Hess A, Franke H D 1979. Response of mouse intestine to 14 MeV neutrons. International Journal of Radiation Biology 35:63

# 5

*C.S. Potten J.H. Hendry*

# The micro-colony assay in mouse small intestine

## HISTORICAL BACKGROUND

In 1967 Withers described an experimental approach for measuring the radiosensitivity of epidermal cells in mouse skin. This involved high radiation doses to ensure that surviving clonogenic cells could, by a series of clonal expansionary cell divisions, produce regenerative foci (clones or colonies) that by 2–3 weeks were large enough to be seen by the naked eye as small nodules in an otherwise denuded epithelium (see Ch. 7). A year later he had adapted the technique for use in the small intestine (Withers & Elkind, 1968; see Ch. 4). Briefly, this macro-colony approach in either tissue is technically limiting for the following reasons: (1) relatively few animals can be handled owing to the rather complex operational procedures and (2) only a limited range of high doses can be used. In both tissues about 2 weeks is required for the development of the macro-colonies and it was quickly realised that the colonies were visible using microscopic techniques at much earlier times, which solved many of the technical difficulties connected with the macro-colony assay (Withers & Elkind, 1970). Early points to be demonstrated using the micro-colony assay were that it provided measurements at lower doses than those possible with the macro-colony assay (Withers & Elkind, 1970; Hornsey, 1973), and that the reciprocal of the slopes of the survival curves, the $D_o$, was about 1.1 Gy with X-rays for both the micro- and macrocolony assay (1.09 and 1.07 Gy for irradiations in air respectively (Withers & Elkind, 1970; see Ch. 4) and 0.92 Gy for both assays with neutrons (Hornsey, 1973). The micro-colony technique has now been widely used, and 26 of the published examples have been reviewed recently (Potten et al, 1983). This review illustrated the variability observed for the position of the survival curves (up to a 10-fold difference in survival at 14 Gy), the fact that the overall average $D_o$ value was between 1.2 and 1.4 Gy, and that the data are presented in various forms. Part of the inter-laboratory variability is possibly due to the use of different mouse strains, slight technical differences and differences induced at the scoring stage.

## DESCRIPTION OF THE TECHNIQUE AND METHODOLOGY

The mice are irradiated usually to the whole-body (partial body may result in slightly different intestinal radiosensitivity as noted for colony — see below) with X or γ rays. They can be lightly anaesthetised for irradiation, or held in an approved restraining device which has adequate ventilation. This is usually a box or a series of tubes made out of tissue-equivalent material, which is perforated and fitted with adjustable closures. The dose-rate is important, and for acute exposures is generally above 2 Gy per minute so that the total exposure time is less than 10 minutes. Ideally the animals should be in a healthy unstressed condition since chronic infection, a variable intestinal fauna, particularly the presence of intestinal parasites and stress from noise and other environmental factors, may all influence the response of the animal. Irradiations in comparative experiments should be performed at a standard time of day, to minimise the effects of circadian variations (survival curves obtained at different times of the day show differences in both $D_o$ and n (Hendry, 1975; Potten et al, 1983)). After irradiation the animals are kept in their usual environment, and killed on day 3 to 3½. This time should be standardised for all experiments, and the intestines should be removed and placed in fixative as soon as possible to minimise post-mortem tissue degradation. A fixed portion of the small intestine can be chosen, e.g. jejunum or ileum, but in practice there is little difference in the response of different portions. The intestine can be sectioned without removing the contents, but some authors prefer to wash out the intestine first, with saline or fixative. The intestines are then processed for histology as outlined below. Segments of intestine must be accurately orientated to ensure that true cross sections are obtained. A convenient method of achieving this is to cut 10 small segments of intestine that are collected together and bound in a bundle with surgical tape (3M micropore surgical tape no. 1530). The tape can be pressed firmly around the 10 pieces that are carefully positioned into a pile (like a pile of logs). The edges of

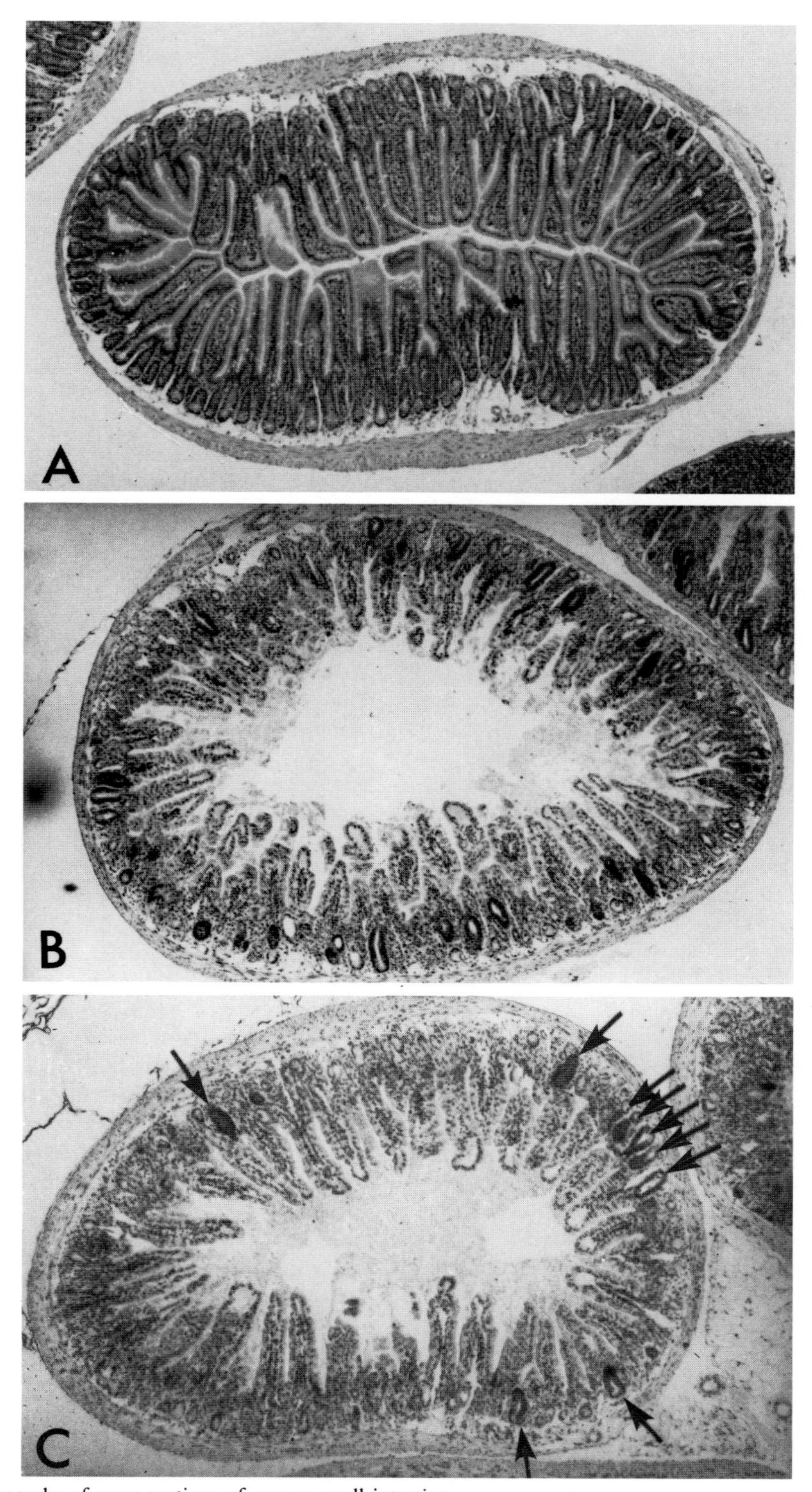


**Fig. 5.1** Photomicrographs of cross sections of mouse small intestine.
**A** Unirradiated controls (about 118 crypts per circumference on average)
**B** 3 days after 12 Gy (about 16 crypts per circumference on average)
**C** 3 days after 13 Gy (about 4 crypts per circumference on average) All × 53. These particular circumferences tend to have more crypts per circumference than the average and were chosen for clarity. Surviving crypts are marked by arrows in panel C.

this large manageable bundle can then be carefully trimmed with a scalpel blade and easily embedded. The tape and all 10 pieces can then be sectioned and processed according to standard histological procedures, providing 10 good cross sections per animal. Indeed, up to four such bundles (40 cross sections) can be embedded into one block and cut together. Thus, an entire experimental group of 4 animals can result in one microscope slide with 40 cross sections (or a group of 6 mice could make 2 slides each with 30 cross sections). Handling of many samples and orientation are both greatly facilitated by this approach. It also provides enough cross sections to count at least 8 per animal, since the occasional cross section might be damaged in sectioning or contain a large Peyer's patch and hence not be representative of the bulk of the small intestine in terms of the number of crypts or colonies per circumference. In our experience the alcohol-based fixatives, particularly Carnoy's [alcohol: chloroform: acetic (or proprionic) acid in ratio 6:3:1] are preferable to the formalin-based fixatives, and haematoxylin and eosin staining is effective in displaying surviving colonies which commonly exhibit a slight chromophilic characteristic.

Figures 5.1 and 5.2 illustrate the appearance of cross sections of the small intestine (in this case ileum) in unirradiated mice and in mice 3–3½ days after irradiation (typically 8–16 Gy). Gastrointestinal mortality does not usually occur earlier than the 3rd day after irradiation but rapidly becomes evident between the 3rd and 7th day after irradiation because the mouse LD50/7-days is about 10–11 Gy. Sterilised crypts continue to empty of cells for a period of about 3 days while crypts containing surviving clonogenic cells take 2–4 days to reform a structure resembling a crypt with a lumen, Paneth cells at its base, and numerous dividing cells. Paneth cells are radio-resistant and hence frequently mark the site of a sterilised crypt for several days. Thus it is clear that the optimum time for sampling and observing clones is on the 3rd or 4th day post-irradiation, in which case few animals succumb to gastrointestinal death, and whole-body irradiations can be used. Since the size of the crypt depends on the rate of cell division, which itself may depend on the treatment used or the state of the animals, the size will vary with time after treatment. The size will influence the probability of a given crypt or colony being seen in any given section and hence the size should be recorded or checked if comparisons are to be made on samples at different times after treatment or at a constant time after different treatments (Potten et al, 1981).

The number of colonies has to be expressed relative to some length or area measurement of intestine. The simplest approach is to express the number of colonies relative to a unit of length of intestine, with a convenient unit of length being a transverse cross section of the intestine (crypts per circumference). Measurements on whole mounts are possible (Khokhar & Potten 1980) but are not widely used. In this case the size of the colonies does not influence the counts of colonies per unit area. This approach provides similar survival curves to those generated from measurements of crypts-per-unit-length (crypts per circumference).

## THE CONSTITUTION OF A COLONY, AND PROBLEMS ASSOCIATED WITH ITS MEASUREMENT

There are a number of practical problems associated with this technique namely:

1. What constitutes a colony in terms of the numbers of cells, the cell types and their staining characteristics?
2. Do all colonies observed at 3 days really represent clonogenic-cell-derived (clonogen-derived) and also clonogen-containing colonies?
3. What is the statistical significance of colony size on the dose-response relationship (survival curves)?
4. How many colonies and intestinal circumferences should be counted?

### Colony definition

Withers & Elkind (1970) defined a surviving colony as one demonstrating the presence of 10 or more well-stained cells in a cluster in the section. This has been adopted by most users. Colonies with less than 10 cells could represent large colonies sectioned tangentially or small (slow-growing) colonies sectioned diametrically. Usually the cut-off is defined as 10 or more healthy-looking chromophilic non-Paneth cells, which would exclude remnants of dying crypts which tend to be less chromophilic, and would also exclude groups of Paneth cells that mark the position of sterilised crypts. There is some doubt whether all crypts as defined above represent growing colonies. This has been checked by pre-labelling with tritiated thymidine and scoring only crypts that contain labelled cells (Burholt et al, 1975; Boarder & Blackett, 1976). This generally gives results that differ little from those obtained without label (Potten, unpublished data), except possibly in some instances where drugs or other treatments have been used (Boarder & Blackett, 1976).

One relevant question is whether all colonies on day three ultimately proceed to re-epithelialise the mucosa with an epithelium that contains further clonogenic cells with a long-term survival capacity. Crypt survival does not change significant when it is assessed on days 3, 4 or 5, after the increase in size of the regenerating crypts (see below) is taken into account (Hendry et al, 1983).

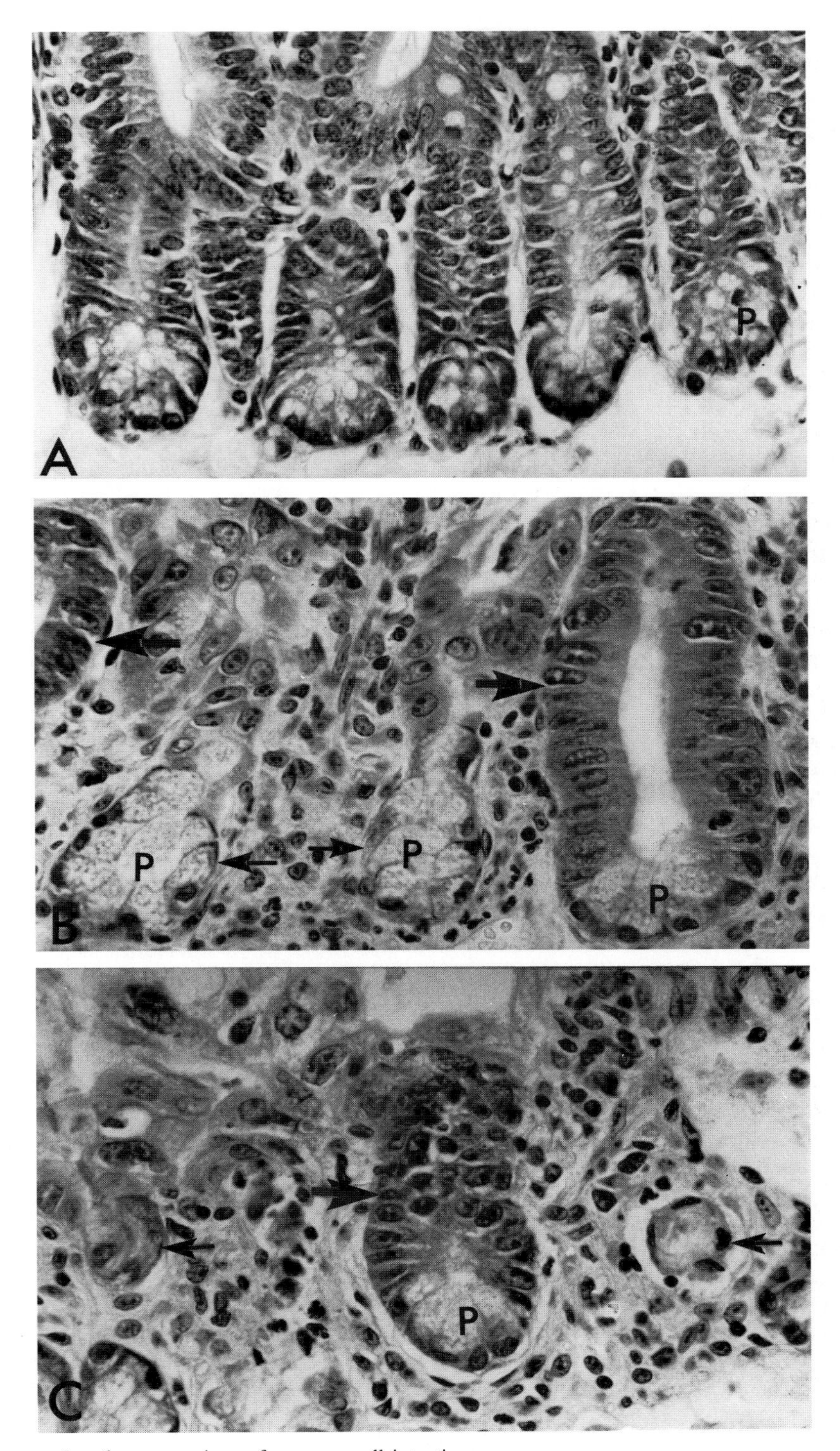


**Fig. 5.2** Photomicrographs of cross sections of mouse small intestine.
**A** Unirradiated control (about 118 crypts per circumference)
**B** 3 days after 12 Gy (about 16 crypts per circumference)
**C** 3 days after 15 Gy (about 1 crypt per circumference)
Surviving crypts (colonies) are marked by large arrows. Remnants of dying crypts, mainly Paneth cells (P) are marked by small arrows. All × 530.

This indicates that there is no net decrease due to the early appearance and subsequent disappearance of clones derived from early committed transit cells. Also, the data obtained by scoring colonies on day 3 appear to be part of the same overall survival curve as the data obtained by counting colonies on the 13–14th day (Withers & Elkind, 1970; Hornsey, 1973; reviewed by Potten et al, 1983 and Ch. 4). This is in spite of the fact that at times greater than day 7 there may be a division and multiplications of crypts (crypt budding) (Cairnie & Millen, 1975). However any new crypts must remain close close together and be presumably within the macrocolonies.

## Clonal origin of micro-colonies

The regenerative foci develop a shape that is, at least initially, similar to a normal crypt. These structures have been assumed by most investigators to be derived from single surviving cells. Eventually, (1) they may spread to cover a large area of epithelium which then contains all the differentiated cell types that characterise the intestine, (2) they are capable of providing further regenerative foci after a second, or several further, radiation exposures, and (3) they sustain the epithelium for the rest of the life of the animal (i.e. are capable of a large number of cell divisions, including the re-establishment of further cells capable of regeneration and self-maintenance ). These observations have led to the usage of the terms cryptogenic cells, or clonogenic cells to describe the cells involved in the formation of the regenerative foci. These imply that the crypt-like foci are indeed clones derived from a single surviving cell. This has yet to be proved for this system and hence it might be better to use the terms micro-colony, macro-colony, or crypt, forming units (MFU or clonogens). A crypt could possibly survive if (1) a single epithelial stem cell survives (a true clone), (2) if a single columnar epithelial stem cell together with a goblet cell precursor and a Paneth cell precursor survive, or (3) a group of columnar epithelial cells of a certain critical number survives (not a true clone). The differences between these possibilities cannot be distinguished at present but all could be covered by the term MFU or clonogen.

## Measurement problems

### *Crypt size*

If regenerating crypts (colonies) are on average larger or smaller than the crypts in unirradiated animals with which they are to be compared, e.g. as a surviving fraction, then there will be a greater or lesser probability respectively that the treated crypts will appear in any random section. This leads to the suggestion that a correction factor should be applied to the survival data (Hendry & Potten, 1974). The dimension that is most relevant is in fact the width of the crypt in the plane at *right angles* to that seen in a transverse section of the gut (i.e. the plane parallel to the longitudinal axis of the gut). The change in size can be seen dramatically when total cellularity is considered (Potten & Hendry, 1975). The necessity for a correction factor has been discussed in further detail elsewhere (Potten et al, 1981) and it was concluded that in many cases the correction can be ignored. However it may be important if:

1. times post-irradiation longer than about 3 days are studied
2. certain other treatments are used, e.g. some drugs where cell regeneration rates may differ
3. very thick sections are used
4. treatments that might affect the length of the gut are employed
5. the data are to be interpreted to extract detailed sensitivity parameters for the target cells (e.g. the extrapolation number is dependent on the correction factor).

All of these are particularly important when the data are expressed as surviving fractions and hence compared with unirradiated crypts. The correction factor (CF) is determined by measuring the components in the formula

$$CF = \frac{D_c - C_c + S}{D_t - C_t + S} \times \frac{L_t}{L_c}$$

where c = control, t = treated, D = diameter of the crypts in the section perpendicular to the section in which the crypts are counted, C = thickness of the two 'end-caps' of crypts that are not scored i.e. the segments containing less than 10 cells, S = section thickness, L = total length of the small intestine. The observed number of crypts per treated circumference is then multiplied by CF. In much of the early work (e.g. Hendry & Potten, 1974) an average value of CF was applied over the whole dose range, as the values showed no strong dependence on dose. More detailed studies over a wider range of dose have now shown that a separate value of CF should be applied to the data for each dose.

### *Number of sections to be counted*

In control animals the numbers of crypts per circumference among different sections from the same mouse are expected to vary very little from the mean value. Variations would be expected due to constrictions or distentions in the circumference before fixation, or differential shrinkage of the material *during* fixation. The mean values of crypts per circumference in controls range from about 90 to 160 in the literature with a mean of about 125.

After depletion, the distribution of crypts per circumference among different sections from the same mouse, or from different mice, would be expected approximately to follow Poisson statistics. In this ideal case the mean (M) should equal the variance which is the square of the standard deviation (SD). The standard error of the mean (SE) is ($SD/\sqrt{N}$), where N is the number of sections counted. Hence, if a given degree of accuracy of the mean is chosen, the number of sections required to be scored can be computed. If the mean is required to have a standard error of 5 per cent then $0.05M = SD = \sqrt{M}$, i.e. $N = 1/(M \times 0.0025)$. Thus, for values for the mean (M) of 50, 10, 5 and 1, the respective total number of sections to be counted would be 8, 16, 40, 80 and 400. With a standard error of 10 per cent, the number of sections would be reduced by a factor of 4. In practice, the distribution of counts can be checked to see whether these expectations are realised. An analysis by the present authors of one series of counts at various doses showed that the distribution of the mean counts between mice was very close to a Poisson distribution. However, the distribution of counts from different sections from the same mouse showed a consistently greater spread than expected. This indicates additional heterogeneity in the data, which could be due for example to differential shrinkage of the material during fixation of different parts of the intestine. Among different groups of 10 sections per mouse, the variance was 1.5 to 4 times the expected value (the mean). This indicates that the number of sections calculated based on the Poisson distribution should be regarded as a minimum number. Also, that the number of sections counted per mouse should be greater than the number of mice per group to equalize the contributions to the total standard error of inter-section and inter-mouse variation.

*Spatial distribution of surviving crypts*

Most investigators have used the assay after *irradiation*, (as an example of a cytotoxic agent), and they count the number of colonies (crypts) in the whole circumferences of transverse sections of the intestine and assume these to be representative of the whole region. The few sections which include part of a Peyer's patch are usually excluded as the intestine is often distorted in this region. With other cytotoxic agents where the delivery of dose to the intestine is not very homogeneous, this averaging procedure around a circumference can be misleading and has been modified. For example, with *heat* it has been noticed that crypts survive preferentially on the mesenteric side, probably mainly because cooling due to blood flow is better near the mesentery (Merino et la, 1978; Hume et al, 1979). In this case counts can be made in a 1 mm segment of the intestine at the mesentery as well as diameterically opposite to it (Hume et al, 1979). There are about 20 crypts per mm in controls and hence about 6 times as many sections would have to be scored for the same accuracy as obtained by scoring whole circumferences containing about 120 crypts.

With *cytotoxic drugs* there is usually a differential survival from one side of the intestine to the other (Moore & Maunda, 1983), which can be very marked, and this depends on the route of administration and the particular drug in use. With radiation, a differential has also been observed leading to a three-fold higher level of survival on the antimesentric side after 16 Gy. The reason for this is unknown, and experiments provided no evidence that it is due to slight hypoxia. On the mesenteric side, the distribution of surviving crypts after radiation is random (Hendry et al, 1983). This was concluded after using a 'runs' test and a test of the frequency of runs of 1,2,3 etc. surviving crypts assuming a Poisson distribution. Although Peyers-patch regions are normally excluded in the scoring procedures, it is interesting that the level of crypt survival after radiation was higher in this region, by a factor of 10 after 16 Gy compared to an average whole circumference (Moore & Maunda, 1983). Again, no explanation was found and there was no evidence that it was due to hypoxia.

## PRESENTATION OF DATA

It is quite feasible for some experimental purposes to compare two mean values of crypts per circumference, these being perhaps at two dose levels or at the same dose but with a modifying agent used with one group. However, it is more common to use a range of dose, and to construct a dose-response relationship. This allows further interpretation in terms of the detailed response of clonogens.

Various authors have presented their data in different forms. The relationship between these is shown in Figure 5.3, where:

A = the number of clonogens per crypt
B = the number of crypts per circumference in controls
C = the number of crypts per circumference after treatment.

The simplest presentation is to give the absolute number of crypts per circumference on a logarithmic scale versus dose (e.g. Hanson et al, 1979). This is similar to the ordinate labelled 1 in Figure 5.3, but in this particular example the ordinate is *not* truly logarithmic except for low levels of survival, because it is drawn so that the values correspond directly with those on the ordinates drawn on the right hand side of Figure 5.3. It is more common to present the data as surviving fractions S = C of the initial number B of crypts per circumference in controls (see ordinate 2 in Fig. 5.3). This implies that the regenerative foci are the same as unirradiated crypts in all respects including size. The latter is approximately

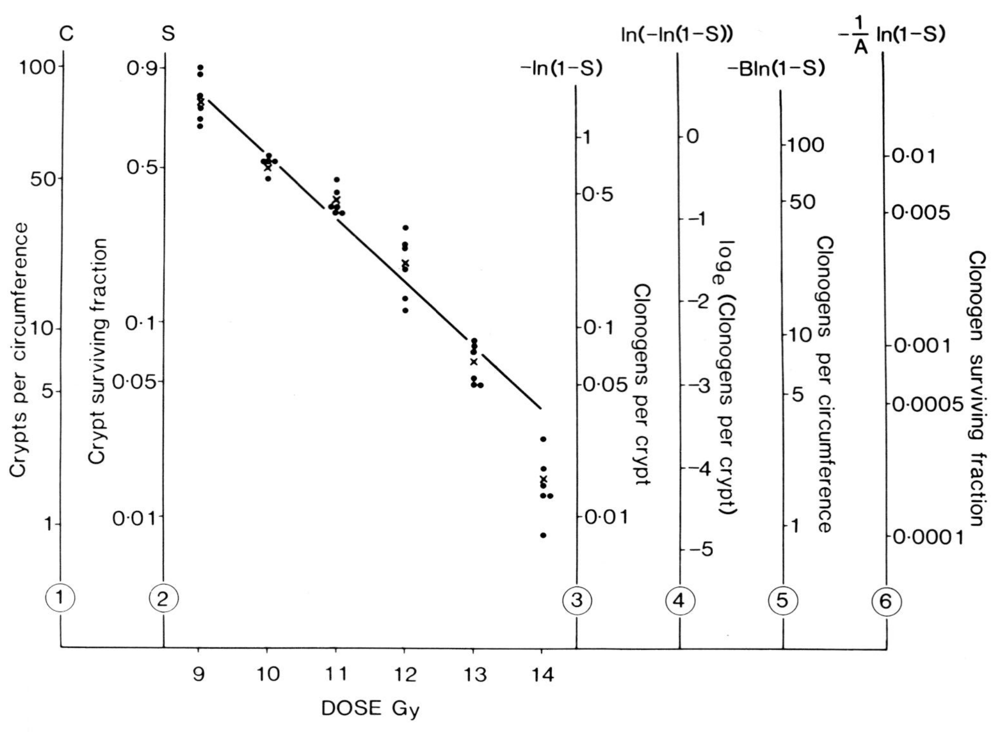


**Fig. 5.3** Schematic relationship of crypt survival to clonogen survival, using different methods of data presentation found in the literature.
A = number of clonogens per crypt
B = number of crypts per circumference in controls
C = number of crypts per circumference after irradiation
$S = \frac{C}{B}$ = surviving fraction of crypts.
Data points taken from Hendry et al (1983), using $B6D2F_1$ mice and acute doses of $^{137}Cs$ $\gamma$-rays. Each solid circle represents the mean value from 6 sections taken from a single mouse. Each cross represents the mean level of survival over the six mice at one dose. For explanation of the different methods of data presentation, see text. Note that ordinates 3,5 and 6 are logarithmic, ordinate 4 is linear, and ordinates 1 and 2 tend to logarithmic scales only at low levels of survival (at high levels of survival they are drawn simply so that the values correspond with values on the other ordinates).

true on the third day after irradiation, and at other times a suitable correction can be applied (discussed above).

The other alternative methods of data presentation shown on the right hand side of Figure 5.3 are based on the assumption that the distribution of counts is Poisson. This can be shown to be approximately true using reasonable sized samples employing several mice and several sections per mouse. In this case the fraction of crypts ablated (1-S) should be exp(-M) where M is the mean number of clonogens surviving per (original) crypt. Hence M = [-1n(1-S)]. In this case M could be plotted on a longarithmic scale against dose (ordinate 3 in Fig. 5.3), similar to a conventional cell survival curve. (It should be noted that the mean number of clonogens per *surviving* crypt (M′) is greater than M by a factor (1/S) i.e. M′ = M/S). An alternative version of ordinate 3 (Fig. 5.3), which involves no further parameters or assumptions, is ordinate 4. This is based on the double-minus-log transformation of the mortality probability (Gilbert, 1974), which is -1n [-1n (1-S)], or 1n [-1n(1-S)] if the signs on the ordinate are reversed. A full description of the derivation of this relationship can be found elsewhere (Lange & Gilbert, 1968; Gilbert, 1974; reviewed in Potten & Hendry, 1983). The relationship corresponds to the log of the number of clonogens surviving per crypt, i.e. 1n M. Hence, instead of plotting M on a log scale (ordinate 3, Fig. 5.3), 1nM is plotted on a *linear* scale. The units on the ordinate are integer values of 1n [-1n(1-S)] which run between 1 and -7 for common values of S between about 0.93 and at least 0.001. A change of one integer on the ordinate corresponds to a change in M by 1/e, i.e. a change in cell survival produced by a dose $D_o$.

Another method of presentation involves the number

of crypts per circumference in controls (B), where the mean number of surviving clonogens per crypt (M) is multiplied by B to give clonogens surviving per circumference (MxB) or [-B.ln(1-S)] (ordinate 5 in Fig. 5.3; Withers & Elkind, 1970). The last method (ordinate 6) involves a knowledge of the initial number of clonogens per crypt (A). If this is known, the mean number of clonogens surviving per crypt (M) can be expressed as a fraction of A so that conventional cell surviving fractions (M/A) or values of [– A.ln(1-S)] can be calculated and plotted on a log sclae versus dose (ordinate 6 in Fig. 5.3; Hornsey, 1973). However, there is still uncertainty about the value for A. The most recent mean value from many published estimates is about 80 (Potten & Hendry, 1983), but the value could be a high as 152 (Thames et al, 1981) or as low as 12 (Zywietz et al, 1979).

The data in any of the schemes of presentation can be fitted mathematically to common model curves for cell survival using methods described in Chapter 1. With data which are approximately exponential with dose, the $D_o$ value can be calculated together with the extrapolation number (E). It should be noted that there is evidence that the curve is gently-bending even at high radiation doses (Masuda et al, 1977), so that the calculated parameters $D_o$ and E should be regarded as average values applying over a particular range of dose. Also, the extrapolation number for ordinates 1–5 in Figure 5.3 represents the total number of subcellular targets per circumference, that is the number of crypts per circumference multiplied by the number of clonogens per crypt and by the number of targets per clonogen. With ordinate 6 the extrapolation number is reduced by the value of A, the number of clonogens per crypt.

A representative set of data is shown in Figure 5.3. This comprises data from 6 sections per mouse taken from each of 6 mice irradiated per dose of $^{137}Cs$ $\gamma$ rays (data included in Hendry et al, 1983). Each solid circle represents the pooled data from a single mouse. The crosses represent the mean level of survival at each dose, i.e. the mean of 36 measurements. It can be seen that the crosses are forming a curve, rather than scattering at random about the fitted line and this is characteristic of many of this type of single-dose data. This raises some doubts about the validity of comparing micro- and macro-colony colony data directly, and it is possible the macro-colonies arise from the more resistant crypts over Peyer's patches (see above; Moore & Maunda, 1983). A continuously-bending curve has been demonstrated more clearly using multi-fractionated doses, where the effect is magnified repeatedly (e.g. Masuda et al, 1977). The effect of this curvature is to increase the standard errors on the parameters which characterise a straight line. The data presented give $D_o = 1.29 \pm 0.12$ Gy, and $E = 1850 \pm 1000$ i.e. a 10 per cent standard error on $D_o$, and an error on E which is commonly not less than 40 per cent of the mean (Potten & Hendry, 1975). For further details of curve fittings see Chapter 1. Reducing the number of mice to 3 per point would not increase the error significantly. However, equal statistical weight is not given to each point because the number of survivors is decreasing according to the surviving fraction at each dose. Hence, the high-dose points have little weight in deciding on the best fit of the curve and this is why they tend to fall below the line at 14 Gy in Figure 5.3. Theoretically, it should be planned to have the same number of survivors at each dose (Ch. 1), as is possible in vitro and with some of the grafting techniques in vivo. With the present assay, the number of sections should be increased as much as is practically possible at the high doses, to increase the accuracy of these points. However, with the line fitting described here, this would not decrease the errors on the parameters, which are determined primarily by systematic errors, i.e. the goodness of fit of large samples to the model (see Ch. 1), but the line would tend to be steeper.

## MODIFICATIONS TO THE MATERIALS AND THEIR CONSEQUENCES

The variation in the number of crypts per circumference (control values) between different mouse strains is very small (Hendry, unpublished data), and is probably due mainly to mouse size. Similarly, the variation in sensitivity between many mouse strains is small. After 12 Gy the level of crypt survival did not differ by more than a factor of 2 among 6 common strains of mice. In $BDF_1$ female mice there was very little effect of age, over a range tested from 6 weeks to 43 weeks, on crypts per circumference in controls.

When the level of crypt survival differs markedly from one region of the circumference to another, as exemplified by the effects of hyperthermia (Hume et al, 1979), it is important to express the crypt surviving fraction in a specified portion of the circumference. This can be orientated with reference to the mesentery if it is carefully cut so as to leave a portion attached to the intestinal wall. Differences in crypt survival in different parts of the circumference and in different proximity to Peyer's patches have also recently been detected after irradiation (Moore & Maunda, 1983).

The clonogens in the small intestine are normally well oxygenated, as evidenced by the lack of sensitization by 1 mg/g body weight of the radiosensitizer misonidazole (Hendry, 1978). In the colon, sensitization of about 10 per cent was reported when mice breathed 95 per cent oxygen instead of air (Hamilton, 1977). Also, the sensitivity of colonic clonogens was lower by about 25 per cent when the colon was irradiated locally compared to whole-body irradiation. Furthermore, the anaesthetic halothane decreased clonogenic sensitivity in the colon by about 10 per cent.

## MODIFICATIONS TO THE METHODS AND THEIR CONSEQUENCES

As mentioned above, the criterion for defining a surviving crypt is usually set at 10 or more clustered chromophilic cells per crypt section, excluding Paneth cells. However, variations in this number can affect the surviving fraction and the shape of the survival curve. If the criterion is set at 15 cells, the survival curve is steeper than the curve measured using 5 or 10 cells (Potten, unpublished data). For example, with 15 cells, survival is lower by a factor of about 2 after 11 Gy and about 3 after 14 Gy, compared to levels of survival measured using 5 or 10 cells. This indicates the importance of standardizing the cut-off limit for surviving crypts. Also, at the higher doses there is a relatively large number of small clones.

Crypts in the small intestine are usually assayed at days 3–4 and in the colon at days 4–5. The slightly longer assay time in the colon is due to the slower rate of regeneration of the crypt cells. At these respective times the size of the crypts is only slightly greater than that of controls and hence the value for the correction factor is not markedly different from 1.0. However, at later times when the crypts are larger, the value for CF will be lower, and on day 4 the CF for crypts in the small intestine can be slightly lower than 0.6 (Moore, quoted in Potten & Hendry, 1983). Hence, unless this correction is made, crypt numbers could apparently be observed to increase by nearly a factor of two. However, when the appropriate corrections are applied, crypt survival in the small intestine does not change markedly between days 3 and 5 (Hendry et al, 1983). One complication in these assessments is the death of animals, which occurs preferentially in mice with low colony counts (Hendry et al, 1983). Hence, this effect must also be taken into account when the assay time is prolonged. Also, by day 7, budding and fission of surviving crypts is occurring (Cairnie & Millen, 1975), and this will also tend to increase the number of crypts.

The technique has been successfully applied to both the colon (Withers &Mason, 1974; Hamilton, 1977) and the glandular stomach (Chen & Withers, 1972) as well as to the small intestine. Although there are small differences, the results obtained for these other regions of the gastrointestinal tract do not differ greatly from those obtained for the small intestine (Masuda et al, 1977), except possibly in the experiments of Hamilton (1977) using colon. The $D_0$ values for small intestine range upwards from 0.85–0.9 Gy (Hendry, 1975; Potten & Hendry, 1983; Boarder & Blackett, 1976; Burholt et al, 1975), with an average value for about 20 estimates of $D_0$ of 1.32 Gy. For colon, the $D_0$ values range from 1.42 to 2.66 Gy while the value for crypts (glands) in the stomach is 1.37 Gy. The total extrapolation number (size of the shoulder on the crypt survival curve) is generally very large (see Fig. 5.3). The extrapolation number is clearly very dependent on an accurate determination of the final slope of the survival curve.

An identical approach has been adppted for determining the survival of crypts in the small intestine exposed to various cytotoxic drugs (Moore, 1979; Moore & Broadbent, 1980). Here, the crypt survival curves commonly have a very small shoulder in comparison with the radiation survival curves. Mechlorethamine hydrochloride (nitrogen mustard), bis-chlorethyl nitrosourea (BCNU), isopropyl methane sulphonate and adriamycin have been studied. These have $D_0$ values of 1.9, 19, 487 and 8.4 mg/kg respectively and low total extrapolation numbers per crypt of 3.0, 176, 1.5 and 1.3 respectively.

The survival of intestinal crypts has also been studied after heat treatment (hyperthermia) (Merino et al, 1978; Hume et al, 1979). This usually involves exteriorisation of a loop of intestine, i.e. techniques similar to those required for the macrocolony assay, but the intestinal temperature can be raised by abdominal inversion in a warmed water bath. The range of temperatures used is common between 37.5 and 45°C. The difference here is that those crypts that are destroyed by the heat treatment disappear very rapidly (within 24 h), hence sampling can be undertaken and an accurate assessment of surviving crypts made one day after treatment. In fact, loss of villi can be detected within a few hours of treatment (Hume et al, 1979).

A very different technical approach was developed by Hagemann & Sigdestad (Hagemann et al, 1971) which is believed to provide results comparable to those generated by the microscopic technique described here (Burholt et al, 1975; Yau & Cairnie, 1979; reviewed by Potten et al, 1983). In this technique the amount of radioactivity in small pieces of intestine (dpm per mg) and in individual isolated crypts (dpm per crypt) is measured using liquid scintillation counting after suitable $^3$H-thymidine labelling and radiation exposure. The number of crypts per mg of tissue can then be calculated from the dpm/mg intestine divided by the dpm/crypt. The technique is discussed in detail elsewhere (Ch. 6). Some authors have shown that in the shoulder region the crypt survival curves using crypts/mg do not differ greatly from those based on crypts/circumference (Burholt et al, 1975; Yau & Cairnie, 1979). The survival curves generated using crypts/mg lie within the 'envelope' of many other published microcolony curves (Potten & Hendry 1983). However, earlier studies using crypts/mg commonly provided rather high $D_0$ values of between 2.5 Gy and 3.3 Gy (Hagemann et al, 1971; Tseng et al, 1978). This is probably partly due to the fact that the crypts/mg technique commonly provides data only on the shoulder of the *crypt* survival curve, where the estimates for the $D_0$ for clonogens would be *expected* to be higher if the clonogen survival curve is indeed continuously bending.

## INFORMATION THAT CAN BE DEDUCED USING THE TECHNIQUE

It is clear from the preceding discussion that the technique can be used to provide basic radiobiological information about mammalian intestinal epithelial cell response. The radio- and drug-sensitivity can be determined and expressed in terms of a conventional $D_0$ and extrapolation number. Combination of drugs and radiation, hyperthermia and radiation etc. can also be easily studied as well as fractionation regimens. The relative biological efficiency (rbe) and oxygen enhancement ratio (oer) can be studied, and the technique is also one of the best biological dosimeters available for assessing normal tissue injury (e.g. Hendry & Greene, 1976).

### Number of clonogenic cells

The shoulder of the *crypt* survival curve has been assumed to be the consequence of the fact that each crypt is composed of many cells, each of which is capable of regenerating the crypt and subsequently the epithelium, i.e. is clonogenic. The crypts are only destroyed when all clonogens per crypt are sterilised. Each crypt contains about 250 cells, of which about 150 are proliferative cells with a rapid cell cycle. There may be a smaller number, about 16 cells, with a slower cycle located at a position in the crypt which indicates that these 16 cells may be the functional stem cells, or include the functional stem cells (for a recent review see Potten & Hendry, 1983). It is generally believed that the Paneth cells and the post-mitotic cells at the top of the crypt are unlikely to be clonogenic. Hence, the maximum possible number of clonogens per crypt is about 170 (150 + 16). One of the questions that can be asked is how many clonogens does an average crypt contain. Does it require 169 cells to be sterilised to leave one final clonogen, the survival of which determines the near exponential portion of the survival curve, or is it 15 out of 16 or some other value? Attempts to answer this have led to differing views which range from estimates of 152 at one extreme (Thames et al, 1981), i.e. numbers similar to the total number of proliferative cells, to 5–15 (Hanson, quoted in Potten & Hendry, 1983; Potten & Hendry, 1975; Zywietz et al, 1979 — all summarised in Potten & Hendry, 1983), i.e. numbers lower than, or roughly equal to, the number of functional stem cells. The average of all estimates is about 80 which is close to many of the individual estimates. Experiments where S-phase specific agents have been administered prior to irradiation during the day suggest that few clonogens are in rapid cell cycle (Hendry et al, 1984) which indicates that the true number of clonogens per crypt *cannot* be as high as 80. Drug survival studies commonly provide estimates between 1 and 6 cells per crypt (Moore, 1979; Moore & Broadbent, 1980).

Estimates of the number of clonogens can be made using several approaches which involve split doses or multifractionated doses. These are described and contrasted elsewhere (Potten & Hendry, 1983). Since the number of clonogens can be estimated and this can be done at various times after depleting the clonogenic population the rate of repopulation (doubling time) for the clonogenic cells can be estimated. This is about 24 hour (Potten & Hendry, 1975), although there are other estimates which are much lower (reviewed by Hendry, 1979).

Unfortunately, since the initial site of regeneration cannot be recognised (i.e. the clonogenic cell itself cannot yet be identified — it can only be inferred from its ability to form a visible clone of many cells), it is impossible to say anything about the location of clonogens within the crypt. Even if very small foci could be identified, their position might not indicate their location prior to exposure to a cytotoxic agent, since the entire crypt organisation is bound to be altered by the level of cellular depletion and hence cells may be found subsequently at positions that they did not occupy originally.

## ACKNOWLEDGEMENTS

The work of the authors is supported by the Cancer Research Campaign (UK).

## REFERENCES

Boarder T A, Blackett N M 1976 The proliferative status of intestinal epithelial clonogenic cells: sensitivity to S phase specific cytotoxic agents. Cell and Tissue Kinetics 9:589

Burholt D R, Hagemann R F, Cooper J W, Schenken L L, Lesher S 1975 Damage and recovery assessment of the mouse jejunum to abdominal X-ray and adriamycin treatment. British Journal of Radiology 48:908

Cairnie A B, Millen B H 1975 Fission of crypts in the small intestine of the irradiated mouse. Cell and Tissue Kinetics 8:189

*The slope of the survival curve*

Perhaps the most serious criticism of the method is not based on the validity of the measurements but upon the usefulness of the results. As was pointed out earlier, this method of crypt assay generates a slope on a survival curve (a measure of radiation sensitivity of the crypt) which is three times less sensitive than that reported using other methods (Withers & Elkind, 1969, 1970). The apparent increased radioresistance implicit in these results was later demonstrated to be due to the fact that only the shoulder of the full survival curve was being measured (Burholt et al, 1975; Yau & Cairnie, 1979). This is further supported by the fact that the Hagemann-Sigdestad method cannot measure survival of crypts below one decade on the survival curve. This shortcoming is because virtually no recognisable crypts are present after these very high radiation doses (>18 Gy). The implications of this limitation will be discussed below.

## APLICATIONS OF THE TECHNIQUES:

This radio-isotope based assay provides a quantitative evaluation of the murine intestinal epithelium either in the normal state or after cytotoxic insult. The most important application of this technique lies in its ability to determine the number of crypts/mg, which is necessary in the construction of crypt survival curves. Hence, applications of this type not only allow quantitation of the intestinal proliferative units but also provide an assay of dose response relationships. In addition to crypts per milligram, this assay can provide values for proliferative cells/mg, villi/mg, crypts/villus ratios and villous cells/mg for the three major subdivision of the small intestine (Hagemann et al, 1970b).

**Information on intestinal crypts (unperturbed)**

We have determined for the mouse intestinal epithelium that, under normal conditions, the crypts per milligram are 855, 892 and 1164 for the duodenum, jejunum and ileum respectively. The values for the entrie small intestine amount to $1.1 \times 10^6$ crypts which correlates well with the values observed by Yau & Cairnie (1979) ($0.933 \times 10^6$ crypts per intestine).

**Information on intestinal crypts (perturbed)**

The surviving fraction of intestinal crypts from irradi-

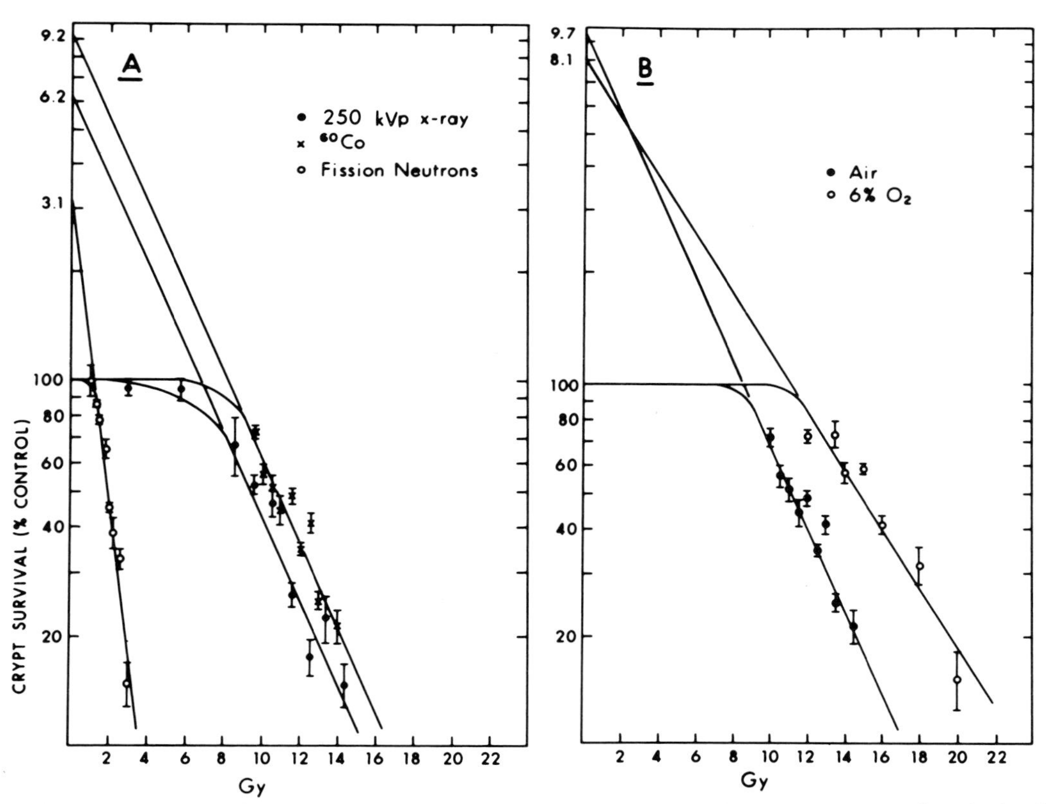


**Fig. 6.5A** Crypt survival curves for three radiation qualities (Sigdestad et al, 1972). **B** Crypt survival curves for mice irradiated with Cobalt-60 gamma rays while breathing either atmospheric air or 6% oxygen (Sigdestad et al, 1973); with kind permission from Academic Press.

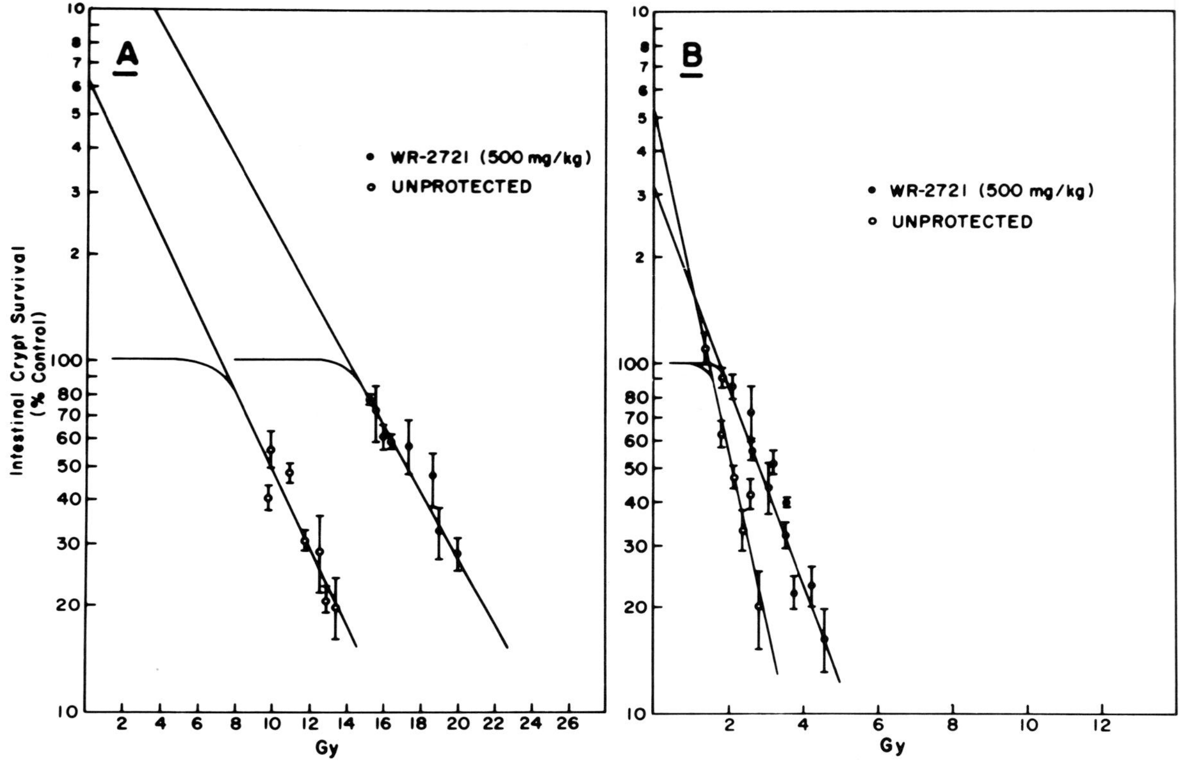


**Fig. 6.6A** Crypt survival curves for mice receiving the radioprotective compound WR-2721 (500 mg/kg) prior to irradiation with 4 MeV X-rays (Sigdestad et al, 1975). **B** Crypt survival curves for mice receiving WR-2721 prior to irradiation with fission neutrons (Sigdestad et al, 1976); with kind permission from Academic Press.

ated animals, as well as the effect of factors which modify the radiation response are of obvious importance. The radiation response has been studied by using radiations with differing LET values, and by altering pre-irradiation environments. Figures 6.5 and 6.6 shows representative crypt survival curves which illustrate the effects of modifying the radiation response by either changing the quality of radiation (Fig.6.5A) or by altering the tissue oxygenation at the time of irradiation (Fig. 5B). Table 6.2 presents the salient parameters defining these curves.

Hagemann et al (1969) were able to demonstrate a relationship between the loss of proliferative intestinal crypts and the median lethal dose needed to produce the gastrointestinal syndrome. It was also shown that crypts/mg and/or crypts/intestine could be correlated with differences in sensitivity between various mouse strains in LD-50(6) using either the Hagemann-Sigdestad method or the Yau & Cairnie modification (Table 6.3).

**Table 6.2**

| Radiation | $D_o$ (Gy) | n | Dq (Gy) | D-30 (Gy) | RBE | LD-50(6) (Gy) | RBE |
|---|---|---|---|---|---|---|---|
| 250 kVp | | | 6.9 | 11.4 | 1.0 | 10.9 | 1.0 |
| X-rays | 3.8 | 6.2 | | | | | |
| Cobalt-60 | 3.8 | 9.2 | 8.3 | 12.9 | 0.8 | 13.5 | 0.8 |
| Fission | | | 1.2 | 2.4 | 4.7 | 2.3 | 4.8 |
| Neutrons | 1.0 | 3.1 | | | | | |

Note: The D-30 is the 30% isoresponse point. This is approximately the LD-50(6) radiation dose

## Information on DNA synthesizing cells per crypt

The number of DNA synthesizing cells on a per crypt or per milligram basis is an important by-product of this method. Hagemann & Lesher (1971) showed that 48 hours (and later) after irradiation with doses from 3–10 Gy the dpm per 3-HTdR labelled nucleus was the same as if the cells were unirradiated. This suggests that the measurement of 3-HTdR activity incorporated into intestinal tissue (either on a per crypt or per mg basis) at three days after radiation is a good measure of the relative numbers of DNA synthesizing cells.

This method nicely demonstrates the compensatory response elicited in the intestinal epithelium after irradiation. The proliferative response is greater than control and appears to increase with increasing dose. The

**Table 6.3**

| Mouse | Crypts/mg | Crypts/intestine | LD-50(6) | Reference |
|---|---|---|---|---|
| C57/J6 | 1005 | $1.1 \times 10^6$ | 12.8 | Hagemann, Sigdestad & Lesher (1970b) |
| Swiss-Webster | – | $0.9 \times 10^6$ | – | Yau & Cairnie (1979) |
| Swiss-Webster | 892 | $1.0 \times 10^6$ | 11.4 | Connor (unpublished data) |

observed response peaks at about 2.0 and 3.5 Gy for fission neutron irradiation and at about 10 Gy with low LET radiations. With higher doses the crypt is unable to evoke the same degree of response, which results in a reduced cellularity. It was noted that even at high doses the cellularity in surviving crypts was not less than that seen in unirradiated control mice.

The value of dpm/mg presented as percent of control likewise represents a measure of the relative number of DNA synthesizing cells per mg of intestinal tissue. The curve obtained when this value is plotted against dose is similar to the response seen in the crypt with a dose dependent increase in proliferative cellularity. With higher doses, however, there is a rapid fall to levels lower than those seen in unirradiated controls. The effect just described has significance in that the surviving crypts, even at high doses, rarely have a cellularity less than that seen in the unirradiated control. This indicates that there is a dose-effect relationship to be found in the compensatory hyperproliferative response observed in the crypts. However, the total cellularity reflects the product of the cellularity per crypt and the fraction of crypts surviving. These two factors are the basis of this method and it indeed measures the loss of intestinal crypts following irradiation.

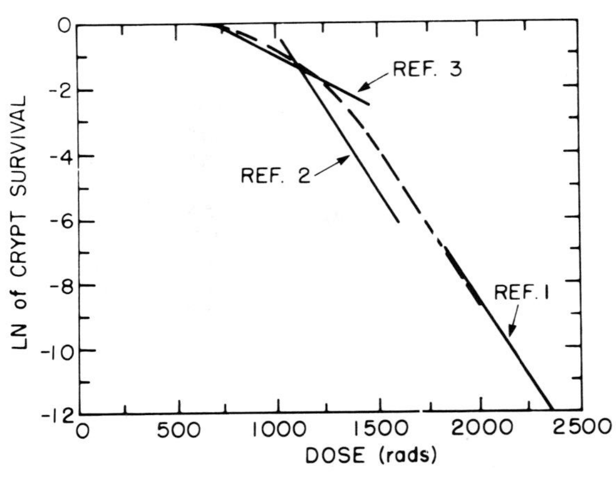


**Fig. 6.7** The natural logarithm (LN) of crypt survival as a function of radiation dose plotted for four different assay systems: Ref. 1, macro-colony assay (Withers & Elkind, 1969); Ref. 2, micro-colony assay (Withers & Elkind, 1970); Ref. 3, crypts/mg assay (Hagemann et al, 1970a); dashed line, crypts/intestine assay (Yau & Cairnie, 1979); with kind permission from Dr Allen Cairnie and Academic Press.

## SUMMARY

It was mentioned above that Yau & Cairnie (1979) slightly modified this isotopic method in order to simplify the technique and to permit an assay of crypt survival at higher radiation doses, i.e. measurement at increased levels of cell kill. When their data are fitted by non-linear regression using a multitarget model with two components, the differences in $D_o$, extrapolation number (n), and stem cells per crypt which occur between results using macro- and micro-colony assays (Withers & Elkind, 1969, 1970) and those using the isotopic crypt survival assay (Hagemann et al, 1970a) may perhaps be explained. In Figure 6.7, four different crypt survival curves are plotted. It may be seen that results using the macro-colony method substantially agree with the apparent exponential portion of Yau & Cairnie's survival curve (dashed line) (1979). This exponential portion is actually continuously bending downward. The conventional micro-colony assay, however, contains response points which appear on both the shoulder region and the exponential region of Yau & Cairnie's composite survival curve. The isotopic method, reviewed in this chapter, provides data points which are entirely on the shoulder region of this composite survival curve. Masuda et al (1977) estimate a $D_o$ of 1.09 Gy, a crypt extrapolation number of 7500, and the number of colony-forming cells as 140 per crypt. Hagemann et al (1971) give values of 3.3 Gy for the $D_o$ and a crypt extrapolation number of 10, which was equivalent to an upper limit for the number of colony-forming cells per crypt. Yau & Cairnie (1979), on the other hand, calculate the $D_o$ as 1.02 Gy, the extrapolation number as 55, and the number of colony-forming cells per crypt as 31. It seems that, given the mathematical formalism of the Yau & Cairnie model, their estimates of the crypt survival curve parameters can be an acceptable compromise between the macrocolony assay, the microcolony assay, and the radioisotopic crypt survival assay. However, this is not meant to imply that any of the methods should be discarded, as long as one remembers to apply the constraints imposed by the interpretation provided by Yau & Cairnie's modification of the crypt dissection method.

It appears that with the above modification of the radioisotopic method, it is indeed measuring the same endpoint as the Withers & Elkind microcolony assay. The choice of the most appropriate assay must be made while keeping in mind the personnel and facilities available to the investigator.

## ACKNOWLEDGMENTS

This study was funded in part by Contract DAMD 17–81-C-1070 from the US Army Medical Research and Development Command, Walter Reed Institute of Research, Washington DC.

## REFERENCES

Burholt D R, Hagemann R F, Cooper J W, Schenken L L, Lesher S 1975 Damage and recovery assessment of the mouse jejunum to abdominal X-ray and adriamycin treatment. British Journal of Radiology 48:908

Cairnie A B, Millen B H 1975 Fission of crypts in the small intestine of the irradiated mouse. Cell and Tissue Kinetics 8:189

Clarke R M 1972 The effect of growth and of fasting on the number of villi and crypts in the small intestine of the albino rat. Journal of Anatomy 112:27

Connor A M, Sigdestad C P 1978 Combined effect of procarbazine and ionizing radiation on mouse jejunal crypts. Experientia 24:368

Hagemann R F, Lesher S 1971 Irradiation of the GI tract: compensatory response of the stomach, jejunum and colon. British Journal of Radiology 44:599

Hagemann R F, Sigdestad C P, Lesher S 1969 Radiation LD50/5 and its relation to surviving intestinal stem cells. International Journal of Radiation Biology 16:291

Hagemann R F, Sigdestad C P, Lesher S 1970(a) A method for quantitation of proliferative intestinal mucosal cells on a weight basis: some values for C57B1/6. Cell and Tissue Kinetics 3:21

Hagemann R F, Sigdestad C P, Lesher S 1970(b) A quantitative description of the intestinal epithelium of the mouse. The American Journal of Anatomy 129:41

Hagemann R F, Sigdestad C P, Lesher S 1971 Intestinal crypt survival and total and per crypt levels of proliferative cellularity following irradiation: Single X-ray exposures. Radiation Research 46:533

Lesher S 1967 Compensatory reactions in intestinal crypt cells after 300 Roentgens of Cobalt-60 gamma irradiation. Radiation Research 32:510

Lesher S, Sacher G A 1968 Effects of age on cell proliferation in mouse duodenal crypts. Experimental Gerontology 3:211

Masuda K, Withers H R, Mason K A, Chen K Y 1977 Single-dose response curves of murine gastrointestinal crypt stem cells. Radiation Research 69:65

Sigdestad C P, Hagemann R F, Lesher S 1970 A new method for determining intestinal cell transit time. Gastroenterology 58:47

Sigdestad C P, Lesher S 1972 Circadian rhythm in the cell cycle time of intestinal proliferative cells. Journal of Interdisciplinary Cycle Research 3:39

Sigdestad C P, Scott R M, Hagemann R F 1972 Intestinal crypt survival: The effect of Cobalt-60, 250 kVcp x-rays and fission neutrons. Radiation Research 52:168

Sigdestad C P, Hagemann R F, Scott R M 1973 The effects of oxygen on intestinal crypt survival in Cobalt-60 irradiated mice. Radiation Research 54:102

Sigdestad C P, Connor A M, Scott R M 1975 The effect of S-2-(aminopropylamino)ethylphosphorothioic acid (WR-2721) on intestinal crypt survival: I 4 MeV X-rays. Radiation Research 62:267

Sigdestad C P, Connor A M, Scott R M 1976 The effect of S-2-(3-aminopropylamino)ethylphosphorothioic acid (WR-2721) on intestinal crypt survival II Fission neutrons. Radiation Research 65:430

Tseng V L, Bybee J W, Osborne J W 1978 Intestinal crypt survival after x-irradiation of the rat small intestine under conditions of radioprotection. Radiation Research 74:129

Wimber D E, Quastler H, Stein O L, Wimber D R 1960 Analysis of tritium incorporation into individual cells by autoradiography of crypt squash preparations. Journal of Biophysical Biochemical Cytology 8:327

Withers H R, Elkind M M 1969 Radiosensitivity and fractionation response of crypt cells of mouse jejunum. Radiation Research 38:598

Withers H R, Elkind M M 1970 Micro-colony survival assay for cells of mouse intestinal mucosa exposed to radiation. International Journal of Radiation Biology 17:261

Yau H C, Cairnie A B 1979 Cell-survival characteristics of intestinal stem cells and crypts of gamma-irradiated mice. Radiation Research 80:92

# Macro-colonies in epidermis

## HISTORICAL BACKGROUND

The radiation response of skin has been studied for almost as long as X-rays have been known to man. It is easy to observe skin damage, and in the early days of the development of X-ray sets, damage to the skin was a common occurrence. When cancer therapy using orthovoltage radiation was introduced the skin was again the limiting normal tissue because it received the highest dose. Initially, arbitrary descriptions were used, e.g. mild, moderate or severe erythema, dry desquamation, moist desquamation or necrosis. Skin tolerance was used as early as 1944 to establish the 'cube root law' for fractionation (Reisner, 1932), and has always been the most abundant source of both clinical and experimental data for evaluation of other fractionation formulae (e.g. Cohen, 1968; Fowler & Stern, 1963; Ellis, 1969).

Many of the first radiobiologists adopted the implicit assumption that early skin damage resulted from the killing of epidermal cells, but it was not until 1967 that this could be validated, as a result of the development of the epidermal macro-colony assay by Withers (1967 a,b,c,). This was the first of the epithelial clonal assays to be developed, but followed the clonal studies with bone marrow (Till & McCulloch, 1961), cartilage (Kember, 1965) and explanted tumour cells (Hewitt & Wilson, 1959). The epidermal assay differed from the two other normal tissue assays in that the size of the initial

**Table 7.1**

| Reference | Animal | Radiation source moat | test | $D_o$ (gray) | Range of test island areas ($mm^2$) | Studies |
|---|---|---|---|---|---|---|
| *Macro-colony assays* | Mice | | | | | |
| Withers 1967a | WHt hybrid | 29 kVp X | 29 kVp X | 1.37 | 0.09–113 | Radiosensitivity |
| Withers, 1967b | WHt hybrid | 29 kVp X | 150 kVp X | 1.40 | 0.09–113 | Oxygenation |
| Withers, 1967c | WHt hybrid | 29 kVp X | 29 kVp X | 1.34 | 0.09–113 | Repair & Repopulation |
| Emery et al, 1970 | SAS/TO | deuterons | 250 kVp X | 1.35 | 0.12–285 | Repair & Repopulation |
| Denekamp et al, 1971 | SAS/TO | deuterons | neutrons | 1.09 | 0.12–120 | Neutron RBE & repair |
| Leith et al, 1971 | CD1 hairless | $^{4}He$ | $^{4}He$ | 0.95 | 3.5–285 | RBEs |
| | | $^{7}Li$ | $^{7}Li$ | – | 31.7 | |
| | | $^{14}C$ | $^{14}C$ | – | 31.7 | |
| Denekamp & Michael, 1972 | WHt & C3H | electrons | electrons | – | 6.6 | Radiosensitizers |
| Denekamp et al, 1974 | WHt inbred | electrons | electrons | 1.45–1.50 | 6.6 | Radiosensitizers |
| Denekamp et al, 1982a | WHt inbred | electrons | electrons | | 6.6 | Radiosensitizers |
| Denekamp et al, 1981 | WHt inbred | electrons | electrons | – | 6.6 | Radio-protectors & oxygen |
| Denekamp et al, 1982b | WHt inbred | electrons | electrons | – | 50 | Radio-protectors & oxygen |
| Denekamp et al 1983 | WHt inbred | electrons | electrons | – | 6.6–50 | Radio-protectors & oxygen |
| Hendry, 1984 | B6D2F1 | None | 300 kVp X | 3.44 | (140) | Comparison with gross reactions |
| Arcangeli et al, 1980 | Human | None | 5.7 MeV X | 4.90 | None | Radiosensitivity (initial slope) |
| *Micro-colony assays* | | | | | | |
| Archambeau et al, 1979 | Yorkshire swine | None | 300 kVp X | 3.37 | None | Single dose histological study |
| Al Barwari & Potten, 1976 | DBA-2 mice | None | 290 kVp X | 2.33 | Whole body | Proliferation |

population could be varied at will, allowing the survival curve to be determined over five or six decades of cell kill. As with other clonal assays a cell is considered to be a survivor if it is capable of proliferating to form a large colony of descendents. However, whereas the in vitro definition is of 50 cells constituting a colony, the macroscopic skin clones are not visible until they exceed 1 mm diameter, at which time they contain at least $10^3 - 10^4$ cells (Withers, 1967a; Emery et al, 1970). Some studies have been performed using histological assessment of the clones, which permits the detection of smaller nodules (Al Barwari & Potten, 1976; Archambeau et al, 1979, Ch. 8).

The epidermal macro-colony assay was used by Withers to study the influence of fractionation, oxygenation and anaesthesia (Withers, 1967 a,b,c). He later went on to develop clonal assays for jejunum, stomach, colon, testis and kidney (see Chs 4, 5, 12, 17). Few workers have adopted the epidermal macro-colony technique in its original form, because it requires specialized equipment and it has a slightly complicated experimental design. However, it does offer advantages over more crude assessments of skin damage in certain situations: Table 7.1 lists the studies that have been performed.

## DESCRIPTION OF THE TECHNIQUE

The essentials of the Withers technique are that islands of skin can be isolated from the surrounding epidermis by a moat of heavily irradiated cells. The islands are subjected to test doses of radiation and scored 2–3 weeks later for macroscopic nodules resulting from the survival of one or more cells in each test island (Fig. 7.1). In order to be able to observe any small clones regrowing in the ulcerated moat, and to have good contact between the shields and the skin during the irradiation of the moat, it is necessary to pluck all hair from the area to be irradiated prior to treatment. The size of the initial population at risk can be varied at will by altering the size of the test islands.

### Choice of animals

Most of the work has been performed on albino mice, either of the TO strain or of the WHT strain (either inbred or as $F_1$ hybrids). Both males and females have been used and mice ranging from 2–15 months of age have been studied, with no obvious age-dependent changes being recorded. Withers (1967 a,b,c) preferred to use female animals because their rear dorsum was flatter and therefore more amenable to uniform irradiation from an incident beam 29 kVp X-rays.

Since bare skin is an essential requirement for these experiments it might seem logical to use hairless mice. These were used by Leith et al (1971) who irradiated their ventral surface. This was also attempted in our early studies, but we found them too susceptible to infection. Instead of healing within 3–4 weeks, the ulcer-

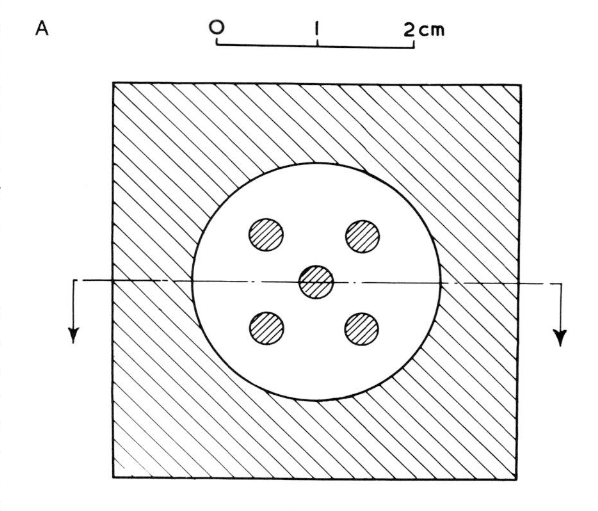


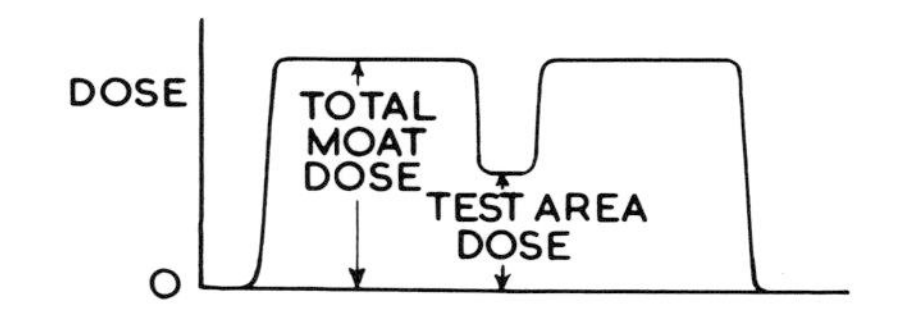


B

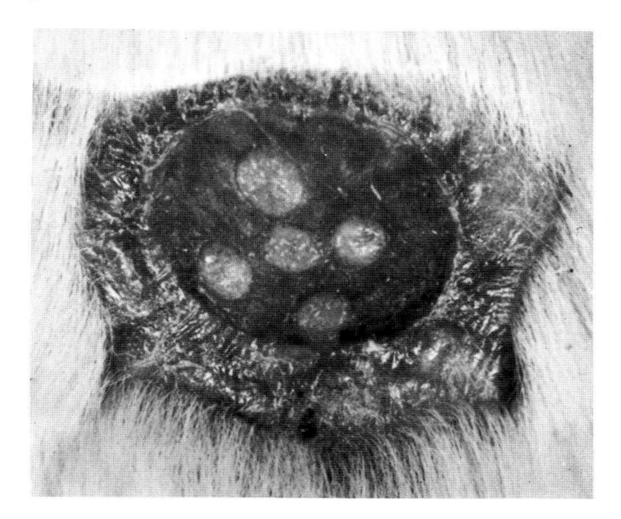

**Fig. 7.1 A** Schematic diagram illustrating method by which five test islands are separated from the surrounding epidermis by a heavily irradiated moat.
**B** Example of five clones regrowing in the irradiated field at about 17 days post irradiation. The original 2.5 cm diameter moat has already contracted because of migration of epithelium from the edges.

ated area often spread beyond its original limits. Pigmented mice have also been used for these experiments, e.g. C3H mice, but the clones are less easy to observe because of the pigment (Denekamp & Michael, 1972).

**Plucking**

The hair is removed from an area of at least 2.5 cm diameter while the mice are anaesthetised. The mouse is held taut across the fingers of one hand while the hairs are carefully removed with the fingers of the other hand, taking care not to damage the skin. With experience this can be a rapid painless procedure, taking 1–2 minutes per mouse. Any animals whose hair follicles are in the growth phase can be detected by the fine down of hairs which are too short to be removed, and these animals should be excluded from the study. In these animals the epidermis lining the hair follicles can extend much deeper (e.g. 500 μm) and it is then difficult to ensure a uniform dosimetry across the whole epidermis. Animals in partial growth phase, i.e. with a small area of the field containing such short hairs, are not routinely excluded from our studies but these are noted so as to determine whether it influences the subsequent response. Chemical depilatory agents were found to be damaging to the epidermis, and allowed the hairs to regrow too quickly, although plastic dressing has been used by some workers as a depilatory agent and the epidermal damage has been quantitated (Hamilton & Potten, 1972).

**Moat dose**

The dose needed to isolate the test islands must be large enough to kill every clonogenic cell in the moat. It can be supralethal, and its upper limit is set by the amount of scatter that can be allowed under the protective shields, and also by the dose that reaches the intestine. In the early studies (Withers, 1967 a,b,c; Emery et al, 1970; Denekamp et al, 1971) the dose to the moat, given whilst the test areas were shielded, was adjusted for each group so that the total dose received in the moat (priming and test dose) was constant (30 Gy). However, the amount of damage to the cells in the test area as a result of scattered radiation underneath the shields was different for the low and high doses used with each shield size, because of the different priming dose to the moats. This may have been part of the reason for the lower apparent radiosensitivity (higher $D_o$) measured using any single shield size than that obtained using the composite from the whole range of irradiated areas (Emery et al, 1970). In more recent experiments the priming dose to the moat has been maintained constant, at doses ranging from 30 to 100 Gy in different experiments (e.g. Denekamp et al, 1974, 1981, 1982; Leith el al, 1971). It is probably more important to maintain a uniform contribution from scattered radiation under the protective shields than to attempt to maintain a constant total dose to the moat.

The radiation used for the production of the moat must be shallow-penetrating to avoid early deaths from intestinal damage. Low energy X-rays, deuterons, heavy ions and relatively low energy electrons have all been used for this purpose, as listed in Table 7.1. All these forms of radiation provide considerable dosimetric problems and expert physics back-up is needed to know the dose-distribution through the critical 0–500 μm of skin. Interfollicular epidermis is usually ⩽ 50 μm in depth, but hair follicles can extend to 200 μm in telogen or even to 500 μm in anagen and clones can regrow equally well from follicular or interfollicular epidermis.

The method of delivering the moat dose is to interpose the appropriately sized shields between the skin and the radiation source. The mice must be deeply anaesthetised for this so that no areas of epidermis are shielded by folds or wrinkles. From one to 16 test islands can be isolated in a 20–30 mm diameter area. In practice five smallish test areas per mouse are often used, with five mice per treatment group (Fig. 7.1).

Withers orginally used shields of four sizes, covering the range shown in Table 7.1. This was expanded to include nine sizes in the studies of Emery et al (1970) but has been confined to a single size in our chemomodification studies. Lead, brass and steel have been used, and ball-bearings were found to be a cheap source of accurately-sized shields for the smaller test areas. The problems of transmission through the edges of the ball, and of scatter under the shield were discussed by Withers (1967a) for 29 kVp X-rays. This problem was minimal with deuterons (Emery et al, 1970; Denekamp et al 1981) but is more important with electrons. The shields must not be allowed to move during the moat irradiation, otherwise the size and precision of the island would be uncertain. This is achieved by pressing them firmly into the back of the mouse, or by attaching them to sellotape which adheres to the plucked skin.

**Test dose**

The test doses must be chosen to give survival of cells in some but not all of the test areas. Both Withers (1967a) and Emery et al (1970) discuss the statistical weight of groups in which 100 per cent, an intermediate proportion, or 0 per cent of the islands regrow. Less accuracy is achieved if all or none of the islands regrow and hence the choice of test doses is very limited. For example, in animals breathing oxygen the response changes from all 3 mm-diameter islands regrowing at 15 Gy to virtually none at 20 Gy (Denekamp et al, 1981, 1982a, 1982b). Considerable prior knowledge is therefore needed about the appropriate dose range if the experiments are to be economical in numbers of mice. If a survival curve is to be obtained over many decades the appropriate range for each shield size must be selected. Similarly if chemosensitization or protection is under study, a fairly accurate estimate of the dose-modifying factor is needed before the experiment can be performed. This is one of the main drawbacks of the technique.

Because of the scattered radiation which may penetrate below the shields it is important to standardize the interval between administering the test and moat doses. The sequence does not appear to be important, but the interval could possibly have a significant effect because of the repair of sublethal injury from the scattered dose. The interval chosen is often not quoted but it is important that it should be standardized. If at least six hours are allowed, full repair of sublethal injury will have occurred, although any changes in sensitivity attributed to synchronous progression of cells could still influence the response. If the irradiations are given in very close sequence, then virtually no repair will occur and all the test areas will respond as if they have suffered the maximum (but equivalent) damage from the scattered dose. This factor will cause varying problems depending upon the type of radiation used for producing the moat, but it is important and needs to be considered in the experimental design.

**Scoring the macro-colonies**

The irradiated skin shows no significant damage before one week, but in the second week ulceration occurs and by 12–16 days the first clones begin to appear as whitish nodules against the background of the ulcerated moat. Between 12 and 21 days these clones become larger and/or more pronounced as the number of cells increases in each because of the rapid repopulation from survivors. By 21 days the irradiated field may be much smaller and quite distorted because of contraction of the skin and migration in from the unirradiated edge. The relative positions of the surviving test areas usually, but not always, reflect the original pattern of the applied shields. The mice are scored 3–5 times a week from day 12 to day 21. The position of each clone is marked and a different symbol is used for possible, probable or definite clones; possible clones usually progress to be scored as definite at later times. Mice in which all the islands have regrown can be sacrificed prematurely.

The proportion of definite clones appearing before or on day 21 is used as the final score. Figure 7.1 shows an example of 5 clones with an irradiated field at about 17 days after irradiation. It has been shown that 3 per cent extra clones may appear beyond day 21 (Emery et al, 1970) but these are difficult to score because of coalescence of existing clones and diminution of the area because of migration of cells inwards from the edge.

Clones may appear initially as small raised nodules, already containing many heaped layers of epidermal cells, or as pale thin flattened discs, presumably only a few cell layers thick. Histological studies have shown that the epidermal cells spread over the dermis and up and down the depopulated hair follicles (Withers, 1967a; Al Barwari & Potten, 1976). Hair growth occasionally occurs within a re-growing island within 21 days, particularly in groups in which every island regrows, and hence in which many cells have probably survived. Withers (1967a) showed that the growth of a clone was probably associated with the survival of one or more individual clonogenic epidermal cells by the use of Poissonian statistics. The inter-dependence of surviving cells within an area was proposed by Foster et al (1967) but was refuted by the analyses made by Emery et al (1970) and by Withers (1967a). Clones which appear and are scored as definite seldom disappear, except in our experiments when hairless mice were used, in which case infection in the ulcerated skin led to a loss of small clones before they were fully established.

**Presentation of data**

The raw data consist of dose groups in which the percent of surviving islands has been determined. Data from several experiments can be pooled, although there can be a very large variation from one experiment to another (Emery et al, 1970). It is therefore unsafe to use historical controls for any comparisons or to use other people's published curves, for example to derive RBE estimates (Leith et al, 1971). The errors shown are usually the binomial error calculated from the proportion of clones regrowing relative to the total number of islands at risk. The data can be plotted simply as percent regrowing islands versus dose if a single shield has been used. However, if more than one test area has been used the data can readily be converted to cells surviving per $mm^2$ from a knowledge of the shield sizes:

$$S = -\ln(1-f)/A$$

where S = surviving clonogenic cells per area A if f is the fraction of islands showing regrowth. If the initial number of clonogenic cells per unit area is known the surviving fraction can be calculated directly, but usually the data are expressed as survivors per unit area. Figure 7.2 shows the single dose X-ray survival curves obtained by Withers (1967a) and by Emery et al (1970), compared with the curves obtained for 2 fractions given in 24 hours. The data are remarkably similar, with $D_o$ values of 1.35 and 1.34 Gy. However the repair increment was larger for the SAS/TO mice than for the WHT mice, and furthermore they required approximately 2.5 Gy

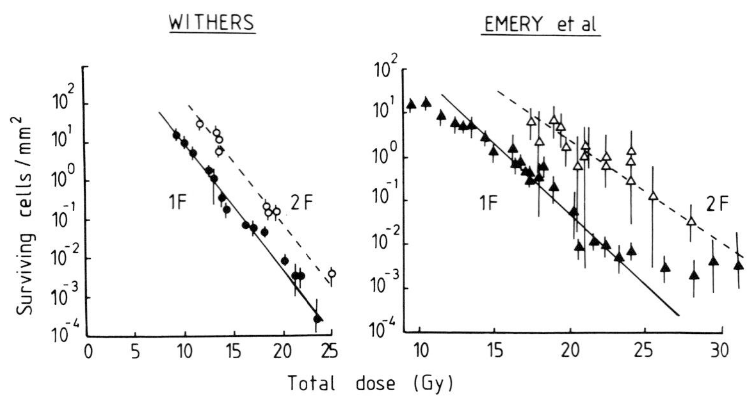


**Fig. 7.2** Survival curves for mouse epithelial cells obtained by irradiating island of various sizes and scoring the percent of regrowing clones within 21 days. Data are shown for single doses (closed symbols) and for two fractions (open symbols) in 24 hours from the experiments of Withers (1967 a, c) and Emery et al (1970). The $D_o$ values are remarkably similar ($\sim$1.35 Gy), but there is a difference in the absolute radiosensitivity and the repair increment ($D_2$–$D_1$).

extra in a single dose to achieve the same endpoint. Thus, the range over which a particular response may be observed can be quite different for different strains of mice. This is nicely illustrated in the review by Potten et al (1983).

When the survival data were fitted by linear regression analysis the quoted values of $D_o$ in Table 7.1 were obtained. However, Figure 7.2 shows that the data are actually sigmoidal, lying below the fitted line at low doses and above it at high doses. Closer examination shows that a shallower slope can be fitted to the individual sets of data obtained with each shield size than that which fits the entire curve. This was discussed at length by Emery et al (1970) and was attributed to heterogeneity of the epidermal cells and the greater influence of losing a clone, or of misinclusion of a migrating edge if the score is close to 100 per cent or 0 per cent than anywhere in the middle of the response range. It may also result from variations in the scatter beneath the shields from the moat dose as mentioned above.

Table 7.1 shows that the $D_o$ values quoted for 250, 150 and 29 kVp X-rays were remarkably similar in the first series of experiments. Arcangeli et al (1980), using colonies regrowing in a patient's skin after irradiation for mammary carcinoma, obtained an estimate of 4.9 Gy for the $D_o$ resulting from multiple fractions per day. This $D_o$ may be assumed to represent the radiosensitivity on the initial portion of a cell survival curve. It is therefore not surprising that it is more than 3 times higher than the values obtained with large single doses. Indeed, it might be expected to be even higher, since the initial slope has been calculated to be four to ten times flatter than the final slope (Denekamp & Harris, 1975; Field et al, 1975). High $D_o$ values have also been obtained for macro-colonies on the mouse tail (Hendry, 1984) and for micro-colonies assessed by histological means in pig skin (Archambeau et al, 1979) and in mouse skin (Al Barwari & Potten, 1976). The reasons for these discrepancies are not understood, but in all three studies the systems do

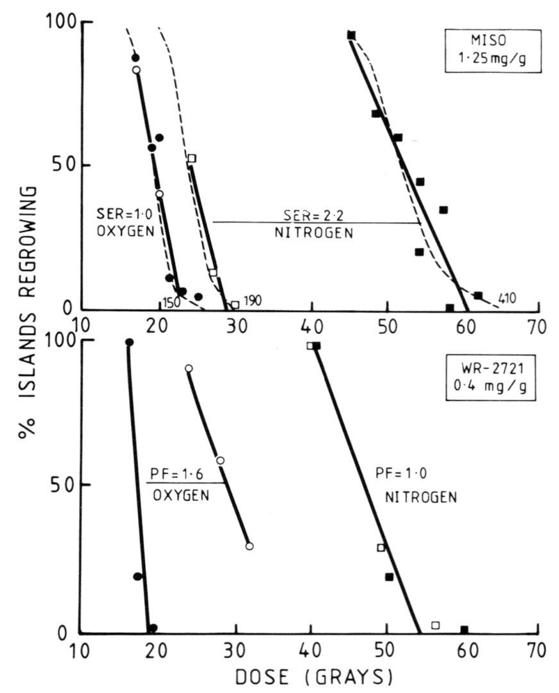


**Fig. 7.3** Clone survival curves obtained after irradiating islands of only one size with graded doses of electrons. Mice irradiated in oxygen (circles) or nitrogen (squares), with or without the addition of the radiosensitizer misonidazole (miso) or the radioprotector WR-2721. **a** Miso sensitized hypoxic skin but had no effect in 100% oxygen. **b** WR-2721 protected mice in oxygen, but not those breathing nitrogen. The solid lines were fitted by eye. The dashed sigmoidal curves in **a** represent computed fits with the values of $D_o$ indicated at the bottom of each curve (in cGy).

not include irradiation of a moat to isolate the test areas. This may be a significant factor.

When a single size of test area is used (e.g. Denekamp et al, 1974, 1982a, 1982b), and the data are plotted against dose, they can be fitted by eye with a straight line, or by computer assuming single-hit kinetics with N targets per area, each with radiosensitivity ($^1/D_o$). N represents the number of clonogenic cells multiplied by the extrapolation number.

Surviving areas per dose D $= 1 - [1 - \exp(-D/D_o)]^n$. The computer generates sigmoidal curves which can be fitted to the data and the corresponding $D_o$ value can then be used for comparing treatments where the actual fraction of islands regrowing does not overlap the two treatment groups. An example of this is shown in Figure 7.3a for mice irradiated in oxygen or nitrogen, with or without the radiosensitizer misonidazole.

## EXAMPLES OF INFORMATION OBTAINABLE

### Fractionation

The macro-colony technique was used at its inception to study the radiosensitivity of cells in vivo compared with cells cultured in vitro and the $D_o$ of 1.35 Gy was considered to be an 'OK value' (Withers, 1967a). It was also used to study the effect of varying the interval between two doses of radiation in order to examine the importance of recovery from sublethal injury and repopulation of surviving cells. Both Withers (1967c) and Emery et al (1970) varied the interval from a few hours to many days and the two sets of data are summarised in Figure 7.4. Many small changes were detectable in the response within the first 24 hours, with an initial peak at 5–6 hours, then a trough and a second peak at 10–12 hours. This was followed by a second trough, attributed to cells entering a more sensitive phase as they progressed through the cell cycle, and finally a steep overall increase in dose recovered which was attributed to repopulation. The doubling time during this phase was estimated to be 22 hours by Withers (1967c) and ~53 hours by Emery et al (1970). Whilst these values both represent an accelerated turnover compared with unirradiated epidermis there is no explanation of the discrepancy between them (Hegazy & Fowler, 1973; Potten et al, 1983). The immediate capacity for a compensatory regenerative response when using this assay in plucked skin contrasts with that observed in unplucked epidermis, e.g. of the foot, in which it is delayed for at least 1 week after large single doses (Denekamp et al, 1969), or after repeated small fractions (Denekamp, 1973; Denekamp et al, 1976). It is important to remember that the assay is in this respect artificial because the skin is already stimulated by the plucking into more rapid proliferation and is not in a steady state of equilibrium at the time of irradiation (Hegazy & Fowler, 1973).

Attempts to use the macro-colony technique to look

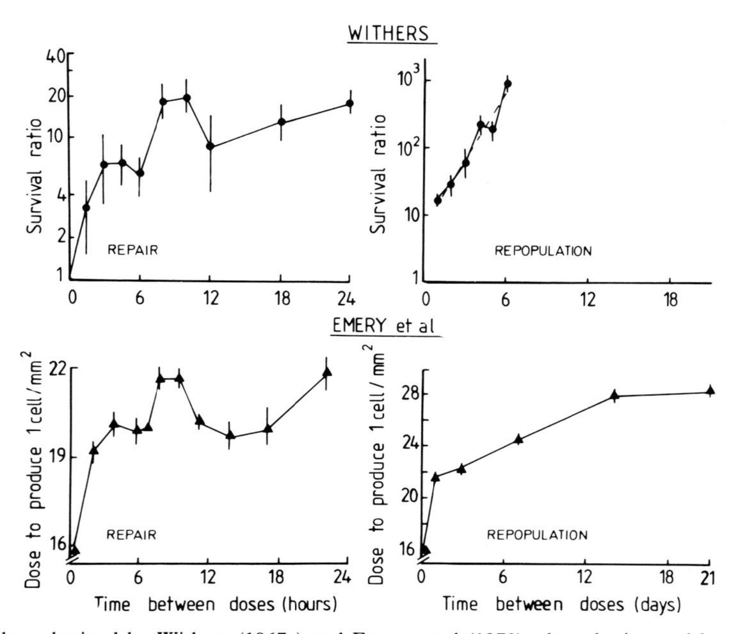


**Fig. 7.4** Fractionation data obtained by Withers (1967c) and Emery et al (1970) when the interval between two doses was increased from 1–24 hours (left hand panels) or for a period of days (right hand panels). The clone data can be expressed in terms of the survival ratios for a fixed dose (upper panels), or, if more complete survival curve data are obtained, as the dose increment in Gy needed to give a constant level of cell survival (lower panels).

at more extensive fractionation schedules proved frustrating (Denekamp & Emery, unpublished). After multiple irradiations (e.g. 5 fractions in 4 days) the dose-response curves obtained were very imprecise and the data were far more scattered than in any of the one or two fraction schedules. The reasons for this were not investigated but they may relate to heterogenity of the cells within each island, from island to island, and from mouse to mouse in terms of the parasynchronous progress of cells around the cell cycle and their repopulation between fractions. Even with two fractions there was much more scatter than with single doses together with a large and significant variation in the $D_o$ values obtained if two fractions were given with a varying interval between them (Emery et al, 1970). The $D_o$ estimates were higher for all intervals longer than 7 hours.

**Relative biological effectiveness of different qualities of radiation**

The macro-colony technique was used to study the RBE of both fast neutrons (Denekamp et al, 1971) and of helium, carbon and lithium ions (Leith et al, 1971). The use of any penetrating radiation makes it necessary that the animals are irradiated tangentially in order to avoid death from damage to the underlying intestine. Whereas Withers was able to use an incident beam of 29 or 150 kVp X-rays, because its shallow penetration permitted the gut to be spared, the use of 250 kVp X-rays or neutrons requires tangential irradiation of the test area as described by Emery et al (1970). Providing a reasonably uniform dose distribution can be achieved across the whole field (i.e. across all the test islands), this is not a technical problem. However, such a tangential irradiation *cannot* be used for isolating the test islands (i.e. for giving the supralethal dose to the moat). Thus such experiments require that two different radiation sources should be used, for the moat and test dose (see Table 7.1).

The RBE estimates for neutrons are very similar to those obtained using average skin reactions and the assay seems to offer no special benefit relative to gross skin reactions for such work. Its only advantage is that it produces a survival curve which is assessable by conventional statistical techniques, whereas skin reactions are arbitrarily allocated and there is some dispute as to whether there is any validity in applying conventional parametric statistical techniques in their analysis (e.g. Leith et al, 1975; Herring, 1980). In practice they give very similar conclusions, e.g. the neutron RBE as a function of dose per fraction for mouse skin clones falls on the same curve as that obtained using gross reactions from either the mouse, rat, pig or human (Field, 1976).

**Oxygenation**

One of the first studies performed with the epidermal macro-colony technique was that in which Withers (1967b) studied the influence of oxygen and anaesthesia. He found that the radiosensitivity of epidermis in air-breathing mice could be enhanced by a factor of 1.2 if they were irradiated in hyperbaric oxygen, or could be reduced by a factor of 2.6 if the skin was made anoxic by means of a suction device, with the mice breathing 8 per cent oxygen instead of air. He concluded that mouse skin was at a uniform and slightly radioprotective oxygen tension and that this was not a result of the anaesthetic used. He calculated the likely oxygen tension in the basal layer of the epidermis to be in the region of 7–17 mm Hg partial pressure. This was lower than might be expected (relative to venous oxygen tensions of 40 mm Hg) but was attributed to the lower affinity of mouse haemoglobin for oxygen than that of human haemoglobin.

The oxygenation of mouse epidermis has recently been of some interest to us in our studies of chemical radioprotectors which act at least partially by oxygen-competitive mechanisms. Using the high electron dose-rates that can be achieved with a van der Graaff generator we have studied the dependence of the radiosensitivity of the skin of the dorsum on the oxygen concentration in the gas surrounding and breathed by the animals. Oxygen concentrations ranging from 100 per cent to 0 per cent (pure nitrogen) have been used, and the dose-reponse curves obtained allow an 'apparent K curve' to be derived as shown in Figure 7.5 (Dene-

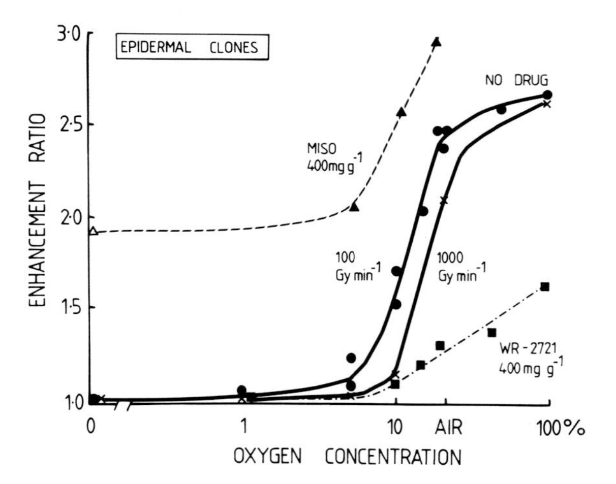


**Fig. 7.5** The oxygen 'K curve' derived from data like those shown in Fig. 7.3. The dose needed to reduce the survival to a constant level has been compared for each oxygen concentration with that in mice breathing nitrogen to calculate an oxygen enhancement ratio. The '$K_{insp}$' value, i.e. the concentration of oxygen giving half the full sensitization is 10–12%, depending upon the electron dose-rate. The addition of misonidazole shifts the curve upwards and to the left, whereas the addition of WR-2721 shifts the curve to the right (i.e. making oxygen less effective).

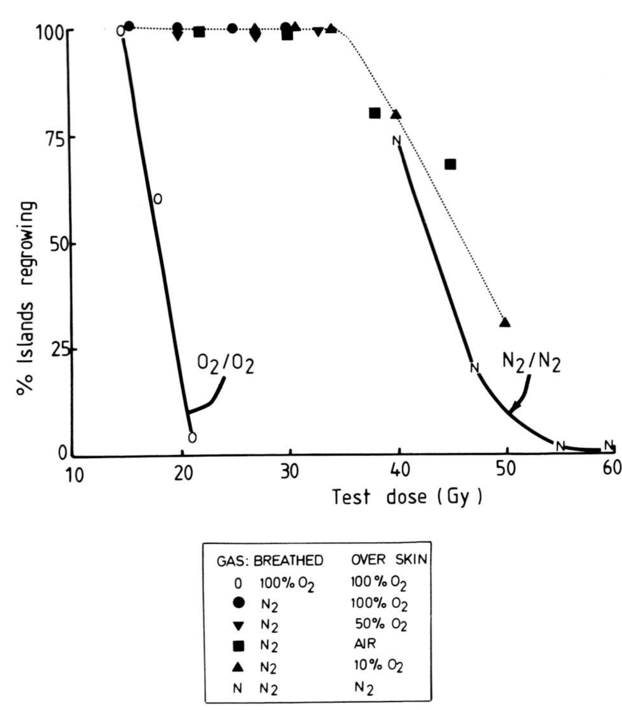


**Fig. 7.6** Data from an experiment in which the oxygen concentration was independently varied in the side of the chamber containing the irradiated skin and in the side containing the mouse's head. The clone survival curves for mice breathing nitrogen were apparently not sensitized if gas mixtures ranging from 10% to 100% oxygen were flowing over the skin (Denekamp, Michael, Rojas & Stewart, unpublished).

kamp et al, 1982b). The '$K_{insp}$ value' or inspired oxygen concentration at which half the oxygen enhancement ratio is observed is dependent upon dose-rate, and can also be modified by chemical agents. The '$K_{insp}$' is about 10 per cent oxygen if the dose-rate for the electrons is 100 Gy min$^{-1}$ and ~12 per cent oxygen if a higher dose-rate of 1000 Gy min$^{-1}$ is used. All these experiments were performed under anaesthetic and it is not possible to comment on how much these 'K values' are influenced by the deep anaesthesia necessary to keep the skin wrinkle-free during irradiation.

It has previously been stated that oxygen can readily diffuse from the gas phase through the thin mouse epidermis on the dorsum to reach the basal cells (Withers, 1967a) or the melanocytes at the base of the hair follicle (Potten & Howard, 1969). However, a recent experiment to test this using the macro-colony assay on the dorsal skin indicated that it is mainly the gas breathed by the mice, rather than the gas flowing over the skin surface, that determines the radiosensitivity of the epidermal clonogenic cells (Fig. 7.6). This agrees well with the oxygen diffusion measurements recently published by Evans et al (1981) which indicated that keratin is a very effective barrier to oxygen diffusion from the surrounding gas phase. The discrepancy with the early results is not understood and requires further elucidation.

**Chemical modifiers of radiation response**

The epidermal macro-colony technique has been used extensively in the last decade at the Gray Laboratory to study radiosensitizers and radio protectors. The ease of inducing radiobiological hypoxia in the skin and the steep dose-response curves obtained have made it an ideal assay for comparing different radiosensitizers in terms of their in vivo potential. In Figure 7.3 some of the data orginally obtained with misonidazole (miso) are shown (Denekamp et al, 1974). These experiments demonstrated that at drug doses that could be tolerated by the mice (they must survive 21 days for this experiment), sensitization could be achieved in mice made artificially hypoxic by breathing nitrogen, but not in those breathing oxygen. Figure 7.3 shows that 1.25 mg g$^{-1}$ miso sensitized the skin of mice breathing nitrogen almost as much as if they were breathing 100 per cent oxygen. Subsequent studies of skin reactions in patients showed a close agreement with these clone experiments when the radiosensitizing efficiency of two drugs on hypoxic skin was compared (Dische & Zanelli, 1976; Fowler et al, 1976).

No particular technical complications were encountered in using the macrocolony assay for sensitizer experiments. The moat and test doses were given in immediate succession, so that the drug was present (or absent) for both irradiations. Because electron-affinic sensitizers are ineffective on well oxygenated cells, and because the moat dose was administered to mice breathing oxygen, any modifying effect of the drug on the scattered radiation received by the test islands during the moat irradiation was unimportant. However, when radioprotectors were used as modifying agents the modifying effect on the moat dose scattered beneath the shields was a major problem (Denekamp et al, 1981, 1982b, 1983). The dose of 30 Gy routinely given to the moat in these experiments is more than adequate to kill all the epidermal cells in the moat. Thus protection of the moat itself (e.g. by large doses of WR-2721) did not give rise to clones growing from surviving cells within the moat. But it did mean that the effect of the scattered dose under the shield was different for mice treated without drug or with radioprotector, and was also dependent upon the dose of drug used. This resulted in completely false estimates of the protection factors in the first experiment performed with WR-2721 (unpublished data).

Since WR-2721 is completely ineffective as a protector if the mice are breathing nitrogen (Fig. 7.3b), it would in principle be possible to avoid these false estimates by giving the moat doses under anoxic conditions for all radioprotector experiments. However, this seems unwise since other protectors may not be ineffective in hypoxic mice and it would not obviate the problem if combinations of sensitizers and protectors were being used to study their interaction. It is therefore recommended that for such drug studies the moat dose is given without drug some hours in advance, so that there is no possibility of modifying the starting number of cells in the test areas accidently by the agents that are being studied. This is an important technical point. The only disadvantages of the two separate series of irradiations are:

1. the mice are subjected to two doses of anaesthetic instead of one
2. it increases the overall time of the experimental irradiations.

However, it has the advantage that full repair of sublethal injury from the scattered dose under the shields will occur and hence this scattered dose will have less effect on the test islands.

Figures 7.3 and 7.5 demonstrate the kind of data that have been obtained with the epidermal clone assay in the study of radiomodifying chemicals. Because of the ease with which the oxygen tension in the skin can be manipulated, and because the skin oxygenation appears to be homogeneous, the assay offers the possibility of studying the influence of oxygen tensions on the modification of radiosensitivity by various chemicals. This is only possible with electron irradiation because of the high dose-rates and hence short exposure times for breathing the requisite low oxygen concentrations. No such experiments have been attempted with any other systems in vivo and it therefore forms an important bridge between studies in vitro and other clonal or functional normal-tissue assays.

## Relationship of the clonal assay to gross skin reactions

The relationship between killing of colony forming units and loss of tissue function has been demonstrated for bone marrow (Robinson, 1968), for small intestine (Hornsey, 1973) and for skin of the mouse tail (Hendry, 1984). Figure 7.7 shows some of the macro-colony data for dorsal epidermis compared with the doses needed to give an average gross skin reaction of 1.0 on the foot of the same strain of mice. In each case, the dose needed to give this degree of gross tissue malfunction corresponds to between 0.1 and 0.01 cells surviving per square millimetere. This can be taken as an indication that erythema and acute desquamation of the foot are directly related to the fraction of epidermal cells surviving after irradiation. It is clear in certain circumstances, particularly in pigs, that this may be modified however by damage to the dermal elements supplying nourishment to the epidermis (Archambeau et al, 1979).

## Nature of the clonogenic cells

Potten & Hendry (1973), Al Barwari & Potten (1976) and more recently Potten et al (1983) have stressed the discrepancy between the number of cells surviving in the epidermis (per mm$^2$) and the initial number of basal epidermal cells. They have postualted that the cells which can give rise to a macro-colony may be only those which act as stem cells at the centre of an epidermal proliferative unit. The epidermis of mice (and other mammals) shows a remarkable organisation into hexagonal units, corresponding in area to the overlying squames (Mackenzie, 1970; Christophers, 1971). These

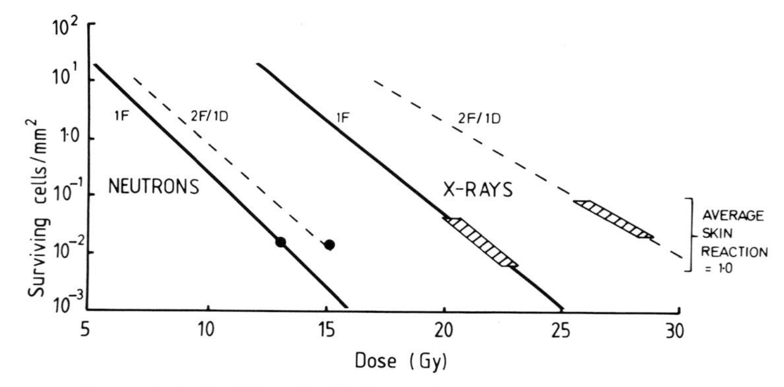


**Fig. 7.7** Diagram to compare the survival of epidermal cells (assayed by the macro-colony technique) with the gross reactions scored as erythema and desquamation over the period 8–30 days. Survival curves are shown for single doses and two fractions/24 hours of X-rays and neutrons. The points from single experiments and the hatched regions from and two combined experiments represent the doses needed to give an average skin reaction of 1.0 on the foot of the same strain of mice. Clone data from Withers (1967 a,c) and Emery et al (1970). Skin reaction data (shaded area) from Denekamp et al (1966, 1969). The skin reaction data all fall within the same decade of surviving fraction i.e. approx $10^{-1}$ to $10^{-2}$ surviving cells/mm$^2$.

units consist of approximately ten cells and if it is assumed that only one of these is a clonogenic stem cell then the discrepancy between the predicted initial number of clonogenic cells per $mm^2$ and the observed number almost disappears. It has been calculated that there are approximately $10^5$ colony-forming units per square cm in dorsal mouse skin and only about $3 \times 10^3$ in mouse tail (Hendry, 1984). Whatever the precise relationship between the surviving fraction and the gross skin damage it is apparent that several thousands of clonogenic cells can be killed with no visible damage being expressed in terms of dry or moist desquamation. Only one or a few clonogenic cells per square cm need to survive in order to prevent desquamation, and these can presumably repopulate the whole irradiated area before the functional cells have been lost.

## SUMMARY

The epidermal macro-colony assay has advantages over most other in vivo clonal assays in that a wide range of survival levels can be investigated by varying the size of the test islands, and the oxygenation of skin can be readily manipulated. It is a rapid assay ($\leq$ 21 days) but requires considerable expertise, good physics back-up and a reasonable prior knowledge of the likely dose-range of interest before it can be applied. For certain types of experiment it offers a considerable advantage over cruder methods of assaying skin damage, but in general the results from the two types of assay are in good agreement.

## REFERENCES

Al Barwari S E, Potten C S 1976 Regeneration and dose-response characteristics of irradiated mouse dorsal epidermal cells. International Journal of Radiation Biology 30:201

Arcangeli G, Mauro F, Nervi C, Withers H R 1980 Dose survival relationship for epithelial cells of human skin after multifraction irradiation: evaluation by a quantitative method in vivo. International Journal of Radiation Oncology Biology and Physics 6:841

Archambeau J O, Bennett G W, Abata T J, Brenneis H J 1979 Response of swine skin to acute single exposures of X-rays: quantification of the epidermal cell changes. Radiation Research 79:299

Christophers E 1971 Cellular architecture of the stratum corneum. Journal of Investigative Dermatology 56:165

Cohen L 1968 Theoretical iso-survival formulae for fractionated radiation therapy. British Journal of Radiology 41:522

Denekamp J 1973 Changes in the rate of repopulation during multifraction irradiation of mouse skin. British Journal of Radiology 46:381

Denekamp J, Harris S R 1975 The response of mouse skin to multiple small doses of radiation. In: Alper T (ed) Proceedings of 6th L H Gray Conference. Wiley, Chichester, p 342

Denekamp J, Michael B D 1972 Preferential sensitization of hypoxic cells to radiation in vivo. Nature New Biology 239:21

Denekamp J, Ball M M, Fowler J F 1969 Recovery and repopulation in mouse skin as a function of time after X-irradiation. Radiation Research 37:361

Denekamp J, Emery E W, Field S B 1971 Response of mouse epidermal cells to single and divided doses of fast neutrons. Radiation Research 45:80

Denekamp J, Michael B D, Harris S R 1974 Hypoxic cell radiosensitizers: Comparative tests of some electron affinic compounds using epidermal cell survival in vivo. Radiation Research 60:119

Denekamp J, Rojas A, Stewart F A 1983 Is radioprotection by WR-2721 restricted to normal tissues? In: Nygaard O, Simic M (eds) Proceedings of the First Conference on Radioprotectors and Anticarcinogens, Gaithersburg, Maryland. Plenum Press, New York, p 655

Denekamp J, Stewart F A, Douglas B G 1976 Changes in the proliferation rate of mouse epidermis after irradiation: continuous labelling studies. Cell and Tissue Kinetics 9:19

Denekamp J, Michael B D, Rojas A, Stewart F A 1981 Correspondence: thiol radioprotection in vivo: the critical role of tissue oxygen concentration. British Journal of Radiology 54:1112

Denekamp J, Michael B D, Rojas A, Stewart F A 1982b Radioprotection of mouse skin by WR-2721: the critical influence of oxygen tension. International Journal of radiation Oncology Biology and Physics 8:531

Denekamp J, Fowler J F, Kragt K, Parnell C J, Field S B 1966 Recovery and repopulation in mouse skin after irradiation with cyclotron neutrons as compared with 250 kV X-rays or 15 MeV electrons. Radiation Research 29:71

Denekamp J, Michael B D, Minchinton A I, Smithen C E, Stewart F A, Stratford M R L, Terry N H A 1982a Comparative studies of hypoxic-cell radiosensitization using artificially hypoxic skin in vivo. British Journal of Cancer 45:247

Dische S, Zanelli G D 1976 Skin reaction — a quantitative system for measurement of radiosensitization in man. Clinical Radiology 27:145

Ellis F 1969 Dose, time and fractionation: a clinical hypothesis. Clinical Radiology 20:1

Emery E W, Denekamp J, Ball M M, Field S B 1970 Survival of mouse skin epithelial cells following single and divided doses of X-rays. Radiation Research 41:450

Evans N T S, Naylor P F D, Rowlinson G 1981 Diffusion of oxygen through the mouse ear. British Journal of Dermatology 105:45

Field S B 1976 An historical survey of radiobiology and radiotherapy with fast neutrons. Current Topics in Radiation Research Quarterly 11:1

Field S B, Morris C, Denekamp J, Fowler J F 1975 The response of mouse skin to fractionated X-rays. European Journal of Cancer 11:291

Foster C J, Hope C S, Orr J S 1967 Dose survival relationship for irradiation of epithelial cells of mouse skin. British Journal of Radiology 40:479

Fowler J F, Stern B E 1963 Dose-time relationships in radiotherapy and the validity of cell survival curve models.

British Journal of Radiology 36:163
Fowler J F, Adams G E, Denekamp J 1976 Radiosensitizers of hypoxic cells in solid tumours. Cancer treatment Reviews 3:227
Hamilton E, Potten C S 1972 Influence of hair plucking on the turnover time of the epidermal basal layer. Cell and Tissue Kinetics 5:505
Hegazy M A H, Fowler J F 1973 Cell population kinetics of plucked and unplucked mouse skin 1. unirradiated skin. Cell and Tissue Kinetics 6:17
Hendry J H 1984 Correlation of the dose response relationships for epidermal colony-forming units, skin-reactions and healing in the X-irradiated mouse tail. British Journal of Radiology (in press)
Herring D F 1980 Methods for extracting dose-response curves from radiation therapy data: I A unified approach. International Journal of Radiation Oncology Biology and Physics 6:225
Hewitt H B, Wilson C W 1959 A survival curve for mammalian leukaemia cells irradiated in vivo (implications for the treatment of mouse leukaemia by whole body irradiation). British Journal of Cancer 13:69
Hornsey S 1973 The effectiveness of fast neutrons compared with low LET radiation on cell survival measured in the mouse jejunum. Radiation Research 55:58
Kember N F 1965 An in vivo cell survival system based on the recovery of rat growth cartilage from radiation injury. Nature 207:501
Leith J T, Schilling W A, Welch G P 1971 Survival of mouse skin epithelial cells after heavy particle irradiation. International Journal of Radiation Biology 19:603
Leith J T, Schilling W A, Lyman J T, Howard J 1975 Comparison of skin responses of mice after single or fractionated exposure to cyclotron accelerated helium ions and 230 kV X-irradiation. Radiation Research 62:195
Mackenzie I C 1970 Relationship between mitosis and the ordered structure of the stratum corneum in mouse epidermis. Nature 226:653
Potten C S, Hendry J H 1973 Clonogenic cells and stem cells in epidermis. International Journal of Radiation Biology 24:537
Potten C S, Howard A 1969 Radiation depigmentation of mouse hair: the influence of local tissue oxygen tension of radiosensitivity. Radiation Research 38:65
Potten C S, Hendry J H, Al Barwari S E 1983 A cellular analysis of radiation injury in epidermis. In: Potten C S, Hendry J H (eds) Cytotoxic insult to tissue. Churchill Livingston, Edinburgh, p 153
Reisner A 1932 Der Hauterythemverlaus bei fractionierter Vereabfolgung grosser strahlenmengen. Fortschritte auf dem Gebiete der Röntgenstrahlen 45:293
Robinson C V 1968 Relationship between animal and stem cell dose-survival curves. Radiation Research 35:318
Till J E, McCulloch E A 1961 A direct measurement of the radiation sensitivity of normal mouse bone marrow. Radiation Research 14:213
Withers H R 1967a The dose survival relationship for irradiation of epithelial cells of mouse skin. British Journal of Radiology 40:187
Withers H R 1967b The effect of oxygen and anaesthesia on radiosensitivity in vivo of epithelial cells of mouse skin. British Journal of Radiology 40:335
Withers H R 1967c Recovery and repopulation in vivo by mouse skin epithelial cells during fractionated irradiation. Radiation Research 32:227

# 8

*C.S. Potten S.E. Al-Barwari*

# Micro-colonies in mouse epidermis

## HISTORICAL BACKGROUND

In 1967 Withers described a technique whereby surviving epidermal cells that possessed a regenerative capacity could be identified by the nodules or colonies that they produced on a heavily-irradiated dermis (see Ch. 7). The nodules grew rapidly with a doubling time of about 22 h so that by the 10th day they could be seen with the naked eye. Providing a rather restrictive range of doses was employed, a cell survival curve could be determined (at less than about 9 Gy the colonies coalesced and at more than about 23 Gy there would be no colonies in the test areas). The technique is technically quite difficult to execute. A range of different sized shields was used within a sterilised field of irradiated skin. The shield size was altered so that very few or no survivors would be expected and the mean number of surviving cells per unit area was deduced using Poisson statistics. The other slight limitation was that much of the animal had to be shielded or protected so that it survived the necessary 2–3 weeks after irradiation. This was achieved partially by using small fields and partially by using low energy X-rays that penetrated only into the dermis.

A very similar technique was later used to determine the number of macroscopically visible colonies present after various doses in the small intestine (Withers & Elkind, 1969) (see Ch. 4). It was then realised that the colonies visible to the naked eye on days 10–14 after irradiation also should be detectable microscopically at much earlier times (3–4 days) as micro-colonies (Withers & Elkind, 1970) (see Ch. 5). A similar approach was then adopted for epidermis by using sheets of epidermis prepared 3–4 days after irradiation and viewed under the microscope. The colonies were identified by prelabelling the mouse with tritiated thymidine and looking for clusters of rapidly cycling cells which hence contained many labelled cells (Al-Barwari & Potten, 1976).

## DESCRIPTION OF THE TECHNIQUE

As with the micro-colony technique in gut, the animals can be irradiated with $\gamma$ rays or X-rays to the whole body (7.5–27 Gy) since they will be sacrificed on the 3rd or 4th day, before the gastrointestinal syndrome causes death. For the macro-colony technique, the animals were prepared by plucking the hairs about 12–24 hours prior to irradiation to stimulate the epidermis. This may provide an essential trigger for the early and rapid proliferation of the colony-forming cells, and is probably also preferable in the micro-colony technique. Without this stimulation the colony-forming cells may begin proliferation only after cellular depletion of the basal layer which may take many days. If the animals are not plucked the micro-colonies are not detected until the 14th day (Keech, 1982). 40 minutes before sacrifice the mice were injected intraperitoneally with tritiated thymidine (approximately 1–2 $\mu$ Ci/g body weight, 5 Ci/mmol). A large piece of skin (approximately 2 × 3 cm) was removed from the back of the mouse from which the hairs if not plucked had been previously clipped with electric clippers. The skin was stretched flat and the muscle and fat layers were stuck to a piece of paper which was then pinned onto a cork board. The remaining hair stubble was removed using a commercial depilatory cream. This usually takes about five minutes and is aided by scraping and respreading the cream once or twice during the five minutes. The skin, still with its attached paper, was washed carefully to remove any remaining cream and the skin and paper were placed in 0.5 per cent acetic acid at room temperature for 1–3 hours before being placed in the refrigerator overnight. The following day the acetic acid was carefully drained off and was replaced by Carnoy fixative for 20–30 minutes. This was then replaced by 70 per cent ethanol in which the skin can be stored until required for separation of epidermal sheets. This was conducted in 70 per cent ethanol under a dissecting microscope using fine watchmaker's forceps or fine dissecting needles. Pieces of epidermis were gently teased from the skin, placed in 70 per cent ethanol on a gelatin-subbed microscope slide (slides dipped in 0.5 g Gelatin plus 0.05 g Chrome Alum in 100 ml water), and then carefully spread out with the cornified cell surface downwards (towards the glass) and the basal layer ex-

**Table 8.2** Criteria used to define surviving colonies based on the size of the clusters of labelled cells

| Days after irradiation with 12 Gy | Minimum number of labelled nuclei per cluster | Colonies or clusters/mm² (± se) |
|---|---|---|
| 2 | 9 | 0.62 ± 0.07 |
| 3 | 17 | 0.56 ± 0.07 |
| 4 | 24 | 0.54 ± 0.05 |
| 5 | 35 | 0.51 ± 0.03 |
| 6 | 50 | 0.52 ± 0.11 |

Taken from Al-Barwari & Potten (1976)

the position of the 3–4 peaks in the cluster size distributions remained roughly constant with time while the maximum on the distributions increased steadily. However, by taking into account several different sets of data a somewhat arbitrary series of threshold values could be derived for defining a surviving cluster or colony (Table 8.2). These took into account the cluster-size distributions, the growth curves of the clusters and the fact that survival levels would be expected to remain constant with time after irradiation. These threshold values excluded any satellite clusters.

The third day after irradiation was selected as optimum because (1) the data for earlier times are more likely to be contaminated with abortive colonies, and (2) the satellite and follicle-derived colonies are less apparent on day 3 than at later times.

**Presentation of data**

Using a criterion of 17 or more heavily-labelled nuclei (about 32 total nuclei) for clusters (or colonies) derived from cluster- or colony-forming cells a survival curve was obtained (Fig. 8.3). A minimum of five mice was used at each of the doses studied. 40 fields, each of approximately 1 mm², were scored from each mouse. Thus, in practice, between 197 and 589 mm² of epidermis were scored at each dose. Over the range of doses this meant approximately 1400 colonies for the lowest dose and about one colony for the highest dose were scored per group of animals. This survival curve has a $D_o$ value of 2.33 ± 0.11 Gy and an extrapolation number of 123 cells/mm² as determined by the computer programme of Gilbert (1969).

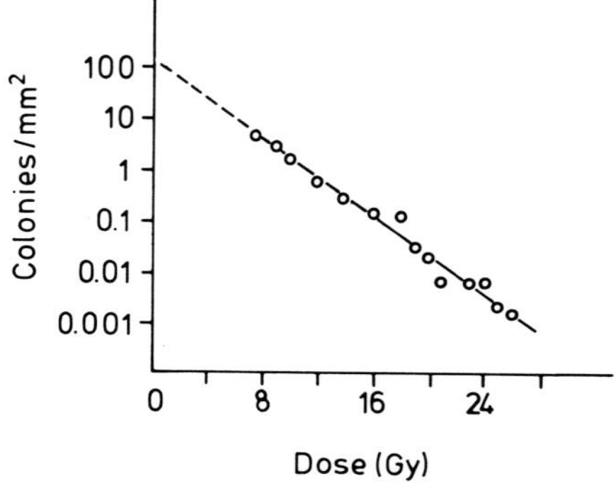


**Fig. 8.3** Survival curve for micro-colonies in skin on the back of DBA-2 male mice. The mice were plucked 20 h prior to irradiation (290 kVp X-rays, 1.5 Gy/min). They were given oxygen five minutes before and during irradiation. The curve was fitted by computer (Gilbert, 1969) and has a $D_o$ of 2.3 Gy and extrapolates to 123 cells/mm².

**Additional data obtained**

Using this approach, survival curves have been also obtained for unplucked animals which were breathing oxygen during irradiation (Fig. 8.4, curve B), animals plucked immediately after irradiation which were breathing air during irradiation (Fig. 8.4, curve D), and for hairless animals which were breathing oxygen during irradiation (Fig. 8.4, curve C). The parameters describing these curves are listed in Table 8.3. These curves appear to differ but the data points (not shown) probably scatter about a common line with a $D_o$ of about 2.5 Gy which would extrapolate to a value of about 335 cells/mm². The only other published information for unplucked DBA-2 mice quotes a $D_0$ value of 4.4 Gy and an extrapolation number of 4.6 cells/mm² (Keech, 1982). This curve is even shallower than curve D in Figure 8.4. These data were obtained by counting clusters 13–14

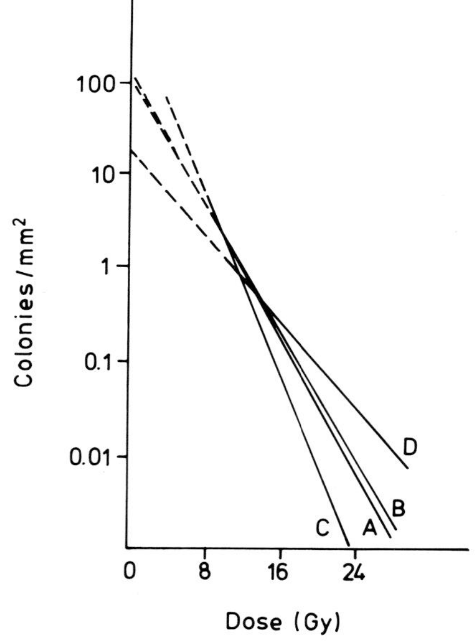


**Fig. 8.4** Survival curves obtained for DBA-2 and hairless mice under different plucking conditions and levels of oxygenation. For an explanation of the curves and the parameters that define them as determined by computer see Table 8.3. Data from Al-Barwari (1978).

**Table 8.3** Micro-colony survival for mice X-irradiated under various conditions

| Mouse strain | Plucking | Oxygenation* | Quality, dose rate (Gy/min) | $D_0$ (Gy) | Extrapolation number (cells/mm²) | Curve in Fig. 8.4 |
|---|---|---|---|---|---|---|
| DBA-2 | 20 h before irradiation | 95% $O_2$ | 290 kVp, 1.5 | 2.3 | 123 | A |
| DBA-2 | Unplucked | 95% $O_2$ | 290 kVp, 1.5 | 2.4 | 100 | B |
| DBA-2 | Immediately after irradiation | Air | 290 kVp, 1.5 | 3.6 | 19 | D |
| Hairless | Unplucked | 95% $O_2$ | 300 kVp, 2.9 | 1.8 | 1100 | C |

Curve A published in Al-Barwari & Potten (1978)
Curves B-C presented in Al-Barwari (1978)
* Gas breathed and over skin

days after irradiation i.e. at a time when macro-colonies might be expected. The author state that clusters could not be observed at earlier times. The causes of the differences remain obscure.

## LIMITATIONS OF THE TECHNIQUE

Besides the limitations outlined above there are others associated with this technique:

1. The survival curves appear to differ from those obtained using the macro-colony approach (Fig. 8.5). However, it is possible that all the data actually scatter about a single common line which would have a $D_o$ closer to 1.35 Gy than the 2.3 Gy reported for the micro-colony assay (see Potten, 1978).

2. The technique is difficult to use in a routine way due largely to the difficulty in defining the characteristics of the surviving colonies.

3. There is no proof that the colonies are in fact clones derived from single cells although this appears at present to be a reasonable assumption.

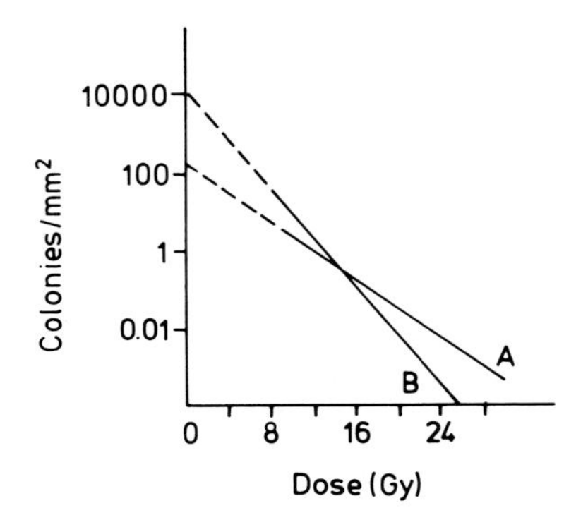


**Fig. 8.5** Comparison of the micro-colony survival curve (A) published in Al-Barwari & Potten (1976) and the macro-colony curve (B) published by Withers (1967).

4. It is unclear why the curves extrapolate to such low numbers of cells/mm² when the macro-colony curves extrapolate to numbers that are similar to the actual number of basal cells/mm², i.e. about 14 000. It might be expected that the extrapolation number is related to the product of the extrapolation numbers for an individual cell and the number of colony-forming cells per mm². The micro-colony curves do in fact extrapolate to numbers that are close to the actual number of hair follicles per mm² (about 40, see Potten, 1983) which is interesting since many of the colonies seem to be associated with, or at least to lie adjacent to, the hair follicle canals (Table 8.4).

5. There appears to be a difficulty in relating the growth characteristics of the micro-colonies to those of the macro-colonies. In the latter case the colony-forming cells divide rapidly to produce a visible lump or nodule that clearly has considerable depth. The micro-colonies, in contrast, appear to spread laterally very rapidly by forming numerous satellite colonies. The explanation for these differences in growth remains obscure.

**Table 8.4** The proportion of clusters of labelled cells that encompass, or are adjacent to, hair follicle canals

| Dose (Gy) | Time after irradiation (d) | % clusters associated with follicles |
|---|---|---|
| 8 | 3 | 53.1 |
| | 4 | 48.4 |
| 12 | 3 | 70.3 |
| | 4 | 71.5 |
| 16 | 3 | 83.7 |
| | 4 | 76.1 |
| 18 | 3 | 74.7 |
| | 4 | 81.7 |
| 20 | 3 | 77.4 |
| | 4 | 74.8 |

Data taken from Al-Barwari (1978)
Clusters were defined as associated if one or more labelled cells were immediately adjacent to a hair follicle

6. It is unclear whether the micro-colonies are derived from interfollicular epidermal or intrafollicular stem cells, some mixture of both, or early transit cells. It is also unclear how and where the macro-colony-forming cells are situated spatially and hierarchically, and what their relationship is to the micro-colony-forming cells.

## REFERENCES


Al-Barwari S E 1978 Cell and population kinetics in the irradiated skin. Thesis, University of Manchester, Manchester

Al-Barwari S E, Potten C S 1976 Regeneration and dose-response characteristics of irradiated mouse dorsal epidermal cells. International Journal of Radiation Biology 30:201

Gilbert C W 1969 Computer programmes for fitting puck and probit survival curves. International Journal of Radiation Biology 16:323

Keech M L 1982 Measurement of micro-colony survival in unplucked mouse skin. British Journal of Radiology 55:941 (abs)

Potten C S 1978 The cellular and tissue response of skin to single doses of ionising radiation. Current Topics in Radiation Research 13:1

Potten C S 1983 Stem cells in epidermis from the back of the mouse. In: Potten C S (ed) Stem cells: their identification and characterisation. Churchill Livingstone, Edinburgh, p 200

Withers H R 1967 The dose-survival relationship for irradiation of epithelial cells of mouse skin. British Journal of Radiology 40:187

Withers H R, Elkind M M 1969 Radiosensitivity and fractionation response of crypt cells of mouse jejunum. Radiation 38:598

Withers H R, Elkind M M 1970 Microcolony survival assay for cells of mouse intestinal mucosa exposed to radiation. International Journal of Radiation Biology 17:261

# 9

*C.S. Potten*

# Hair follicle survival

## HISTORICAL BACKGROUND

Three months after the description by Roentgen of his production of X-rays such rays were used experimentally for the location of a foreign object in the head. This was preliminary to locating a bullet in the head of a child. The subject used for the experiment lost the hair in the exposed area within a period of 21 days (Daniel, 1896). This was the first recorded case of radiation-induced epilation.

In the succeeding 87 years epilation has been studied sporadically with the production of rather variable and imprecise data. Epilation is clearly the consequence of radiation damage to the cells of the hair-producing organs, the hair follicles. The hair follicles when actively producing hair are structures containing several thousand cells many of which are dividing rapidly (see Fig. 9.1 and Ch. 12 for more details). In the mouse they divide with cell cycle times as short as any renewing cell populations (i.e. the cycle duration is about 12 hours; Cattaneo et al, 1961; Griem, 1966; Fry et al, 1968). The consequence of this division activity is that hair is produced at a rate of up to 1 mm/day. The hair contains a record of the proliferative history of the follicle cells. Any interruption or reduction in hair production may be apparent as a reduction in hair diameter — this is the basis of the hair dysplasia end-point first noted in 1906 by Williams but elaborated by others, notably Griem & Malkinson (1967, 1969). The upper limit for the doses that can be used with this end-point is the dose (about 10 Gy) resulting in such a thinning of the hair that it falls out — the threshold epilation dose for growing hair follicles. Hair is not grown continuously by the follicles. When a mature hair is produced the follicle 'switches off' proliferatively, shrinks in size and enters a dormant resting phase or telogen, in contrast to the growth phase or anagen. The resting follicle tends to be much smaller, contains only a few hundred cells, none of which is dividing, synthesising DNA or passing through the cell cycle (see Fig. 9.1). Hair growth cycles tend to be synchronised in male mice with the first cycle terminating on the 21st day after birth and the second beginning on about the 28th day and terminating on the 49th day (7 weeks) (Silver et al, 1969). A new hair growth cycle can be initiated at any time by the act of plucking the hair from a resting follicle. In female mice the later hair growth cycles become somewhat erratic and desynchronised as a consequence of the influence of the varying levels of sex hormones. In males they show a greater tendency to remain synchronised.

The hair follicles are formed by infolds of the epidermis during late embryonic and early post-natal development after which their numbers do not change (Claxton, 1967). Once destroyed there is no indication that they can be replaced or re-initiated from the epidermis. There is a specialised region of dermis, the dermal papilla, which plays an important role in organising the follicle in successive waves of hair growth (Oliver, 1971). Should the follicle germ and the dermal papilla be separated the follicle will fail to function and grow a hair. For additional details on follicle structure and hair growth see Chapter 10.

Thus, the hair follicle can be regarded as a closed proliferative unit that contains many rapidly dividing cells in anagen and a few quiescent germ cells in telogen, the structure being organised to some extent by the dermal papilla. If all germinal (stem) cells are killed or separated from the dermal papilla the follicle 'dies', while on the other hand, if one or more germinal cells survive the follicle may be capable of repopulation, reorganisation and regrowth of a hair, i.e. the surviving germinal cells may express some regenerative or colony forming capacity. The clonal origin of these reformed colonies has not been investigated. In practice the sensitivity of the follicles is clearly very dependent on their position within the hair growth cycle. This may be in part due to differences in (1) the sensitivity between quiescent and cycling cells, (2) the closeness of the association of the dermal papilla and the germinal cells, and (3) the number of germinal cells per follicle. These factors together with the use of a wide range of different assay procedures probably account for the very variable results obtained for follicle survival. These points will be discussed in greater detail later in this paper.

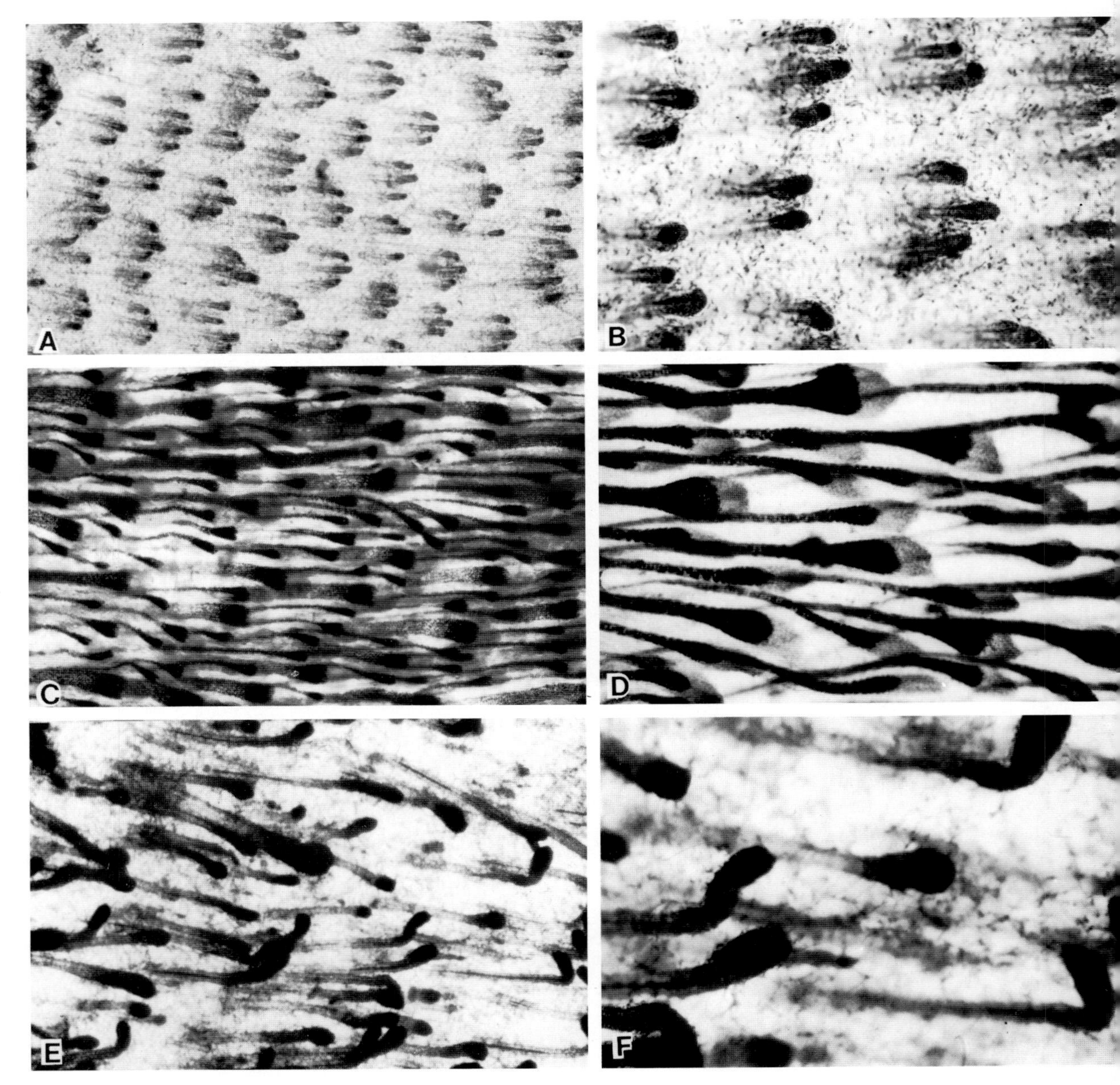


**Fig. 9.1** Whole mounts of skin with the muscle and fat layers removed by a one hour treatment with 5N HC1 after fixation in Carnoy's fixative. The pieces of skin are lightly stained with haematoxylin. **A** and **B** Normal telogen (resting) hair follicles. These are normally arranged in rows or small clusters of three or more follicles. **C** and **D** Normal anagen (growing) hair follicles. These vary in size and contain a core of intensely active melanocytes and a regular linear deposition of pigment in the forming hair. **E** and **F** Anagen follicles about 10 days into the second post-irradiation hair growth cycle. These mice received 10 Gy soon after plucking the hair to initiate new hair growth. They were replucked 28 days later and killed 10 days after the second plucking. The follicles are very irregular in size and are reduced in numbers. A, C and E x 33, B, D and F x 111.

## DESCRIPTION OF THE TECHNIQUES

One of the major difficulties in assessing the value of this technique and in comparing the published data is that the method of assay, the time of assay and the stage of the hair growth cycle at the time of irradiation all tend to vary considerably. Basically the technique involves either irradiating whole animals with soft X-rays, or selected parts of the animal with $\gamma$-rays or hard X-rays. Damage to the follicles is assessed by either the presence or absence of hair as viewed from the surface, or the presence or absence of hair follicles as viewed from the underside of skin removed from the animal (Fig. 9.1). If damage is assessed early after irradiation, epilation is being assayed, while in contrast, assessment at later times measures follicle regeneration and function. The longer the post-irradiation sampling time, the more stringent is the assay for colony regeneration. If the skin is in telogen at the time of irradiation and particularly if the hair is plucked immediately before or after irradiation to initiate a new hair growth cycle, then the follicles or skin can be assayed as early as 10–12 days post-irradiation when they should be in mid-anagen. Some workers, however, have allowed this post-irradiation hair cycle to terminate and have looked at the second post-irradiation hair cycle, in most cases following a second plucking. This clearly is testing the ability of surviving follicle germ cells to reorganise and repopulate a follicle, grow a hair, re-enter telogen and respond to another stimulus to reorganise another phase of growth: a fairly stringent test of regenerative ability.

Irradiation of telogen follicles in skin is best achieved by selecting male mice of the age 7–8 weeks at which age the majority of the follicles will be synchronously in telogen (Silver et al, 1969). Alternatively, the test area can be plucked to initiate a new cycle and the animals irradiated 21 days later when the follicles will have again reached telogen. Irradiation of skin with growing follicles is best achieved by plucking 7–8 week old male mice and exposing them to radiation 10–12 days later when the follicles will be in mid-anagen. Other stages of anagen can be treated by selecting other post-plucking times.

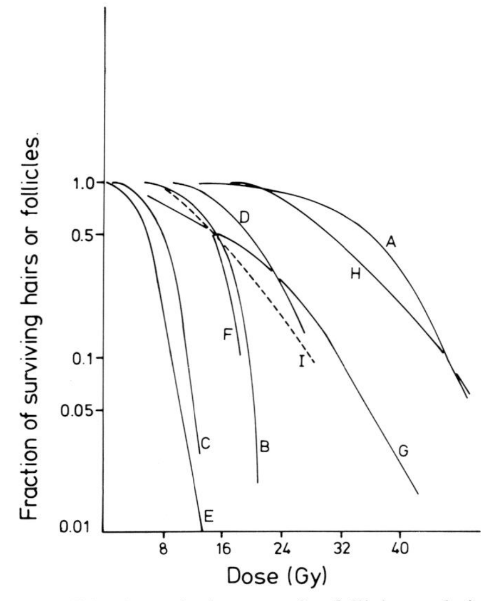


**Fig. 9.2** Published survival curves for follicles or hairs. Details for each curve are given in Table 9.1.

## SCORING PROCEDURES

Basically two approaches have been adopted. The first involves counting the hairs in a defined area. This generally requires the animal to be killed and samples taken for study under a low power microscope but in principle it could be also a non-invasive in situ assay. The presence of a hair at some long time interval after irradiation (times equivalent to one or more hair growth cycles) can be taken as an indication of the functional integrity of the hair follicle. The second approach involves counting the number of hair follicles per unit area by viewing the under-surface of the skin under the microscope (see Fig. 9.1). The difficulty here is in defining what constitutes a surviving hair follicle. The best assay is to score only large growing hair follicles. However, even here some subjective criterion on the size has to be applied. One problem is that a functionally dead follicle may persist as a peg of epithelial tissue possibly even complete with functioning sebaceous glands (e.g. it may merely have lost contact with the dermal papilla).

## DATA PRESENTATION

There have been only a limited number of published survival curves using these approaches. They are summarised in Figure 9.2 and Table 9.1. As can be seen there is an enormous range in sensitive expressed by these various reports. The anagen curve of Griem et al 1979 (curve C) may more correctly be thought of as an epilation assay. This may also to some extent be true for their telogen curve, and that of Dubravsky et al (1976) for early anagen, since all of these are assayed at early times post-irradiation. The most stringent tests for long-term survival were applied in the case of the telogen experiments of Burns et al (1968) and Hendry et al 1980 (curves A and G respectively). These together with the data for tail represent the most resistant survival curves. The following general statements can be made:

1. Early post-irradiation sampling times tend to generate data to the left on figure 9.2 (i.e. data exibiting greater sensitivity). These may be more accurately measuring cell killing and epilation. These studies tend to have $D_0$ values in the range 1.35–1.9 Gy i.e. data more in accord with other mammalian cell survival data in vivo, in particular those for interfollicular epidermal colony-forming cells.
2. Anagen follicles appear to be more resistant than telogen follicles though the data are not conclusive on this point.
3. Follicles in tails tend to be more resistant than those in the body skin.

**Table 9.1** Hair and hair follicle survival in experimental animals

| Animal | Age (days) | Hair cycle when irradiated | Radiation | Time of scoring post-irradiation (days) | Scoring | Fig. 9.2 curve | Threshold or $D_q$ (Gy) | $D_{50}$ (Gy) | $D_o$ (Gy) | Reference |
|---|---|---|---|---|---|---|---|---|---|---|
| Rabbits NZW | – | Telogen (around eye) | $^{60}$Co $\gamma$ rays | ~100 | Subjective index and Hairs on photographs | I | 8 | 16 | – | Cox et al, 1981 |
| Rats CD ♂ | 58 | Telogen | 0.7 MeV electrons | 560 | Follicles in whole mounts | A | 16–20 | 36 | – | Burns et al, 1968 |
| Mice CF ♀ | 84 | Telogen (preplucked) | 45 kVp X-rays | 36 | Hairs on photographs | B | 8–14<br>2–5 | ~15<br>7.5 | 1.35<br>1.35 | Griem et al, 1979 |
| " | " | Anagen (14 days post-plucking) | " | 7 | " | C | | | | |
| " | " | Telogen | " | 2 hair cycles (>42 days) | " | D12– | 17 | 19 | 4.9 | Griem et al, 1973 |
| " | " | Anagen (14 days) | " | ~7 | " | E | 2–5 | 6 | 1.7 | Griem et al, 1973 |
| C3H ♂ | 84 | Early anagen (3 days post-plucking) | 50 kVp X-rays | 17–19 | Hairs/area | F | – | ~15 | 1.9 | Dubravsky et al, 1976 |
| BDF1 ♂ | 84 | Telogen | 300 kVp X-rays | 2nd hair cycle (~47 days) | Follicles/area | G | ~2 | 16 | 5.4 | Hendry et al, 1980 |
| | | Telogen (tail) | | 84 | | H | 18 | 30 | – | Hendry et al, 1980 |

$D_{50}$ = dose to give 50% survival

4. The level of oxygenation at various points in the follicle, particular those on the tail, is important in determining their survival.

5. There are three quite similar sets of data for telogen follicles (using long-term survival criteria), one for the hairs around the eyes of rabbits and two for mice. These have $D_{50}$ values of 16–19 Gy and possible $D_o$ values of about 5 Gy. However, it should be realised that the curves probably represent little more than the shoulder on the hair follicle survival curves.

6. One set of data for telogen follicles in rats suggests that these are more resistant than in the mouse.

## LIMITATIONS OF THE TECHNIQUE

The major limitations of the technique are as follows:

1. The inherent variability suggested by the experiments conducted so far.

2. The difficulty in defining the optimum conditions for irradiation and particularly those for scoring (the time of scoring, the post-irradiation plucking schedule, the selection of follicle or hair counting procedures etc.).

3. The difficulty in defining the clonal or multicellular basis for follicle survival and the uncertain influence exerted by the dermal papilla.

Some of the reasons for the variability in the results obtained are as follows:

a. The oxygen concentration in the tissue surrounding the hair follicles and in the hair follicles themselves and the surrounding (external) ambient gas may vary.
b. The end-points used have not been standardised: some may score the hairs shed, some those that remain, others the follicles that remain intact, while yet others score those follicles which regrow a new hair. Many of these involve scoring at different times after irradiation.
c. Hair follicles in adult mice may contain more than one hair. Hence, follicle and hair counts may differ.
d. There are different types of hair growing from different sized follicles. Each type may have a different sensitivity determined largely by the number of target cells. The proportions of the various follicle types may vary from mouse strain to strain and from species to species.
e. Follicles in different stages of the their growth cycle contain very different numbers of cells at differing stages of the cell cycle and both differences in number and cell cycle phase will influence the overall follicular sensitivity.
f. Damage to one follicle may result in the loss of the hair but the follicle may reorganise to regrow another hair, or alternatively the follicle may be sterilised and be incapable of further hair growth. It is conceivable that it may continue to contain a hair for some time even though the follicle may be sterilised.
g. During the post-irradiation reorganistion of the epithelium it is possible that new follicles form from epidermal cells or sebaceous gland cells although this is thought to be unlikely. If it does occur this reorganisation would be expected to take a considerable time.

## REFERENCES


Burns F J, Albert R E, Heimbach R D 1968 The RBE for skin tumors and hair follicle damage in the rat following irradiation with alpha particles and electrons. Radiation Research 36:225

Cattaneo S M, Quastler H, Sherman F G 1961 Proliferative cycle in the growing hair follicle of the mouse. Nature 190:923

Claxton J H 1967 The initiation and development of the hair follicle population in tabby mice. Genetical Research Cambridge 10:161

Cox A B, Keng P C, Glass N L, Lett J T 1981 Effects of heavy ions on rabbit tissues: alopecia. International Journal of Radiation Biology 40:645

Daniel J 1896 The X-rays. Science 3:562

Dubravsky N, Hunter N, Withers H R 1976 The effect of precooling on the radiation sensitivity of the proliferating hair follicle. Radiation Research 65:481

Fry R J M, Kessler D, Kisieleski W E, Weber C L, Griem M L, Malkinson F D 1968 A method for the study of cell proliferation of the hair follicle. Argonne National Laboratory Biology and Medical Research Division Annual Report ANL 7535:77

Griem M L 1966 Use of multiple biopsies for the study of the cell cycle of the mouse hair follicle. Nature 210:213

Griem M L, Malkinson F D 1967 Some studies on the effects of radiation and radiation modifiers on growing hair. Radiation Research 30:431

Griem M L, Malkinson F D 1969 Some effects of radiations and radiation modifiers on growing hair. Progress review. Frontiers of Radiation Therapy and Oncology 4:24

Griem M L, Dimitrievich G S, Lee R M 1979 The effects of X-irradiation and adriamycin on proliferating and nonproliferating hair coat of the mouse. International Journal of Radiation Oncology, Biology and Physics 5:1251

Griem M L, Malkinson F D, Marianovic R, Kessler D 1973 Some studies of the X-ray effects on resting hair cell populations. In: Duplan J F, Chapiro A (eds) Advances in radiation research, biology and medicine 2. Gordon & Breach, New York, p 845

Hendry J H, Edmundson J M, Potten C S 1980 The radiosensitivity of hair follicles in mouse dorsum and tail. Radiation Research 84:87

Oliver R F 1971 The dermal papilla and the development and growth of hair. Journal of the Society for Cosmetic

Chemistry 22:741
Silver A F, Chase H B, Arsenault C T 1969 Early anagen initiated by plucking compared with early spontaneous anagen. In: Montagua W, Dobson R L (eds) Advances in biology of skin, IX. Hair growth. Pergamon, Oxford, p 265
Williams A W 1906 A note on certain appearances of X-rayed hairs. British Journal of Dermatology 19:63

# 10

*C.S. Potten*

# Melanocyte colonies in mouse hair follicles

## DESCRIPTION OF THE BIOLOGICAL MATERIAL

Hair grows from small infolds of the epidermis called hair follicles. Hair growth is a cyclic phenomenon with each hair follicle in humans acting largely independently of all others while in small rodents the follicles of the body often tend to be synchronised in relation to hair production. In the mouse the growth phase lasts 21 days after which there is a shrinkage in the follicle size, particularly in its cellularity, and a complete cessation of cell proliferation. This dormant phase is known as *telogen* while the growth phase which is characterised by a large follicle with many rapidly dividing cells is termed *anagen*. For additional detail of hair follicle structure and hair growth cycle see Chapter 9. There are follicles of different sizes producing the different types of large guard hairs and the predominant small zigzag hairs. Between 60 and 75 per cent of the hairs are the small zigzag underfur hairs. These contain about 20–30 active (melanogenic) melanocytes synthesising melanin in the form of granules (melanosomes) which are injected into, or ingested by, the neighbouring keratinocytes. These then move into the new hair, carrying their pigment with them. Once incorporated the pigment is permanently locked into the hair. The melanocytes are located on a specialised connective tissue peg, the dermal papilla. In the telogen follicle there are no active melanocytes but the follicle contains a small number of precursor (stem) melanocytes or melanoblasts. The actual number of melanoblasts per telogen zigzag follicle has been estimated from split-dose irradiation experiments to be only 1 or 2 (Potten & Chase, 1970). At one time it was thought that these represented melanocytes which had switched off melanogenic activity at the end of the cycle, i.e. it was assumed that all melanocytes were equal in their stem cell potential (division capacity) and all passed through cyclic phases of melanogenic activity during which they underwent few, if any, cell divisions. At the end of the hair growth-cycle many melanocytes were lost into the forming hair, with only one or two remaining as stem cells for the next cycle (Potten, 1972). In the light of some recent experiments this interpretation seems unlikely (Potten, 1982). It is now believed that the few telogen melanoblasts divide early in anagen, replacing their own numbers while also generating other cells which divide further to generate the 20–30 melanogenic melanocytes. Once formed, the hair follicles represent a closed system i.e. no cells enter or leave the follicle. Thus, when melanoblasts are killed the follicle will be deficient in pigment for the rest of the life of the animal. In rodents, with their synchronised hair follicles, the effects of the hair growth cycle on the radioresponse can be controlled and studied. This is facilitated by the fact that the hair growth cycle in telogen follicles can be initiated, and hence controlled, by the simple procedure of plucking the hair from the follicles.

## HISTORICAL BACKGROUND

One of the early radiobiological observations made by Coolidge (1925) and then Hance & Murphy (1926) was that the hair on rabbits' ears, or the abdomens of black mice, turned white after X-irradiation. This phenomenon was reported intermittently over the next 20 years until the fairly extensive studies of Chase which began in about 1949 and continued until about 1963. These experiments received a significant impetus from the development firstly of nuclear fission and secondly of high altitude air travel and space flights (i.e. possible exposure to increased levels of cosmic rays). This earlier work was all based on studies on hair removed (plucked), from the animal. The incidence of hairs with normal, intermediate (mosaic) or without pigmentation (greyed or white) was recorded. Chase's experiments demonstrated several features:

1. The level of greying was strongly influenced by the stage in the hair growth cycle of the follicles.
2. The maximum effect (greatest sensitivity) was observed when the hair follicles were in the resting state with no melanogenically-active melanocytes (Chase & Rauch, 1950).

3. The effect only appeared in hairs grown subsequent to the irradiation.

4. The effect was permanent — a hair never re-established a pigmented state, but mosaic hairs tended to turn white in succeeding hair cycles.

5. The effect has been observed by various authors in guinea pigs, rabbits, hamsters, mice, chickens, pigeons, cats (see Potten, 1967) and in man after radiotherapy treatment. (However, in man it has also been reported that in naturally-greyed hair, which presumably reflects a switching off of melanogenesis rather than a killing of melanocytes, a re-activation of pigmentation can also occur.)

6. The dose-response curves for the fraction of unaffected follicles over the range 3–10 Gy has an extrapolation number of 2.7, a $D_0$ value of about 3 Gy, a dose inactivating 50 per cent of the follicles of about 4.25 Gy and a 'threshold' dose of about 2.8 Gy for telogen follicles (see Potten, 1967 and Figure 10.1)

7. The smaller zigzag-hair follicles are the most sensitive with the sensitivity being strongly influenced by follicle size (Chase & Smith, 1950). Thus, there is a higher resistance, for example of guinea-pig follicles which all are of a large size.

8. The response is sensitive enough to detect differences in the levels of tissue oxygenation (Chase & Hunt, 1959).

9. Using charged ions (neon, oxygen or carbon) it was deduced that the sensitive target cells in telogen follicles were at least 225 microns deep into the skin, i.e. near the follicle base. (Chase et al, 1963).

10. There are no clear indications that changes in sensitivity due to dose-rate and LET effects can be detected.

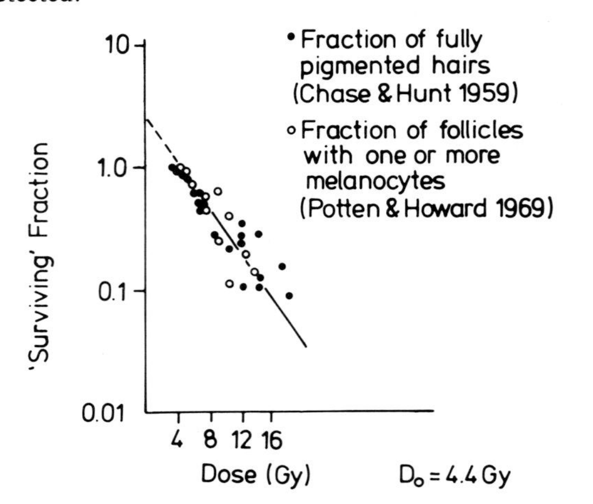


**Fig. 10.1** Dose-response curves for the fraction of fully pigmented hairs (from Chase & Hunt, 1959) and the fraction of follicles with one or more functional melanocytes (from Potten & Howard, 1969). In both instances these fractions have been referred to as the 'surviving' fraction for simplicity and in both cases telogen follicles were irradiated.

One of the difficulties with this end point is that the detection of melanocyte-killing (hair-greying) is dependent on a hair subsequently regrowing from the follicle. If the radiation dose is high enough to sterilise the follicle itself (permanent epilation) then greying cannot be detected.

The hair represents the end product of the proliferative activity of the hair follicle over a preceding period of up to 21 days. Detailed information on the changes in the cells responsible (i.e. the melanoblasts and melanocytes) cannot be gained from observations on the hair but may be gained by observations on the hair follicles and their respective melanocytes.

## DESCRIPTION OF THE TECHNIQUE

As in most experiments on rodent skin, the hair growth cycle has to be carefully controlled. In the initial experiments this was achieved by plucking the hair from the animals under anaesthetic and waiting for one complete hair growth cycle (i.e. 21 days), and then irradiating the telogen follicles. This has to involve some irradiation device designed to irradiate only part of the body since the animals must survive for 12–24 days after irradiation, i.e. until the sampling time. The irradiation is preceded or immediately followed by another plucking to restimulate the follicles into another growth phase. This first post-irradiation growth cycle could be analysed by sampling in mid-anagen (about 12 days after plucking), or alternatively, the animals could be re-plucked at the end of the post-irradiation growth cycle and analysed in mid-anagen of the second post-irradiation cycle. For most studies the latter was adopted. One of the problems here is that sampling in the first or second post-irradiation hair growth cycle tests different aspects of the colony-forming ability of the pigment-cell precursors (the ability to form some functional melanocytes possibly involving a few cell divisions in one case, and the ability to pass through a complete cycle of melanocyte production and stem-cell replacement in the other case). The latter is clearly the stricter criteria for melanoblast function. However, a second problem is that the duration of the post-irradiation hair growth cycle is dependent on radiation-dose and after high doses it is very difficult to know when to repluck the hairs.

The preparatory plucking before irradiation may be unnecessary if care is taken to use male mice about 7 weeks old. Adult male mice tend to have more uniform hair growth-cycles than adult females where the cyclic levels of the sex-hormones influence the hair growth-cycle. At 7 weeks the male animals are in their second post-natal telogen phase which usually lasts for at least a week.

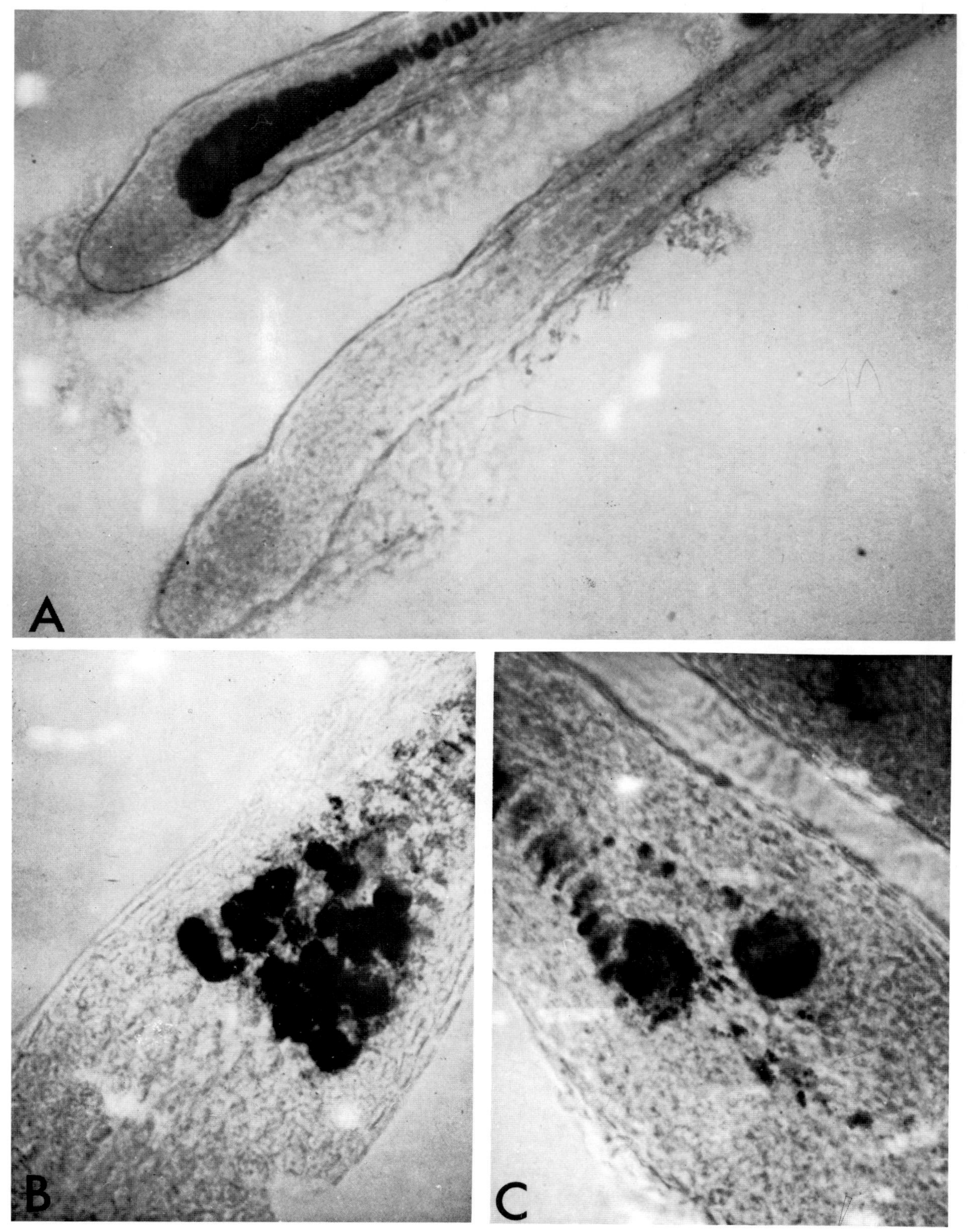


**Fig. 10.2** Whole-mount preparations of isolated mouse hair follicles (lightly stained or unstained phase contrast pictures). **A** Fully pigmented and unpigmented 'white' follicle. **B** Squashed normal (unirradiated) follicle showing about 20 melanocytes. **C** Follicle with two remaining functional melanocytes. Taken from Potten (1967).

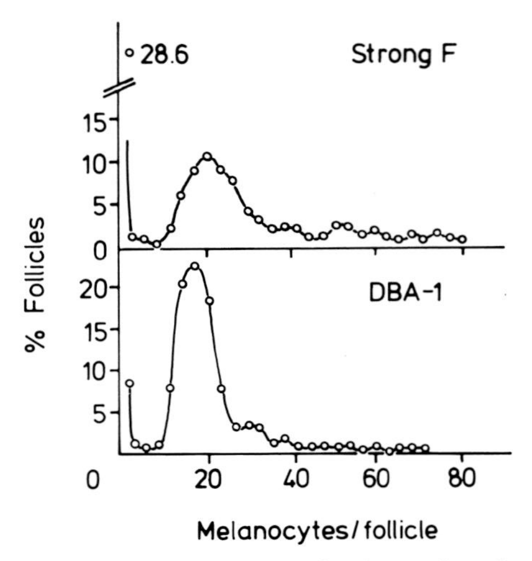


**Fig. 10.3** Frequency polygons (%) for the number of melanocytes per hair follicle for two inbred strains of mice.

At the time of sampling, large areas of anagen skin are fixed in Carnoy's fixative for 20–30 min. The skin is then washed and stored in 70 per cent ethanol. A small sample from the centre of the irradiation field is then removed and hydrated and hydrolysed in 1N HCL at 60°C for 6–7 minutes. After washing, the muscle and fat layers are removed and the remaining tissue is dehydrated and cleared in terpineol overnight. This causes the follicles to become quite brittle so that they can be snapped off the tissue sample and 100–300 can be mounted on a microscope slide. These slides can be scored under low magnification for the fraction of follicles containing one or more functional (melanogenic) melanocytes (see Figs 10.1, 10.2) — scoring procedure 1. This provides data which can be assumed to be very similar to the fraction of mosaics plus pigmented hairs (or one minus the fraction of totally white hairs). The dose-response curves for follicles scored in this way are very similar to those for the fraction of fully-pigmented hairs (Potten, 1967) (Fig. 10.1) since the fraction of mosaics is small and the follicle assay is conducted at a time when most mosaics will have turned white.

An elaboration of the technique involved carefully applying a pressure to a warmed slide thus slightly squashing the hair follicles. When this was done the number of individual melanocytes could be estimated (see Fig. 10.2) and melanocyte colony-size distributions could be determined (Fig. 10.3) — scoring procedure 2.

## CONSTITUTION OF A COLONY

The normal distributions of melanocytes in the hair follicles of two strains of mice are shown in Figure 10.3. The large peak is believed to be the consequence of the predominance of the small zigzag hair follicles, and the tail to the right the consequence of the various categories of the larger guard-hair follicles. The modal values are 17 and 20 melanocytes per follicle for the two strains with arithmetic mean values of 18.4 and 16.3 for DBA-1 and Strong F respectively. There are normally relatively few follicles with less than 10 melanocytes.

Early work had assumed that the telogen follicle contained few (3 or 4) melanocyte precursors (Chase, 1951). This was subsequently reduced to 1 or 2 per follicle by split-dose experiments (Potten & Chase, 1970), supported by other more recent experiments (Potten, 1982). Hence, the colony of 16–20 melanocytes must arise from divisions, early in anagen, of the 1 or 2 melanoblasts and colonies cannot strictly be interpreted as clones derived from single cells although this may in practice be the case in many follicles.

If the fraction of follicles with 1 or more functional melanocytes is regarded as containing 'surviving' melanoblasts (procedure 1) a dose-response curve is obtained which can be defined by the following parameters; n = 2.2 (1.7–2.8), $D_o$ = 3.5Gy (3.1–4.0) for Strong F, and n = 2.4 (1.4–4.0), $D_o$ = 5.1Gy (3.9–7.3) for DBA-1 (95 per cent confidence intervals in brackets). When the presence of 10 or more functional melanocytes (Fig. 10.3) is used as the criterion defining melanoblast

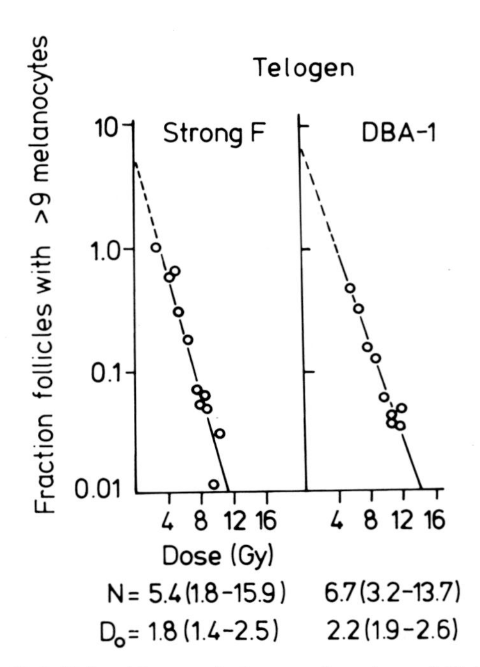


**Fig. 10.4** Melanoblast survival curves for telogen follicles in two inbred strains of mice. Survival of melanoblasts is defined by their ability to produce colonies of 10 or more functional melanocytes in the second hair growth cycle post-irradiation.

survival (scoring procedure 2) the parameters are as follows: n = 5.4 (1.8–15.9), $D_o$ = 1.8Gy (1.4–2.5), and n = 6.7 (3.2–13.7), $D_o$ = 2.2Gy (1.9–2.6), for Strong F and DBA-1 respectively (see Fig. 10.4). Raising the threshold further, up to about 12 melanocytes per follicle, has relatively little effect. Raising it even further clearly eliminates many normal follicle colonies (Fig. 10.3). Thus, the presence of 10 or more functional melanocytes has been adopted as the criterion defining melanoblast survival. Full distributions of melanocytes per follicle following a range of doses have been presented in the literature (Potten, 1968).

## DATA PRESENTATION

Figure 10.4 illustrates the typical dose response curves obtained using this approach. The parameters defining these curves, which were fitted by a computer programme devised by Pike & Alper (1964), are presented above.

## INFORMATION DEDUCED USING THE TECHNIQUE

Sections with a thickness of 0.5 $\mu$m showed that after irradiation the follicles lack completely not only functional melanocytes but also amelanotic melanocytes (Potten & Chase, 1970) which indicates that melanoblasts have been sterilised rather than having their melanogenic potential switched off. Split-dose experiments demonstrated that the number of melanoblasts per telogen follicle is not more than one or two (Potten & Chase, 1970). In one experiment a recovery factor of 6.6 was obtained which was almost exactly the same as the extrapolation number (6.7) for the single-dose response curve suggesting that the follicles contain no more than 1 melanoblast. In a second experiment the extrapolation number of a two-dose survival curve was 2.7 compared with an extrapolation number of 6.8 for a single-dose curve suggesting that the follicles contain 2.5 melanoblasts. Since both estimates are based on studies where all the follicle types were analysed and since it seems reasonable to assume that the smallest zigzag follicles contain the fewest melanoblasts, then these smallest follicles are likely to contain only a single melanoblast. Hence, the colonies scored are in most cases clones derived from a single cell.

Using the less stringent criterion for 'survivors' (follicles with one or more functional melanocytes — procedure 1), the differences in sensitivity between various mouse strains has been investigated (Potten, 1969). Using a common extrapolation number of 2.47 the survival curves for six different coat colour genotypes (including 3 different inbred strains — Strong F, DBA-1 and C57B1/J) have $D_o$ values ranging from 3.3 to 5.6Gy (independent analysis of each curve showed that n varied between 2.0–3.7 with the $D_o$ values varying between 3.7 and 5.8 Gy). Using the same type of analysis the variation in sensitivity throughout the hair growth cycle was studied (Potten, 1970). Although there is some variation between strains when telogen follicles are irradiated (maximum differences are by a factor of about 1.6) the anagen follicles are between 1.2 and 11 times more resistant. The anagen follicles in one strain (Strong F) differ only slightly in sensitivity from telogen follicles while anagen follicles in DBA-1 and C57B1 are 8–11 times more resistant (Potten, 1970). Using the criterion of 10 or more functional melanocytes per follicle to define survival (procedure 2) the dose-response curves become 1.6 times more resistant three days after plucking and 2.5 times more resistant at the end of the melanoblast division-activity at five days after plucking. The reasons for these changes remain unclear at present.

Using procedure 1 for defining 'survivors', the influence of external and inhaled oxygen tension has been studied (Potten & Howard, 1969). These studies stemmed from earlier work by Chase & Hunt (1959). The experiments showed that by clamping the vascular supply and surrounding the skin with nitrogen the follicles could be rendered about 2.9 times more resistant (2.3–2.8 times, if the more stringent criteria of 10 or more melanocytes is used, i.e. procedure 2). The sensitivity using procedure 1 could be increased by a factor of 1.2–1.3 by replacing the surrounding and inhaled air by pure oxygen. The results also demonstrated that under conditions where the vascular supply of oxygen was eliminated (clamping) significant amounts of oxygen could diffuse into the follicles from the external gaseous environment.

## REFERENCES

Chase H B 1949 Greying of hair. 1. Effects produced by single doses of X-rays on mice. Journal of Morphology 84:57

Chase H B 1951 Number of entities inactivated by X-rays in greying of hair. Science 113:714

Chase H B, Hunt J 1959 In: Gordon M (ed) Pigment cell biology. Academic Press, New York p 537

Chase H B, Rauch H 1950 Greying hair. II. Response of individual hairs in mice to variations in X-radiation. Journal of Morphology 87:381

Chase H B, Smith V W 1950 X-ray effects on mouse pigmentation as related to melanoblast distribution. Zoologica 35:24 (abs)

Chase H B, Straile W E, Arsenault C 1963 Evidence for indirect effects of radiations of heavy ions and electrons on hair depigmentation. Annals of the New York Academy of Sciences 100:390

Coolidge W D 1925 High voltage cathode rays outside the generating tube. Science 62:441

Hance R T, Murphy J B 1926 Studies on X-ray effects. XV. The prevention of pigment formation in the hair follicles of colored mice with high voltage X-ray. Journal of Experimental Medicine 44:339

Pike M C, Alper T 1964 A method for determining dose-modification factors. British Journal of Radiology 37:458

Potten C S 1967 The radiation inactivation of follicular melanocytes in mice. Thesis, University of London, p 374

Potten C S 1968 Radiation depigmentation of mouse hair: a study of follicular melanocyte populations. Cells and Tissue Kinetics 1:178

Potten C S 1969 Radiation depigmentation of mouse hair: effects of mouse strain. British Journal of Dermatology 81:289

Potten C S 1970 Radiation depigmentation of mouse hair: effect of the hair growth cycle on the sensitivity. The Journal of Investigative Dermatology 55:410

Potten C S 1972 The X-irradiation of melanocyte precursor cells in resting hair follicles. In: Riley V (ed) Pigmentation: its genesis and biologic control. Appleton-Century-Crofts, New York

Potten C S 1982 Sensitivity of folicular melanoblasts in newborn mouse skin to tritiated thymidine: Evidence for a long term retention of label. Experientia 38:1464

Potten C S, Chase H B 1970 Radiation Depigmentation of mouse hair: Split-dose experiments and melanocyte precursors (amelanotic melanoblasts) in the resting hair follicle. Radiation Research 42:119

Potten C S, Howard A 1969 Radiation depigmentation of mouse hair: the influence of local tissue oxygen tension on radiosensitivity. Radiation Research 38:65

# 11

*R. Dover*

# Clonal survival assay for human epidermal keratinocytes in vitro

## HISTORICAL BACKGROUND

It has been possible for many years to propagate in vitro several cell types from a variety of species, both from normal and diseased tissue. In vitro systems allow the investigator to manipulate the cells to an extent which would often be impossible in vivo, and many experiments which, for practical or ethical reasons are impossible in vivo, can be carried out in vitro.

The criteria which are required for the assay of reproductive survival in vitro are:

1. The cells should continue to grow clonally at low density.
2. The cultures should be free from contamination by other cell types.

If the studies are to be used to deduce information for cells in vivo:

3. The cultures should display some features typical of the tissue in vivo.
4. The cultures should be genetically stable and normal.

There are now many systems for the maintenance of epidermal cells in culture. Most do not satisfy the criteria outlined above. To be useful for survival assays one should be able to grow the cells at low density so that there is a clear separation between the colonies. This will ensure that any colonies derived from single cells can be identified. Many of the available culture methods require a large number of cells to be plated, resulting in cultures in which one cannot identify individual clones. In these, if the cells are plated out at low densities using conditions the same as for high plating densities the epidermal cells do not usually grow. The methods available at present for growth of keratinocytes at low density are detailed in Table 11.1.

One of the major problems usually associated with the cultivation of epithelial cells in vitro is that mesenchymal cells — 'fibroblasts' often contaminate and eventually overgrow the cultures. Unless one is working with

**Table 11.1** Epidermal keratinocyte culture systems which proliferate at low cell density

| Species | Cell isolation | Culture conditions | Reference |
|---|---|---|---|
| Human | Dermis trimmed. Serial trypsinization | DMEM 20% FCS 0.4 μg/ml HC. Plated with 3T3 feeder layer 37°C | Rheinwald & Green, 1975 |
| Human | As above | Medium MCDB 151 1 mg/ml dialized FCS. Low $Ca^{++}$. High adenine 37°C | Peehl & Ham, 1980 |
| Human | Floated on trypsin 18–24 h 4°C | Collagen substrate M199 + 10% FCS. 0.17 $mMCa^{++}$(low) monolayer. 0.4 $mMCa^{++}$ (normal) stratified 36.5°C | Hawley-Nelson et al, 1980 |
| Human | Dermis trimmed. Trypsin/EDTA for 30 min | 'Defined' media. EGF, transferrin, insulin, ethanolamine, phosphoethanolamine, trace elements. Low $Ca^{++}$ 37°C | Tsao et al, 1982 |
| Mouse | Trypsin floatation overnight at 4°C | Dermal fibroblast conditioned. Eagles medium 8% FCS. 0.02 mM $Ca.^{++}$ Killed fibroblast 'substratum' | Yuspa et al, 1981 |

a tissue free from mesenchyme (e.g. cornea or lens) the preparation of a homogeneous population of epithelial cells may be difficult. Several methods have been reported selectively to remove or inhibit the growth of fibroblasts. These include the use of cis-hydroxyproline (Kao & Prockop, 1977, 1980; Liotta et al, 1978) or D-valine (in place of L-valine, Gilbert & Migeon, 1975) in the culture medium. Epidermal cells have also been grown on a dead dermal substratum (Freeman et al, 1976) to inhibit fibroblast growth. Edwards et al (1980) produced monoclonal antibodies to kill fibroblasts selectively. Density gradient centrifugation has also been used to separate epidermal cells from fibroblasts (Fusenig & Worst, 1975; Marcelo et al, 1978).

The epidermis itself consists of several cell types, keratinocytes, melanocytes, Langerhans cells and Merkel cells. The evidence available indicates that the minority constituents either never attach in culture or fail to proliferate. Ideally one would prefer to have only keratinocytes in the culture but all the systems for growth of normal mammalian epidermis probably suffer *to some degree* from contamination by other cell types, most notably fibroblasts. The possibility that such cells may interact in some way with the keratinocytes to affect their responses cannot be completely ruled out, but in most systems the detectable levels of contamination are very low and therefore probably have little effect.

If one wishes to analyse regenerative ability the model chosen must be capable of regenerating to produce a normal tissue. None of the available culture methods produce an epidermis identical to that found in vivo. Typically those cells directly attached to the substratum, the 'basal cells' are not cuboidal but flattened. Further, a stratum corneum is rarely found.

Probably the most popular method for the culture of clonal human keratinocytes is that described by Rheinwald & Green (1975). The methods described by Peehl & Ham (1980), and Tsao et al (1982) (see Table 11.1), both use the techniques described by Rheinwald & Green to produce the primary cultures. The method described by Hawley-Nelson et al (1980) is not claimed to produce clonal growth but as cells can be plated at low densities they might produce colonies from single cells. The biochemistry of differentiation of the keratinocyte has been extensively studied by Green and co-workers (for review see Green, 1980). These studies revealed that cultured human keratinocytes synthesize all keratins except those with the highest molecular weights. A more recent change in culture technique (Fuchs & Green, 1981) was to use a medium free of vitamin A. Under these conditions formation of a stratum corneum is reported and keratin species of higher molecular weight can be detected. Human keratinocytes maintained in culture by the methods of Rheinwald and Green produce the best available model for epidermis. The only culture technique for murine epidermal cells growing at clonal densities is that described by Yuspa et al (1981). In these cultures the level of calcium present in the medium is controlled. At low calcium levels monolayer cultures are produced and the cells still produce enucleated squame-like cells but do not stratify. If the calcium level is raised to normal, desmosomes form, the cultures stratify and the balance of cell production is towards differentiation not self-renewal. This means that the cultures senesce at normal levels of calcium. Obviously this is not as satisfactory as a 'steady-state' renewing system, where cell loss due to differentiation is balanced by cell production. Another method for the culture of murine keratinocyte is that described by Fischer et al (1980), which produces plating efficiencies similar to the values reported by Yuspa et al (1981). The method is not claimed to be clonal but, again, it is possible that it may be. Both these methods could be of use in studies of clonal regeneration but would require further characterization.

## DESCRIPTION OF THE TECHNIQUES

### Analysis of the survival of human keratinocytes

*The Rheinwald & Green method*

The Rheinwald & Green (1975) method involves the use of a lethally-irradiated 'feeder layer' of 3T3 cells. Feeder layers are reproductively-sterilized but metabolically-active cells used to support the growth of other viable cells plated at low density. The feeder approach has been used for radiation survival studies (Fisher & Puck, 1956). Here, lethally-irradiated HeLa cells were used to support the growth of irradiated HeLa cells at low levels of survival.

The method for human keratinocytes is novel in that the feeder layer is not of the same type of cell as the ones requiring support. Of several lines tested, the 3T3 line of Todaro & Green (1963) proved the most useful. Human dermal fibroblasts can also support colony initiation and growth but the 3T3 line is more effective. The 3T3 line was derived from the embryos of Swiss albino mice, but the exact tissue origin of the cells is unknown as the whole embryos were minced to produce the cell suspension.

Rheinwald & Green used neonate human foreskin as the source of keratinocytes but skin from other sites can also be used. Skin from younger donors has a higher plating-efficiency. The skin is prepared by trimming away as much subcutaneous fat and dermal material as possible and is then minced with scissors and subjected to a series of trypsinizations of 30 minutes duration, using 0.25 per cent trypsin. Rheinwald (1980) advocates the use of a trypsin: collagenase mixture, particularly for isolating keratinocytes from adult skin. After a period of enzyme treatment the fragments are allowed to settle, the supernatant is removed and collected, fresh enzyme solution is added and the process continued. When disaggregation of the tissue is complete the cells are spun down and resuspended in fresh medium. The medium originally used by Rheinwald & Green and in the experiments described here was Dulbecco's modified Eagles medium (DMEM) with 20 per cent fetal calf serum (FCS) and 0.4 μg/ml hydrocortisone. After counting, the keratinocytes are plated into a culture vessel already containing $2 \times 10^4$ 3T3 cells/cm$^2$. The 3T3 cells are able to inhibit the growth of any contaminating fibroblasts as well as in some way to modify the culture environment to allow growth of keratinocytes. Fibroblast contamination is a potential problem, particularly when some media additives are used. This problem can easily be overcome by treating the culture with 0.02 per cent EDTA for 30 seconds follwed by vigorous pipetting, when both the 3T3 cells and the fibroblasts will detach, leaving the keratinocytes behind (Sun & Green, 1976). Freshly irradiated 3T3 cells can then be added, and once colonies are well established they will continue growth in the absence of a feeder layer (Rheinwald, 1980).

Several modifications can be made to the basic method. These changes include the addition of Epidermal Growth Factor (EGF) (Rheinwald & Green, 1977) and Cholera toxin (Green, 1978). These factors when used together, produce more colonies which are faster-growing and which can be maintained in culture for longer. The addition of transferrin, insulin and tri-iodothyronine reduces the serum requirement to 10 per cent (Watt & Green, 1981). The use of serum free of vitamin-A may be useful for some studies (Fuchs & Green, (1981).

*Experimental details*

The techniques used have been described (Dover & Potten, 1983a). Human foreskins from patients aged between 2 and 10 years of age were used in the present study. Cells were isolated by four serial, 30 minute trypsinations (0.25% w/v Sigma type III). Freshly isolated cells were washed, resuspended in medium (DMEM, 20% FCS with 0.4 μg/ml hydrocortisone) and counted. Cells to be irradiated as primary suspensions were irradiated at this point, before being plated with $3 \times 10^5$ lethally-irradiated 3T3 cells into Falcon 25 cm$^2$ culture flasks and then incubated at 37°C in a humidified 95 per cent air, 5 per cent $CO_2$ atmosphere. The medium was changed twice weekly.

3T3 cells were reproductively sterilized by a dose of 60 Gy (from a linear accelerator). Cell suspensions of primary keratinocytes held at 4°C were irradiated with gamma rays from a $^{137}$Caesium source (dose rate 5 Gy/min). Alternatively, cells were irradiated with X-rays using a Pantak unit operated at 290 KV and 10 mA (dose rate 1.5 Gy/min).

Cultures were fixed and stained ten days after plating, and colonies containing 50 or more cells were counted using a low power microscope. The data were analyzed by computer (see later). The survival curve for primary cell cultures irradiated with gamma rays is shown in Figure 11.1. The survival curve for cells in confluent primary cultures, which were removed from the flask with trypsin/EDTA and then irradiated with X-rays, is shown in Figure 11.2.

Survival was assayed in these experiments by scoring the reduction in plating efficiency (PE) attributable to the dose of radiation given. The PE of unirradiated cells in these experiements was between 0.09 and 0.71 per cent for primary cultures. This is very low, but it must be remembered that the original suspension contains cells from all strata of the epidermis, of which only the basal cells would be expected to be capable of cell division, only a few of which may be clonogenic. The cell suspension will also contain non-keratinocytes such as melanocytes and Langerhans cells. Thus PE cannot reflect how many potentially clonogenic cells actual initiate colonies in vitro. Despite this, relative changes in PE can be used accurately to measure killing of the clonogenic cells. The PE can vary for cells isolated from two different speci-

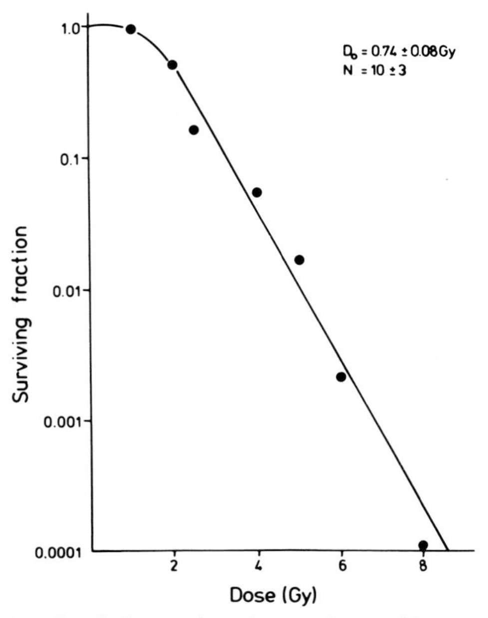


**Fig. 11.1** Survival curve for primary cultures of human keratinocytes irradiated with gamma rays. If cholera toxin is added to the cultures exactly the same shaped curve is produced.

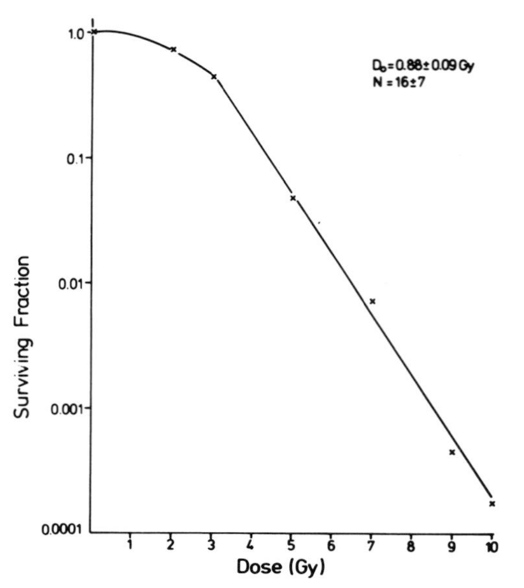


**Fig. 11.2** Survival curve for human keratinocytes at first passage irradiated with X-rays in suspension prior to plating.

mens of skin of the same age and from the same site, but the reduction in PE after a given dose of radiation was found to be remarkably constant between samples. Pilot experiments were carried out using a wide range of doses in order to obtain estimates of the radiosensitivity of the cells. By repeating these experiments several times using different concentrations of cells for each dose, the most appropriate combination of plating-density and dose was selected. The combination was chosen to produce sufficient colonies per flask to keep the total number of flasks to a manageable level whilst minimizing colony fusion before the cultures were scored. Combinations of plating density and dose could also be selected to produce equal numbers of surviving colonies per flask at different doses. At low levels of survival the flasks contained many dead cells which were not removed until the first medium change (3–5 days). These did not appear to affect survival in any way. This was demonstrated by adding large numbers of freshly-isolated keratinocytes $(1-5) \times 10^6$, which had been sterilized with a dose of 60 Gy, to flasks in experiments measuring survival. Survival was identical for a given dose and plating density, irrespective of whether or not extra dead cells were added.

It is difficult to give 'typical' numbers of colonies which would be expected because of the large differences in PE. However, some indication can be obtained if one assumes a PE of 0.4 per cent (which is in the mid-range of the observed values) then one would need to plate $2.5 \times 10^4$ keratinocytes to obtain 100 colonies. From Figure 11.1, it can be seen that a dose of 3 Gy reduces survival, and hence the observed PE, by a factor of five. So in this example one would need to plate five times as many cells (i.e. $1.25 \times 10^5$) to obtain 100 surviving colonies after 3 Gy. Hence, having established a range of PE values for unirradiated cells, the range of the numbers of cells which should be plated after a dose of radiation can be estimated. It is critical to have control cultures which can be counted accurately, with minimal colony fusion. Thus one should plate control cultures at a variety of densities ($5 \times 10^3$–$10^5$ cells/flask in the present study). As long as fusion of colonies at the higher plating densities is negligible, all the control plates should give the same (or very similar) estimates of PE.

As there was little variation in survival between skin samples and as the number of primary cells available was limited, the data from several experiments were pooled. The survival curve in Figure 11.1 is derived from a total of over $2 \times 10^8$ cells plated (over 20 specimens of skin). The data were analyzed using Gilbert's (1969) computer programme. The data were entered by calculating the total number of cells plated for each dose and the total number of survivors at each dose, from all experiments. The programme then produces the line of best fit to the data (see Ch. 1).

### The definition of a colony

The value of 50 cells was chosen as a reasonable criterion for colony formation (see Ch. 1). Assuming exponential growth the clonogenic cell will have undergone six divisions. However in these cultures it is improbable that all the cells produced remain in the proliferative compartment as some differentiate and become post-mitotic. Colonies become multi-layered at early stages of growth. In unirradiated cultures four to ten days post-plating labelling indices of 30 per cent are common and the cycle time is about 14 hours (Dover & Potten, 1983b). No data are available to quantitate cell loss from the upper layers of cultures at this stage of growth, but the probability remains that when a colony of 50 cells is observed that cell loss has occurred and the clonogenic cell may have divided more than six times. Despite this probable underestimation of proliferation the 50 cells/colony criteria may still overestimate clonogenicity in that not all colonies may have true long-term self-maintaining ability. If a larger colony size was chosen one would need to score survival at times later than 10 days, in which case it would probably be necessary to reduce the number of survivors per flask to prevent colony fusion.

## MODIFICATIONS TO THE MATERIALS AND THEIR CONSEQUENCES

At present the only site from which skin has been assayed for keratinocyte radiosensitivity is foreskin. This is largely because we wished initially to study primary cultures and foreskins are relatively frequently available. Skin from other sites and from patients of other ages could be used after subculture to produce sufficient cells. The only other species for which a clonal-density culture technique is available is mouse (Yuspa et al, 1981) but this has not been used to determine a survival curve.

## MODIFICATIONS TO THE METHODS AND THEIR CONSEQUENCES

Pilot experiments were carried out where the cells were irradiated in suspension at either 4°C or 37°C, and survival was found to be equal at both temperatures. 4°C was chosen for routine use as it was easier to maintain during transport to the radiation source. The effects of temperature on human keratinocyte survival could easily be studied using this system. In some experiments, intact skin specimens were irradiated and then immediately subjected to trypsinization. The survival of these cells was identical to that of cells isolated from an unirradiated piece of the same skin specimen and irradiated when in suspension. This suggests that the isolation procedure does not affect the intrinsic radiosensitivity of the cells.

As mentioned earlier, there are several factors which can be added to the culture medium to enhance growth. Cholera toxin (CT) raises the internal cAMP levels and increases the PE, rate of growth and longevity of keratinocytes in culture (Green, 1978). Without cholera toxin there are often many small abortive colonies, whereas with cholera toxin the number of such colonies is greatly reduced. The abortive colonies may be derived from non-clonogenic differentiated (transit) cells 'running down' through a last few divisions. Cholera toxin retains a larger proportion of cells within the cultures at small cell diameters and the cells with small diameter have a high proliferative ability (Green, 1980). Thus cholera toxin may increase the size of the transit compartment. In experiments where primary cell suspensions were irradiated with gamma rays and grown in medium supplemented with $10^{-9}$M cholera toxin, a survival curve was produced identical to that shown in Figure 11.1. Although CT increased the plating efficiency in all cases, compared with cells from the same sample grown in its absence, the relative survival was unaffected. Two possible interpretations are:

1. CT aids the retention of early transit cells and permits a higher than usual self-replicative capacity, i.e. 'short term clonogenicity', these cells having a similar radiation response to the true clonogenic cells.
2. CT increases the size of the clonogenic population or enhances the PE of a population of clonogenic cells.

The effects of EGF or combinations of CT and EGF have not been studied.

## INFORMATION WHICH CAN BE DEDUCED USING THE TECHNIQUE

Basic radiobiological parameters such as the $D_0$ and extrapolation number can be derived using the culture system. The curves produced for human keratinocytes have shoulder regions larger than those reported for human skin fibroblasts (see Ch. 19).

The extrapolation numbers of 10–16 (Figs 11.1 and 11.2) are of the same order as the recovery factors deduced from epidermal clonogenic cells (data reviewed by Potten et al, 1983). However, the sensitivity ($D_0$) is greater for clonogenic cells in vitro than in vivo, and the reasons for this have yet to be elucidated. It would be expected that the response to X-rays would be slightly greater than for $\gamma$-rays (Figs 11.1 and 11.2) but the reverse is observed here.

The methods used here could easily be adapted to study the effects of hyperthermia, split-dosage or additives to the medium. Survival after drug treatment could also be studied. The technique of Rheinwald & Green can be used to grow a variety of epithelial tissues, e.g. cornea, conjunctiva (Sun & Green, 1977) cervix (Stanley & Parkinson, 1979), oral mucosa (Taichman et al, 1979), and oesophagus and vagina (Watt & Green, 1982). Diseased tissue may also be used. Baden et al (1981) reported that psoriatic keratinocytes could be cultured if CT was used and Wu & Rheinwald (1981) have cultured human squamous cell carcinomas.

The major drawback to the system is the requirement for 3T3 feeders cells. This can be overcome by the use of 3T3 conditioned medium and 3T3 'conditioned' culture vessels (see Rheinwald, 1980). The presence of 3T3 cells could be a major problem when studying drug-responses as it could be difficult to distinguish between the metabolism of the drug by the 3T3 cells and by the keratinocytes. It is simple to treat the keratinocytes in the absence of 3T3's, which then could be added after treatment or they can be dispensed with altogether in the case of large colonies. Table 11.1 includes some culture systems which do not require 3T3 feeders cells and if these media (Peehl & Ham, 1980; Tsao et al, 1982) become commercially available the use of such systems would become simpler. However, for studies of the radiation response of keratinocytes, the presence of 3T3 cells is acceptable. The potential of epithelial culture systems for the study of regeneration and drug and radiation response remains relatively unexploited.

REFERENCES

Baden H P, Kubilus J, MacDonald M J 1981 Normal and psoriatic keratinocytes compared in culture. Journal of Investigative Dermatology 76:53

Dover R, Potten C S 1983a Radiosensitivity of normal human epidermal cells in culture. International Journal of Radiation Biology 43:681

Dover R, Potten C S 1983b Cell cycle kinetics of cultured human epidermal keratinocytes. Journal of Investigative Dermatology 80:423

Edwards P A W, Easty D M, Fostel C S 1980 Selective culture of epitheloid cells from a human squamous carcinoma using a monoclonal antibody to kill fibroblasts. Cell Biology International Reports 4:917

Fisher H W, Puck T T 1956 On the functions of X-irradiated 'feeder' cells in supporting growth of single mammalian cells. Proceedings of the National Academy of Sciences USA 42:900

Fischer S M, Viaje A, Harris K L, Miller D R, Bohrman J S, Slaga T J 1980 Improved conditions for murine epidermal cell cultures. In Vitro 16:180

Freeman A E, Igel H J, Herman B J, Kleinfeld K L 1976 Growth and characteristation of human skin epithelial cell cultures. In Vitro 12:352

Fuchs E, Green H 1981 Regulation of terminal differentiation of cultured human keratinocytes by vitamin A. Cell 25:617

Fusenig N E, Worst P K M 1975 Mouse epidermal cell cultures. II. Isolation, characterization and cultivation of epidermal cells. Experimental Cell Research 93:443

Gilbert C W 1969 Computer programmes for fitting puck and probit survival curves. International Journal of Radiation Biology 16:323

Gilbert S F, Migeon B R 1975 D-valine as a selective agent for normal human and rodent epithelial cells in culture. Cell 5:11

Green H 1978 Cyclic AMP in relation to proliferation of the epidermal cell: a new view. Cell 15:801

Green H 1980 The keratinocyte as differentiated cell type. The Harvey Lectures 74:101

Hawley-Nelson P, Sullivan J E, Kung M, Henning H, Yuspa S H 1980 Optimized conditions for the growth of human epidermal cells in culture. Journal Investigative Dermatology 75:176

Kao W W Y, Prockop D J 1977 Proline analogue removes fibroblasts from cultured mixed cell populations. Nature 266:63

Kao W W Y, Prockop D J 1980 Can proline analogues be used to prevent fibroblasts from overgrowing cultures of epithelial cells. In: Danes B S (ed) In vitro epithelial and birth defects. Liss, New York, p 53

Liotta L A, Vembu D, Kleinman H K, Martin G R, Boone C 1978 Collagen required for proliferation of cultured connective tissue cells but not their transformed counterparts. Nature 272:622

Marcelo C L, Kim Y G, Kaine J L, Voorhees J J 1978 Stratification, specialization, and proliferation of primary keratinocyte cultures. Journal of Cell Biology 79:356

Potten C S, Hendry J H, Al-Barwari S E 1983 A cellular analysis of radiation injury in epidermis. In: Potten C S, Hendry J H (eds) Cytotoxic insult to tissues: effects on cell lineages. Churchill Livingstone, Edinburgh, p 153

Peehl D M, Ham R G 1980 Clonal growth of human keratinocytes with small amounts of dialyzed serum. In Vitro 16:526

Rheinwald J G 1980 Serial cultivation of normal human epidermal keratinocytes. In: Harris (et al (eds)) Methods in cell biology, vol 21A. Academic Press, New York, p 229

Rheinwald J G, Green H 1975 Serial cultivation of strains of human epidermal keratinocytes: the formation of keratinizing colonies from single cells. Cell 6:331

Rheinwald J G, Green H 1977 Epidermal growth factor and the multiplication of cultured human epidermal keratinocytes. Nature 265:421

Stanley M A, Parkinson E K 1979 Growth requirements of human cervical epithelial cells in culture. International Journal of Cancer 24:407

Sun T T, Green H 1976 Differentiation of the epidermal keratinocyte in cell culture: formation of the cornified envelope. Cell 9:511

Sun T T, Green H 1977 Cultured epithelial cells of cornea, conjunctiva and skin: Absence of marked intrinsic divergence of their differentiated states. Nature 269:489

Taichman L, Reilly S, Garant P R 1979 In vitro cultivation of human oral keratinocytes. Archives Oral Biology 24:335

Todaro G J, Green H 1963 Quantitative studies of the growth of mouse embryo cells in culture and their development into cell lines. Journal of Cell Biology 17:299

Tsao M C, Walthall B J, Ham R G 1982 Clonal growth of normal human epidermal keratinocytes in a defined medium. Journal of Cell Physiology 110:219

Watt F M, Green H 1981 Involucrin synthesis is correlated with cell size in human epidermal cell cultures. Journal of Cell Biology 90:738

Watt F M, Green H 1982 Regulation by vitamin A of envelope cross-linking in cultured keratinocytes derived from different human epithelia. Molecular Cell Biology 2:1115

Wu Y J, Rheinwald J G 1981 A new small (40 Kd) keratin filament protein made by some cultured human squamous cell carcinomas. Cell 25:627

Yuspa S H, Koehler B, Kulesz-Martin M, Hennings H 1981 Clonal growth of mouse epidermal cells in medium with reduced calcium concentration. Journal of Investigative Dermatology 76:144

# 12

*H.R. Withers*

# Spermatogenic colonies

## HISTORICAL BACKGROUND

The response of the testis to radiation has been studied from the very beginnings of radiobiology. Useful reviews have been made by Mandl (1964) and Bianchi (1983). The classic statement by Bergonie & Tribondeau (1905) to the effect that rapidly proliferating systems were the most radiosensitive (nowadays, we would say radioresponsive) was based on work on the testis by Regaud and his colleagues (e.g. Regaud & Nogier, 1911; Regaud, 1924; Regaud & Ferroux, 1927).

The colony technique for quantifying spermatogenic stem cell survival depends upon counting discrete colonies of epithelium in histological sections of the testis. For identification of discrete colonies of regenerating epithelium it is necessary that the stem cell population be reduced sufficiently that regenerating colonies do not coalesce before the testis is taken for assay. The essential histological change, frequently called focal or lacunar regeneration, was described as early as 1911 by Regaud & Nogier, and in scores of publications thereafter. A literature search on testicular irradiation would have expedited our development of a colony assay for spermatogenic stem cell survival.

## DESCRIPTION OF THE TECHNIQUE

### The kinetics of response of spermatogenic epithelium

The methods used for measuring stem cell survival after cytotoxic insults to the testis are most easily appreciated through an understanding of the proliferation kinetics of spermatogenic epithelium.

A diagrammatic presentation (Meistrich et al, 1978) of the most generally-accepted model of spermatogenesis in the mouse (Huckins, 1971; Oakberg, 1971; Oakberg & Huckins, 1976) is shown in Figure 12.1 together with an indication of the radiosensitivities of the various cell types.

About 40 days elapse between the mitotic division of the stem cell and the consequent appearance of as many as 1000 mature spermatozoa in the seminiferous tubules. (Many cells are 'lost' during spermatogenesis so that the average yield per cell division is less than 2 (Huckins & Oakberg, 1978; Bennett et al, 1971).) Thereafter, the sperm travel for about a week from the testis along the vas deferens to the seminal vesicles, from whence they are sporadically ejaculated. The putative stem cell, the isolated type A spermatogonium ($A_s$ or $A_{is}$) divides on average at least once every cycle of the spermatogenic epithelium, which is 8–9 days in the mouse (Oakberg, 1971; Oakberg & Huckins, 1976; de Rooij, 1982). Therefore, assuming no major effect of irradiation on stem cell proliferation kinetics, regeneration of a focus of spermatogenesis from a surviving stem cell would be underway within 8–9 days of irradiation (Oakberg, 1971; Oakberg & Huckins, 1976). Other types of spermatogonia are highly radiosensitive and, since they are dividing rapidly, are lost within a few days when exposed to the relatively high doses necessary to reduce stem cell survival significantly. On the other hand, spermatocytes, spermatids and spermatozoa are relatively insensitive to radiation and hence remain in the tubules for periods of time consistent with their normal turnover kinetics. In view of these observations, it is apparent from Figure 12.1 that differentiating cells that had reached the spermatocyte stage before irradiation would continue to differentiate and could be detected in the tubules for up to 28 days after exposure to high doses. By that time, there will be essentially no progeny of the radiosensitive differentiating spermatogonia remaining, but the survival of the more resistant stem cells will be evidenced by newly-generated spermatogonia, spermatocytes and spermatids. In the colony assay to be described, 35 days are usually allowed between irradiation and histological assay to ensure complete depletion of pre-existing differentiated cells and to allow adequate regrowth for easy scoring of colonies derived from surviving spermatogenic stem cells. It should be noted, however, that the production and differentiation of spermatogenic cells passes along the length of a tubule in synchronous waves about 2 cm long (Oakberg, 1956). Consequently cross sections at different levels within one focus of regeneration, or

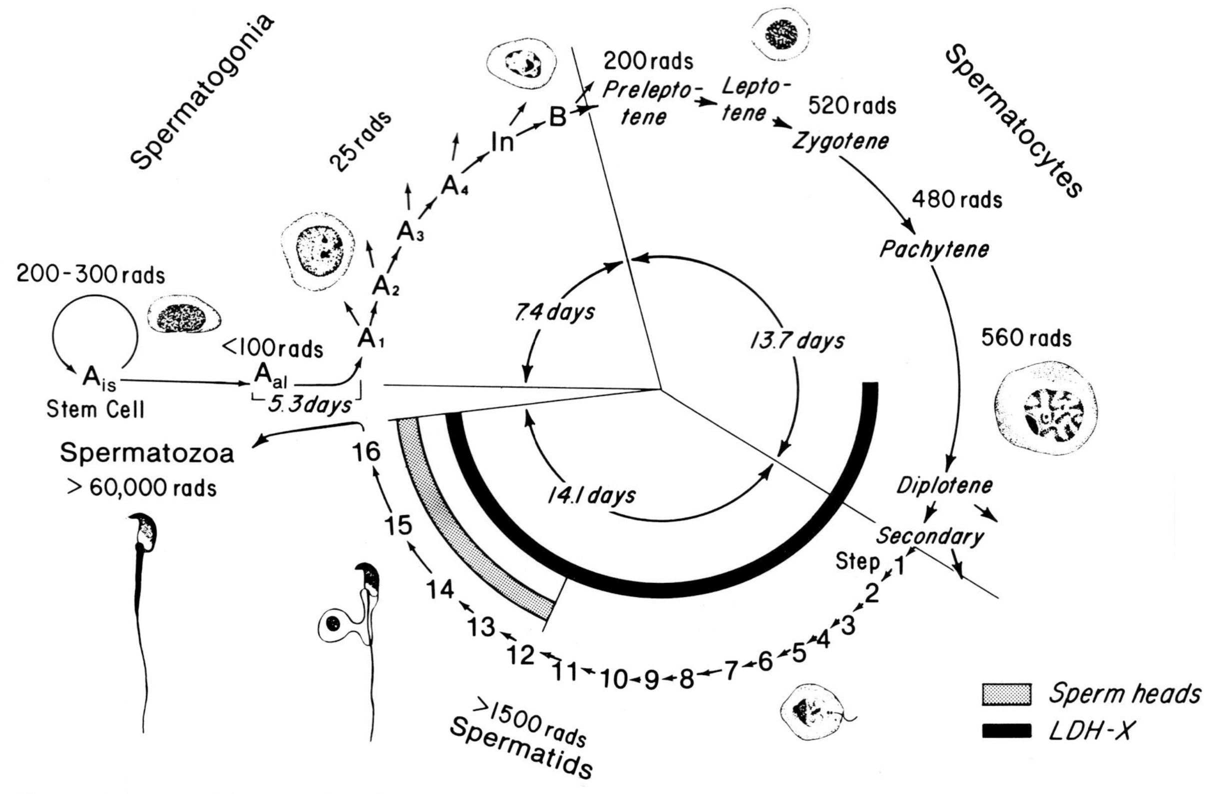


**Fig. 12.1** Diagram of the generation of spermatazoa from a stem cell in the mouse showing the stages of differentiation and their duration, estimates of the dose to reduce the various populations to 50%, and those subpopulations which are assayed by sperm head counts or LDH-X enzyme levels. Note that the total time between stem cell division and migration of the spermatazoa out of the testis is about 40 days. Not shown on the diagram is the cycle time of the stem cell, which is 8–9 days. (From Meistrich et al, 1978, by permission of the authors and Academic Press.)

from different colonies at intervals along a tubule, can show quite different distributions of spermatogenic cell types

The Sertoli cells, which are not spermatogenic and turn over slowly, if at all, show no detectable effect of irradiation during the time involved in the assay of spermatogenic cell survival.

**Details of the technique**

Identification of colonies of spermatogenic cells in histological cross sections of seminiferous tubules as discrete entities regenerating from surviving stem cells requires that a proportion of the tubules contain no colonies (Fig. 12.2). When a tubule cross section is not repopulated, it can be assumed that no stem cells survived in the vicinity: when it contains a focus of repopulation, at least one stem cell survived. To ensure that spermatogenic cells will not have repopulated all tubules and that there will be a minimum of coalescence of adjacent colonies, single doses of gamma-rays greater than about 8 Gy should be given. Such doses reduce stem cell survival to less than 10 per cent (Thames & Withers, 1980) and leave at least 50 per cent of the tubules depopulated of spermatogenic epithelium (Withers et al, 1974) for months after exposure. Since stem cell survival from radiation is random, it is to be expected that when most tubule cross sections show no evidence of surviving stem cells, colonies in the remainder will be derived mainly from single cells. However, some will have resulted from the coincidental, random, survival of two cells in the same area, and even fewer from three cells or more (see p. 109).

35 days after irradiation or other cytotoxic treatment, both testes are removed from the animal for separate evaluation and preserved in a fixative such as Bouin's solution, Carnoy's, Zenker-formal, etc. The thick fibrous capsule (tunica albuginea) of the testis slows penetration of fixative. In the mouse, the irradiated testes are small enough to allow fixation to occur quickly despite the capsule, but if larger specimens are being used, fixation should be facilitated by cutting the capsule near one pole of the organ, well removed from the proposed plane of subsequent sectioning for histological study.

The testis is ovoid and packed tightly with 20–30

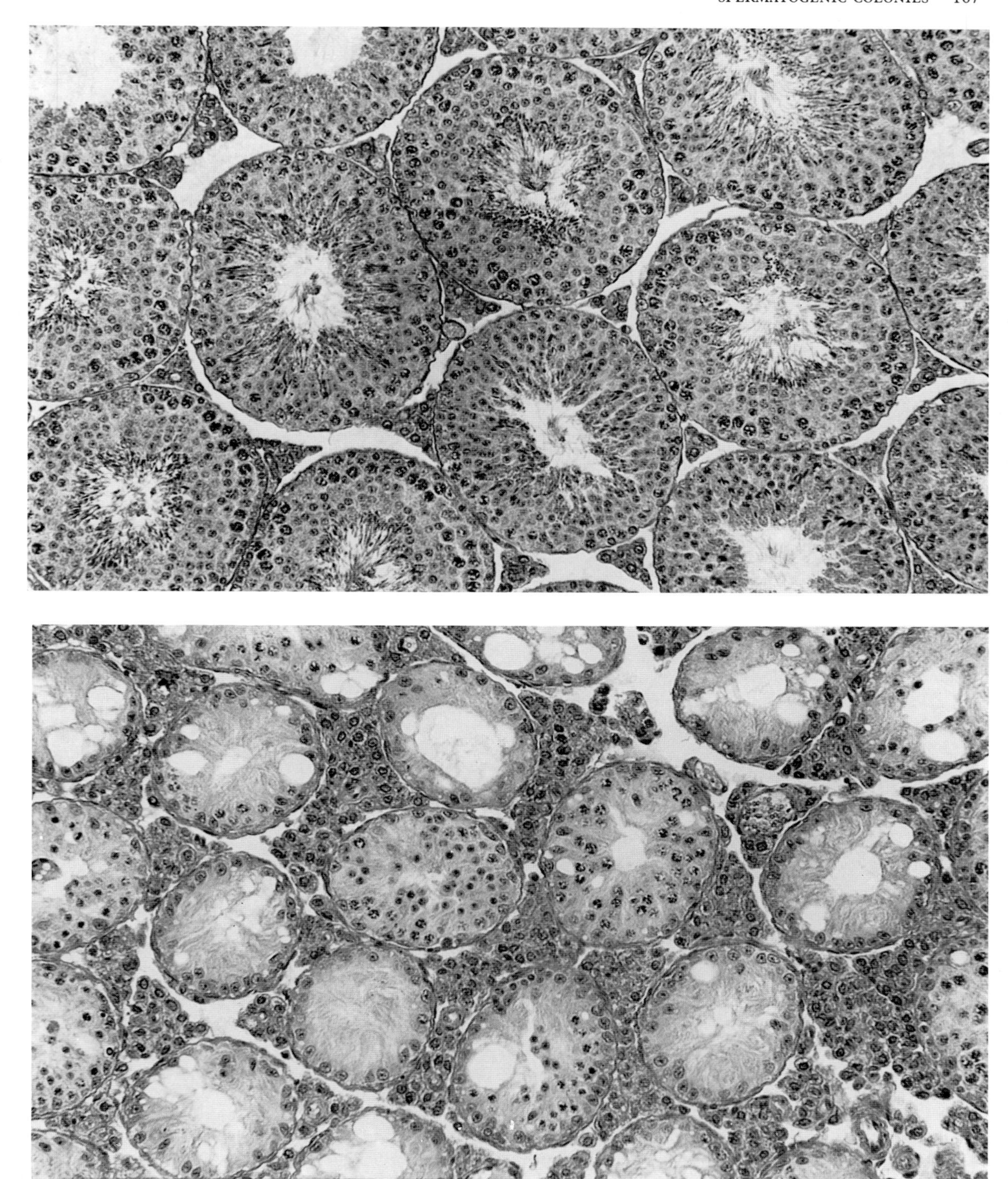

**Fig. 12.2** Photomicrograph (approximately × 190) of a histological section of the normal testis of a mouse (upper panel) and a testis 35 days after exposure to 9 Gy gamma-rays. In the irradiated specimen, a proportion of the tubules are repopulated by spermatogenic epithelium but in others, only Sertoli cells remain.

seminiferous tubules totaling about a metre in length in the mouse. These tubules course back and forth so that, in histological sectioning, the same tubule is cut many times and in various planes relative to its long axis. This complicates scoring colonies of epithelium since the different angles of sectioning make the average length of wall sampled per tubule cross section variable. The most desirable preparation would be one in which all tubules were cut perpendicular to the axis. An exhaustive assessment of the optimal plane of section through the testis has not been reported, but the best appears to be at the equator, transverse to the long axis. Review of slides of 100 testes from one of our experiments showed that the number of tubules per equatorial cross section ranged from 84 to 224, mean = 159 ± 3 (SE). This range reflects the considerable variation in the angle at which testis tubules cross the equator. Deciding whether a section has too many oblique or longitudinal cuts is obviously arbitrary: we reject any slide with fewer than 50 tubule cross sections.

For most purposes, routine hematoxylin and eosin staining is adequate but some investigators prefer to use the more tedious PAS (periodic acid Schiff) reaction and hematoxylin. The thickness of the section is not crucial, varying between 4–10 μm in different laboratories.

Colonies of regenerating spermatogenic epithelium are counted using a microscope, usually at a magnification of 100 times. The criteria for a colony are described later. The total number of tubules can also be counted using the microscope but, if large numbers of slides are to be scored, it is easier and quicker to set up a microprojector to permit scoring on a sheet of white paper using the naked eye and dotting each tubule image as it is counted. We find it convenient to mount the microprojector horizontally on a shelf above the bench and to bring the image to the bench through a prism lens.

## WHAT CONSTITUTES A COLONY?

### Histological considerations

The exact composition of a colony of spermatogenic cells will depend upon the time after irradiation at which the testis is examined, and to some extent, the dose delivered. Colonies of spermatogenic cells originate from a stem cell, and, in their early growth, are composed only of spermatogonia in various stages of differentiation. Spermatocytes do not appear until more than 10 days after the division of a stem cell (Fig. 12.1). Although surviving stem cells have resumed proliferation within about 8 days (Oakberg, 1971) serial sections have shown that after high doses of radiation a normal distribution of the stages of maturation may not be achieved, even by 15 weeks (van den Aardweg et al, 1983). Nevertheless, by 35 days after irradiation, all surviving stem cells have formed a colony (de Ruiter-Bootsma et al, 1976), and, although variable in size (Withers et al, 1974; van den Aardweg et al, 1983), it will contain one or more stem cells, differentiating spermatogonia, spermatocytes, and, in many cases spermatids. As mentioned earlier, because spermatogenesis proceeds in waves along a tubule, the types of cells seen in a 4 μm cross section will vary from colony to colony.

The most readily detected cells are spermatocytes because of their characteristic nuclear appearance. Spermatids are less commonly seen in sections taken even five weeks after irradiation but, when present, also facilitate identification of a colony. The leading edge of a regenerating colony as it expands along and around tubule walls consists of spermatogonia and it is the scoring of these cells that provides most uncertainties in the assay system. Each investigator has to set an arbitrary value for the number of spermatogonia required to score a tubule as being repopulated with cells: we have chosen to require at least four spermatogonia (Withers et al, 1974; Reid et al, 1981) but others are content with one (van den Aardweg, et al, 1982).

### Stem cell content

From multifraction data it is possible to estimate that the number of stem cells 'at risk' for forming a colony detectable in a cross section of a tubule, five weeks after irradiation, ranges between about 3 and 11 in different strains and ages of mice (Thames & Withers, 1980; Bianchi et al, 1983). Although the length of a colony five weeks after high doses of radiation varies greatly, a reasonable mean value lies between 1 and 2 mm (van den Aardweg et al, 1983). If then there are 3–11 stem cells at risk per 1–2 mm of tubule, the total number per testis ( 1 meter of tubules) would lie between 1500 and 11 000. A similar estimate of about 2000 stem cells/testis was made from extrapolation to zero dose of an exponential curve for survival of stem cells per total testis after neutron irradiation (deRuiter-Bootsma et al, 1976).

The number of isolated spermatogonia counted in unirradiated testes varies between 20 000 and 30 000 in different strains of mice (Meistrich et al, 1984). Thus, if a stem cell is defined as one capable of regenerating a persistent colony of spermatogenic epithelium, only a small proportion, less than 30 per cent, of isolated spermatogonia are stem cells. Alternatively, an average of 70 per cent of isolated spermatogonia are so radiosensitive that they do not affect the single-dose radiation response detectably. If the latter were true such exquisite radiosensitivity may contribute to the radiosensitizing effect of redistribution of cells through the cell cycle in multifraction or continuous low dose-rate irradiation.

Not only is the initial number of stem cells per testis low, but those which survive to form a colony after irradiation are slow to regenerate, showing no evidence of it within two weeks (Withers et al, 1974) and requiring as long as five weeks to double in number (Meistrich et al, 1978; van den Aardweg, et al, 1982). A colony of repopulating spermatogenic epithelium evident in histological sections five weeks after irradiation, but commonly containing just one or two stem cells, is obviously the result of a mode of proliferation more characteristic of steady-state than regenerative conditions. Among all normal tissues studied by clonal endpoints, this relative lack of a regenerative response is unique to the testis: but it is probably also a characteristic of the oocyte population which has not been, and probably cannot be cloned.

## PRESENTATION OF DATA

The response of the spermatogenic epithelium of each testis is expressed in terms of the proportion of tubules being repopulated, either as that value itself (repopulation index), or as the number of stem cells per tubule cross section (stem cell survival index) determined from the repopulation index using the rules of random (Poisson) distribution statistics.

The purpose of a Poisson correction is to account for the probability that some colonies represent the coalescence of cells derived from more than one stem cell, a probability that increases as the density of regenerating colonies increases, i.e. at lower doses and, to some extent, as the interval between radiation and assay is increased. It is a reasonable transformation since the survival of colonies throughout the tubules is random (van den Aardweg et al, 1983), and bears a linear relationship to the number of colonies counted in a histological cross section of the organ (deRuiter-Bootsma et al, 1976). The correction is applied to the score from each testis by taking the proportion of tubules containing no colonies and calculating the average number of stem cells surviving per tubule on the assumption of a Poisson distribution: average number of stem cells per tubule = $-\log_e f$, where f is the proportion of tubules without a colony. The conversion of repopulation index to stem cell survival index does not affect the scores at low levels of cell survival, becoming a significant factor only when the repopulation index exceeds about 30 per cent; for example, conversion factors that would be applied to repopulation indices of 30 per cent, 50 per cent and 70 per cent are 1.18, 1.39 and 1.72 respectively. In practice, we design experiments to achieve repopulation indices below 50 per cent to minimize the problems of a multicellular origin of colonies. A single dose survival curve for stem cells is shown in Figure 12.3.

In isoeffect experiments, in which the same level of injury is compared for various modes of insult, it is desirable to use an easily scored level of injury, e.g. a re-

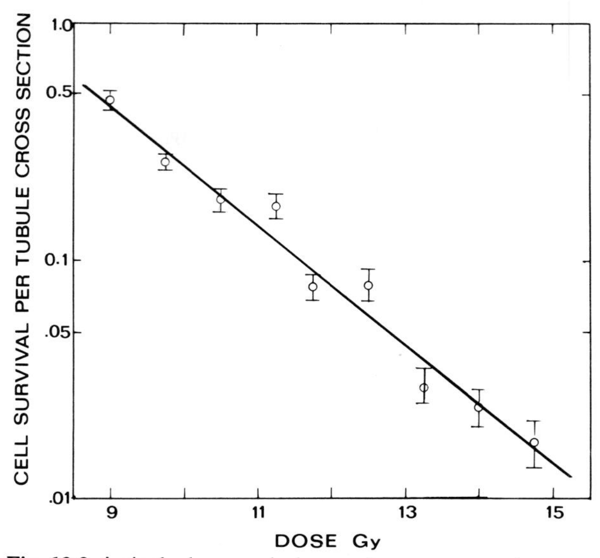


**Fig. 12.3** A single-dose survival curve for spermatogenic stem cells exposed to γ-rays. The data points and their 95% confidence intervals were obtained from the proportions of tubules showing regeneration of spermatogenic epithelium by applying a correction to account for multiplicity of stem cells per regenerative focus. The $D_0$ value and its 95% confidence interval is 1.74 (1.52–2.40) Gy.

population index of 10 per cent. This value is in the middle range of a readily-determined survival curve, and therefore is characterized by less uncertainty from interpolation than values near the extremes of the scorable range.

In order to study the response of the testis to lower doses, a complete survival curve can be reconstructed from multifraction experiments if it is established that each dose in a series produced an equal (logarithmic) cytocidal effect. When 4-hour fractionation intervals were used this condition was probably met, and testicular stem cell survival characteristics at low doses were seen to resemble closely those of jejunal crypt cells (Thames et al, 1981). A single dose curve can also be reconstructed from two fraction data (Hendry, 1979). Lower-dose regions of the survival curve can also be obtained by using a shorter post-irradiation regrowth period, since this allows identification of a larger number of discrete colonies (see p. 110).

While it has been presumed in the above discussion that there is just one population of stem cells which is relatively radioresistant, the existence of a small subpopulation of radiosensitive cells cannot be excluded by the experiments described.

## MODIFICATION OF METHODS

### Anaesthetic

The use of anaesthetic does not appear to modify the

radiosensitivity of spermatogenic stem cells (Suzuki et al, 1977; Bianchi et al, 1984), but it could indirectly affect the radiation dosimetry. In anaesthetized mice the testes descend into the scrotum whereas in mice that are awake, they are usually retracted into the inguinal canals. With $^{60}$Co or $^{137}$Cs gamma-rays it may be necessary to bolus the testes of anaesthetized animals and with all beams it is necessary to determine the dose in the inguinal canal, not the scrotum, if the animals are to be irradiated awake.

**Interval between radiation and assay**

With time after irradiation the colonies expand along the tubule. Although the rate at which the tubule surface is re-covered varies from colony to colony, and with dose, and will be slower initially as the colony grows from its unicellular origin. An average rate of about 30 $\mu$m/day has been measured during weeks 3 to 8 following high doses of radiation (van den Aardweg et al, 1982, 1983). Therefore, even though the absolute number of colonies does not increase (de Ruiter-Bootsma et al, 1976), the probability that a colony will appear in a histological section increases slowly with time. Nevertheless, colony counts per tubule remain a reliable index of total stem cell survival per testis regardless of when the assay is performed between 4 and 20 weeks after irradiation (de Ruiter-Bootsma et al, 1976; van den Aardweg et al, 1982). The effect of prolonging the regrowth time for colonies, at least for intervals longer than 4 weeks, is to shift the survival curve slowly to the right with no change in slope (de Ruiter-Bootsma et al, 1976).

35 days is a convenient time to assay because the colonies are easily identified and the scores reflect clonogenic survival accurately. However, one advantage of using shorter intervals is that responses after lower doses can be measured because larger numbers of colonies can be counted before they coalesce. (Note that early colonies will contain relatively more spermatogonia and fewer spermatocytes.) By scoring at 21, instead of 35, days responses to doses 20–25 per cent lower can be measured (de Ruiter-Bootsma et al, 1976). However, results from regrowth periods of 3 weeks or less may not be strictly comparable with those from longer intervals, because the time at which stem cells make their first division after exposure may vary, at least theoretically, by up to 8 or 9 days. Further, shrinkage of the testis is not complete until 4 weeks, and the absolute rate of increase in length of colonies may still be changing. These effects could alter the colony counts in a dose-dependent manner. Nevertheless, modifying the technique by using shorter regrowth times may be useful in studying responses to lower doses, especially if the aim of the experiments is to achieve iso-survival from different treatment procedures rather than survival curve slopes.

**Methods for assaying the consequences of stem cell survival**

Meistrich et al (1978) developed two useful methods for estimating the consequences of stem cell survival: counts of sperm heads (nuclei from spermatids in the final stages of development) obtained by homogenization of the whole testis by ultrasound, and assays of LDH-X, the X-isozyme of lactic dehydrogenase which is produced only by progeny of stem cells beyond the stage of mid-pachytene spermatocytes, and which is present in the supernatant of the homogenate prepared for sperm head counts. The stages of spermatogenesis at which these assays apply is illustrated in Figure 12.1. From this figure it is apparent that these assays should not be performed until more than 40 days after exposure: in practice they are usually performed at 8 weeks. Such assays are especially useful for assessing responses to lower doses than those necessary for the stem cell assay. Sperm head counts have been used to measure the cytotoxic effects of an injected radioisotope (Mian et al, 1982). Furthermore, they provide a measurement of the 'functional' survival of stem cells since they measure the output of differentiated products of spermatogenesis. Although it had been thought that stem cell differentiation returned to normal after irradiation (Nebel et al, 1960; Oakberg, 1971) a recent report (van den Aardweg et al, 1983) indicates that there is a dose-related interference with production of cells in later stages of differentiation, mainly after the resting spermatocyte stage and especially in the elongation of spermatids. These findings are consistent with the observations of Meistrich (1982), that the survival curve for sperm heads is steeper than that for loss of LDH-X, which in turn is steeper than that for stem cell survival. Also, the complex shapes of survival curves obtained with functional endpoints such as sperm head counts and LDH-X levels (Lu et al, 1980), may result from factors other than just stem cell survival, for example a dose-dependent alteration in differentiation pattern.

Weight loss, commonly measured at 28–30 days after irradiation (Kohn & Kallman, 1954) is an easy quantitative assay that is useful for some purposes, but it is predominately a measure of the killing of differentiating spermatogonia, not a measure of stem cell survival.

**Oxygenation**

The spermatogenic epithelium in the mouse exists at an oxygen tension less than that necessary for maximal radiosensitivity. If irradiations are performed in pure oxygen at 3 atmospheres pressure, the doses required for iso-survival are reduced by about 20 per cent, and a similar reduction can be achieved with misonidazole (Suzuki et al, 1977).

### Hyperthermia

Hyperthermia is effective in killing testis stem cells. Those most affected are furthest from where the veins leave the testis (Reid et al, 1981). This asymmetry of thermal injury in contrast to the more symmetrical radiation injury, presumably reflects a gradient of temperature across the organ similar to that across intestinal wall (Merino et al, 1978; Hume et al, 1979). Another difference between the histological patterns of injury after hyperthermia and radiation is that Sertoli cells are killed by heat and sperm remained 'marooned' in seminiferous tubules, presumably from a lack of Sertoli cell secretions in which to move (Setchell & Waites, 1974).

### Chemotherapeutic drugs

The cytotoxic effect of chemotherapeutic drugs varies markedly. Of 21 drugs tested (Lu & Meistrich, 1979; Meistrich et al, 1982), only two (Adriamycin & Thiotepa) were capable of reducing stem cell survival sufficiently to permit assessment of chemosensitivity using the clonogenic cell survival assay. The effect of some other drugs may be readily assayed by measuring depression of sperm head counts, and their site of action within the differentiating population can be deduced from the time course of the depression (Meistrich, 1982; Meistrich et al, 1982).

Naturally, the interaction of two or more agents (e.g. drugs, hyperthermia, radiation), or their lack of interaction in stem cell killing, can always be established by the clonogenic assay because, even in the absence of any effect of the other agents, the radiation dose can be adjusted to ensure that colonies could be counted.

## CONSEQUENCES OF MODIFYING MATERIALS

### Strain of mouse

The response of spermatogenic cells to single and split doses of radiation varies between mouse strains (Bianchi, 1983; Bianchi et al, 1984; Meistrich et al, 1984). While much of the variation reflects differences between laboratories, e.g. in techniques and criteria for scoring colonies, and in handling of data, real differences do exist. This is shown by variations between strains reported by the same investigator (Meistrich et al, 1984; Bianchi et al, 1984). Such differences are less apparent in other normal tissues.

### Species

The system could be used for measuring radiosensitivity of spermatogenic stem cells in mammals other than mice and rats since histological changes consistent with clonally-derived regenerative foci have been described in several different mammalian species (Regaud & Nogier, 1911; Erikson, 1963; Pitcock, 1964; Courot, 1964), including man (Heller et al, 1968).

### Age

The spermatogonia of prenatal and early postnatal mice show a high radiosensitivity which decreases rapidly within a few days of birth (Erikson & Martin, 1972; Hilscher et al, 1982). There is no significant change in the radiosensitivity of stem cells between 7 and 110 weeks of age although there is an exponential decline in stem cell number, the halving time being about 35 weeks in C3H mice (Suzuki & Withers, 1978).

## INFORMATION THAT CAN BE GAINED USING THE TECHNIQUE

The survival assay for colony forming cells provides an accurate method for measuring the cytotoxic effects on spermatogenic stem cells from various agents, e.g. radiation, chemotherapeutic agents, hyperthermia or other cytotoxins, provided at least 90 per cent of the stem cells are killed. Because regeneration of stem cells is so slow, it provides an excellent model for studying, in vivo, radiobiological problems such as the effect of cell cycle redistribution on tissue responses.

Perhaps more importantly, it provides another biological endpoint with which to study the normal biology of the testis. For example, demonstration that the stem cell, as defined by its ability to regenerate spermatogenic epithelium, was radioresistant, helped confirm that the stem cell is an isolated type A spermatogonium. The endpoint has also helped in a better understanding of the kinetics of spermatogenesis, and it could be a useful tool in studying mechanisms determining steady-state or regenerative proliferation kinetics. Further, since radiation selectively kills spermatogenic epithelium, leaving Sertoli and Leydig cells relatively unaffected, the endpoint could prove valuable in quantitative research into the hormonal and/or nutritional interrelations of the cells involved in the male reproductive system. It could also benefit fertility studies (it has already been possible to correlate infertility with the reduction of stem cell survival to 10%), and conversely, it could be useful in quantitating effects of strategies for male contraception. For many of these purposes the sperm head counting technique can be substituted, especially for lower dose effects. For example, the ratio of sperm counts immediately after radiation to those at some time later (e.g. 56 days for mice, but longer for man) would be a good

Howard Thames and Jolyon Hendry for useful suggestions.

This investigation was supported in part by PHS

# 13

*N.F. Kember*

# The cartilage clone system

## INTRODUCTION

The cartilage clone system has the modest claim to fame that it was one of the earliest in situ systems to be developed — preceding the skin clones of Withers (1967) by a few months. The claim, however, has to be modest in that the author of this chapter is the sole experimenter to publish results using the system. None of the findings obtained have been confirmed either by other workers, or in animals other than the Wistar rat, or in growth plates other than the proximal tibia. It may be conjectured that radio-biologists have not repeated the studies or extended them because it is considered that the growth cartilage is not a typical tissue — but which tissue is 'typical'? The system has some unique features and has the potential for special studies on, for example, the comparative radiobiology of young, mature and senile proliferating populations.

The clones of cells that are seen in cartilage at a few weeks after a dose of the order of 15 to 20 Gy (see

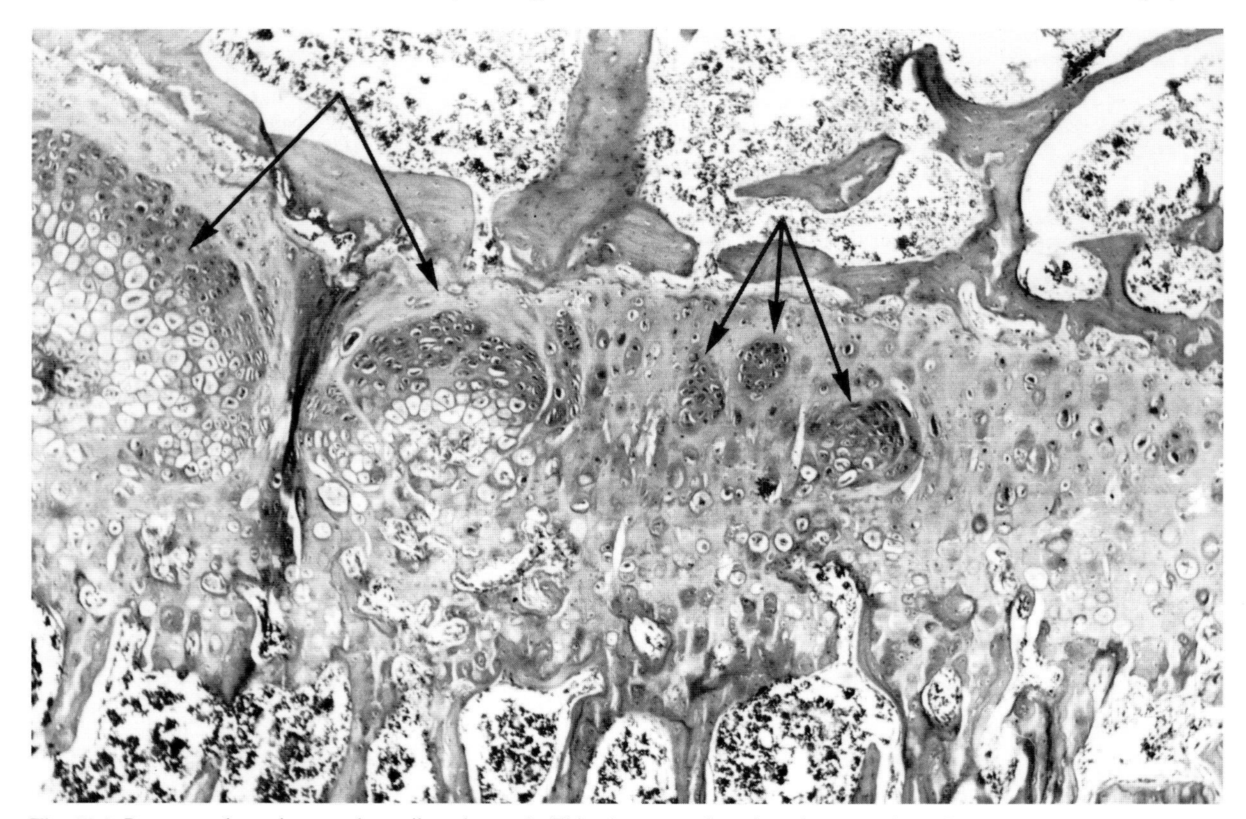

**Fig. 13.1** Recovery clones in growth cartilage (arrows). This view was selected to show a variety of clone appearances. The two clones on the left are well advanced with hypertrophic cells and vascular invasion from the metaphysis. The 3 clones, centre right, could either represent slices from the centre of small clones or 'edge' slices through large ones. The tissue to the right is typical of irradiated cartilage before clones appear. Sagittal section of rat proximal tibia. Stained H & E × 200.

Fig. 13.1) have been observed by a number of workers. Dahl (1936) was, perhaps, the first. He reports 'groupes isogeniques de chondrocytes' at 31 days 'apres l'irradiation, reaction cutanée du degré 7' (after about 13.5 Gy). The effect was also observed by Gall et al (1940) as 'small focal clusters' of actively growing chondrocytes and by a number of other workers in more recent years.

## OUTLINE OF THE SYSTEM

The growth of a long bone is dependent on division of the chondrocytes of the epiphyseal growth plate. The tissue has a characteristic structure with columns of cells which can be divided into a 'reserve cell zone' adjacent to the epiphysis, a proliferation zone, a maturation zone and a hypertrophic zone in which the matrix between the cells is calcified. These hypertrophic cells are invaded by vascular loops from the metaphysis, at a rate which under normal conditions is equal to the cell production rate so that the cartilage columns have a constant length during the active period of growth.

If the cartilage plate is irradiated to about 20 Gy the majority of the proliferating cells die and the architecture of the plate is disrupted. The cell columns disappear leaving only the matrix with a few scattered empty lacunae containing the debris of chondrocytes. However, in some plates and at about 2 to 3 weeks after irradiation clones of actively dividing chondrocytes may be observed (Fig. 13.1). These expand within the damaged matrix. Cells at the metaphyseal side of the clone become hypertrophied and new blood vessels break through the matrix from the metaphysis to re-establish normal endochondral ossification. Eventually these clones repopulate the plate and growth continues as before (Fig. 13.2).

If the number of cells at risk in a growth plate can be estimated and if the number of clones that appear after a given dose of radiation can be counted the system can be used to measure cell survival.

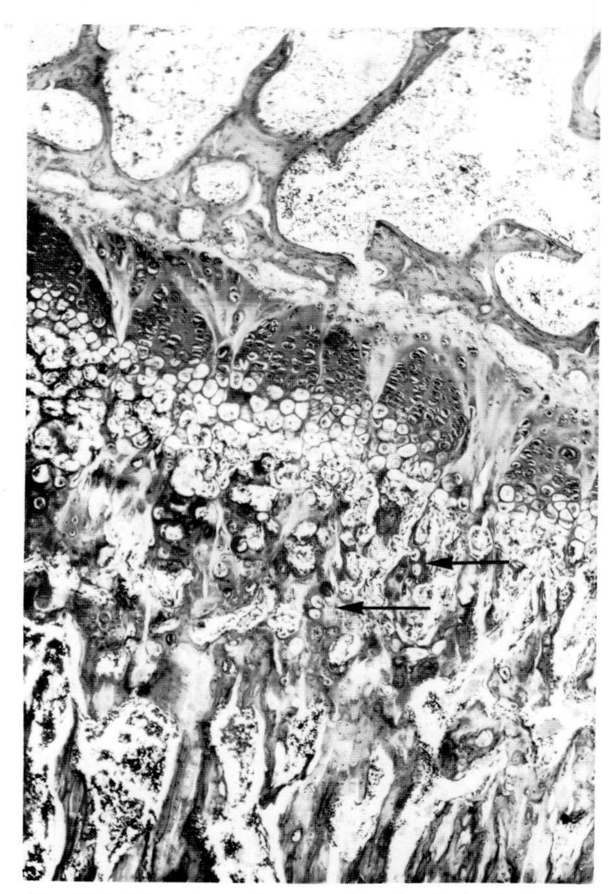

**Fig. 13.2** A later stage of recovery from irradiation in the growth plate. Most of the clones are still distinct but endochondral ossification has resumed across the full width of the plate. In the metaphysis there is evidence of the old cartilage (arrows) below the new primary spongiosa. Sagittal section of rat proximal tibia. Stained H & E × 163.

### The number of cells at risk

All the discussion in this paper refers to the growth plate at the proximal end of the tibia in a 6-week-old Wistar rat. The author has always assumed that only cells in the reserve and proliferation zones are at risk, i.e. that only these cells can give rise to recovery clones. This is not necessarily the case since some recovery from maturing cells may take place following treatment with hydroxyurea (Kember, 1972). It is, however, most unlikely that cells external to the cartilage could migrate through the matrix into the substance of the cartilage plate and start to divide and form clones. This chapter, therefore, will assume that the clones arise in those parts of the cartilage plate where cells may be found labelled at one hour after an injection of tritiated thymidine.

In kinetic studies the number of cells per cartilage column that are in proliferation can be measured from the labelling profile for 100 cells with an accuracy of about plus/minus 20 per cent. In the proximal tibia in rats of 5 to 7 weeks of age counts show that there are 15 to 25 proliferating cells per column. The number of columns per cartilage plate can also be estimated from counts on sections. Thus a count of 200 columns across a section taken from the midline of the bone represents a total of $3.14 \times 100^2$, or 30 000 columns in the complete disc of the growth plate. From these estimates the total cells at risk are (20 × 30 000), i.e. 600 000 cells, in one cartilage plate. An error of plus/minus 50 per cent on this figure does not invalidate the cartilage clone method since surviving fractions are measured at the $10^{-3}$ to $10^{-6}$ levels.

**Clone counting**

Ideally a count of all clones appearing on each plate should be made. It might be possible to count clones in a cartilage plate that had been dissected out following vital staining but only histological methods have been used to date. A complete count would entail a detailed and tedious survey of about 400 serial sections per cartilage plate. A less than ideal method has therefore been adopted which should, however, give an adequately precise count on a sufficient number of replicate cartilage plates per experimental point.

In sagittal sections the clones are seen as roughly circular discs with varying diameters. If the thickness of the section is 't', then a volume $t.\pi.(d/2)^2$ of the total clone is seen on the section. This slice lies somewhere in the spherical volume of the clone which has a diameter 'D' at least as large as the observed diameter 'd'. If it is assumed that the observed diameter is equal to the most probable diameter of a slice taken at random through the sphere i.e. equal to $\pi.D/4$ then the observed disc represents a fraction

$$f = \frac{t.\pi.(d/2)^2}{4/3.\pi.(2.d/\pi)^3}$$

of the total volume of the clone. (This may also be regarded as the probability that the clone arose from a surviving cell in that particular section.) In practice the diameter of each clone in a section is measured using an eyepiece graticule. The sum of the individual f values is then found. The volume of the section can be calculated as a fraction F of the total volume of the cartilage plate. The clone count in that plate is then given by (sum of f)/F. A simple computer programme has been used for the calculation of the total clone count from the input of clone diameters and the breadth of the plate as it appears on that section. The section thickness 't' cancels out in the calculation. The surviving fraction for the particular dose of radiation equals clone count divided by the cells at risk.

In practice 91 serial sections of 10 $\mu$m thickness were cut in the sagittal plane near the centre line of the bone. Every tenth section was collected and stained with H and E for clone counting. The published data are based on mean clone counts for these 10 sections from each bone.

## ASSUMPTIONS IN CLONE COUNTING

The first assumption is that the clones arise from single surviving cells. This may not be justified and some clones may result from 2 or 3 adjacent survivors. This should theoretically overestimate the extrapolation number on the survival curve but is probably a small correction in comparison with the other errors in the system.

What is the criterion for identifying a clone? Figure 13.1 shows that clones may vary considerably in appearance even on the same section. Clones must contain cells that are histologically viable in appearance in order to be counted. In large clones with some hypertrophied cells the criterion for viability is clear but small clones with only 6 or 8 cells visible on the section offer a greater possibility for error in the definition of an active clone. Large clones, however, present a problem when they start to fuse with one another. Such fusion occurs more rapidly at the lower doses and sets an upper limit on the survival fractions that can be estimated by the method. In general clone numbers may be overestimated at high doses when there is a tendency to count any group of cells as a positive clone, and underestimated at low doses when the clones are closely packed. Allied to this error is the fact that the method for calculating the clone numbers gives a greater weighting to small than to large clones.

Clones do not always appear to be equally distributed across the diameter of the cartilage plate. In many sections more clones are seen at the anterior edge than in the centre of the cartilage plate. The most probable cause of this variable distribution is a dose gradient across the bone (Kember, 1967a).

At times later than 25 days some clones, particularly following the higher doses, may show signs of degeneration, i.e. they do not continue to grow or be invaded by blood vessels but appear to 'silt up' with matrix. These late abortive clones will be included as viable in counts carried out at an earlier time.

The optimum period after irradiation for clone counting has been investigated (Kember, 1965). It is necessarily a compromise between waiting for all clones to appear and counting before adjacent clones have time to grow and fuse. If experiments were carried out on single limbs then it would be possible to adjust the time of sacrifice to the optimum for given dose level but most experiments have been designed to give a comparison between treatments on the two limbs in the same animal — i.e. a single dose to one leg and a split dose to the other, so that the compromise delay of 25 days before sacrifice of the rats was selected.

The plot of data (Fig. 13.3) illustrates the wide variation in clone counts that were found in rat bones subjected to the same radiation dose. This inherent variability of the technique makes it of no value in the comparison of radiation regimes that only produce a small difference in effect.

One further factor that affects the numbers of clones that may be found after a given dose of radiation is the general health of the animal. When the number of clones per plate is plotted against the weight gain of the animal during the 25 day post-irradiation period it is seen that there is a strong link between clone count and weight

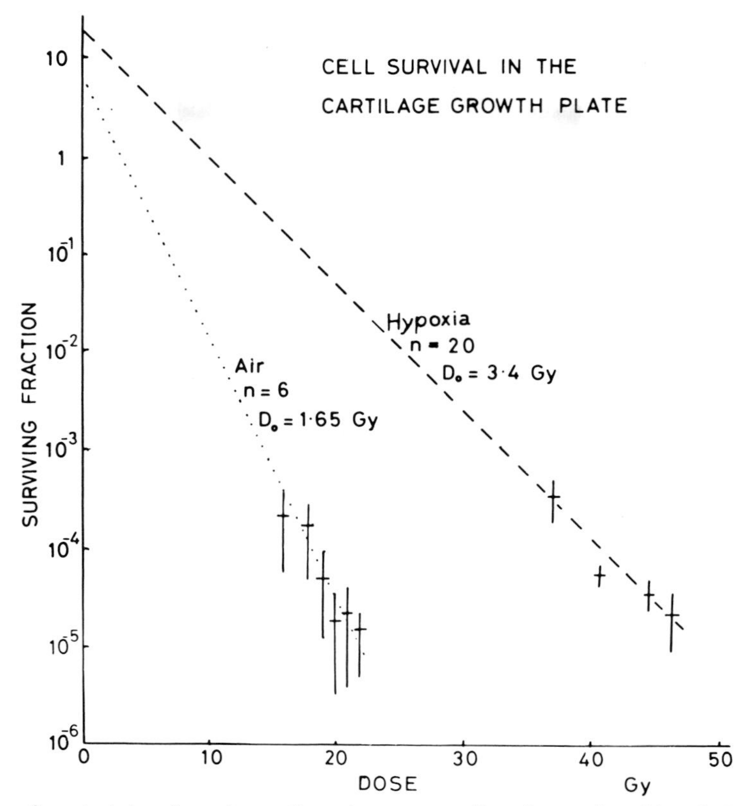


**Fig. 13.3** Standard plot of survival data from the cartilage clone system. Data for rats breathing air (total of 94 animals) and for legs made hypoxic (14 animals) by the application of a tourniquet. Error bars indicate standard deviations.

gain (Kember, 1965). Data from animals showing a weight gain of less than 30 per cent during the 25 days were excluded from the published results.

## DOSIMETRY

For the majority of irradiations the rats were not anaesthetised but were restrained in close-fitting perspex moulds. A 200 kVp X-ray machine was used at a dose rate of about 2 Gy per minute. The animals were shielded with 4 mm lead except for the two hind limbs which protruded from the shield. The legs could be separately covered with lead for differential exposures. In experiments for which the limbs were made hypoxic for irradiation the animals were anaesthetised for the duration of the application of the tourniquet. Doses were measured during trial irradiations with a Baldwin-Farmer ionisation chamber inserted into the mould in the position of the rat tibia. The measured dose provided a calibration for the monitor chamber that was used to check output during the experimental irradiations. In this way good consistency was achieved in the doses given in the various experiments.

In addition to the relative dosimetry it is necessary to consider the variation in dose that may occur within the cartilage plate. Optimum conditions call for an equilibrium thickness of scattering media around the limbs to reduce local variations in dose but the amount of material close to the leg had to be kept to a minimum to reduce scattering under the lead shield into the trunk. Thus the model for dose is roughly that of a small cylinder irradiated in air. The dose variations will therefore include a gradient from anterior to posterior — about 10 per cent and also a change across the plate due to the circular geometery and lack of side scatter. A variation of 10 per cent in an overall dose of 20 Gy is equivalent to a change in dose of over 1 $D_0$ to cells in different parts of the cartilage plate.

## MEASUREMENTS WITH THE CLONE SYSTEM

In considering the appearance of clones it is natural to inquire if more clones start from cells in the reserve zone of the plate than from cells in the proliferation zone. Such a difference could arise because of variation in radiosensitivity or because the reserve cells have a greater potential for repeated division. The question cannot be answered definitively since clones are already 3 or 4 cells in diameter at the time of detection and have therefore expanded beyond their site of origin. An experiment with younger rats which have wider growth plates indicated that clones could originate from surviving cells within any part of the reserve or proliferation zones. The simple fact that each rat has two tibiae makes comparative experimental designs attractive. In split dose experiments a single dose can be given to one leg and the divided dose to the other. Although the system can be used for ($D_2$–$D_1$) experiments it is simpler to give the same total dose to both limbs but split into two fractions for one leg. The ratio of clones (recovery ratio) in the two legs should equal the extrapolation number provided that the survival curves for single and split doses are parallel.

Since it is possible to make one limb hypoxic by tying a tourniquet around the thigh the system can be used for OER measurements. One problem is that under these conditions the doses required to reduce survival to countable levels are high and special care has to be taken to ensure that the scattered dose to the trunk of the rat is kept low. The convenience of the system has made one type of experiment possible — the investigation of recovery under hypoxia. In the published study it was found that when limbs were kept hypoxic no recovery was observed between split doses (Kember, 1967b). Experiments with fast neutrons have also been described by Kember (1969a).

**Survival and overall growth loss**

Because the cell columns within the growth plate constitute an effectively linear growth system it is possible to relate cell survival to the loss of overall bone growth. This can be most conveniently accomplished using a computer model of the system (Kember, 1969b). While general agreement between cell survival levels following a given dose of radiation and the corresponding loss of bone growth can be established there are too many unknown variables (such as possible acceleration of cell division in irradiated cartilage) to draw any precise conclusions about the relationship between cell survival and overall tissue effect.

## CONCLUSIONS

Within the limitations of the assumptions outlined above the cartilage clone system does provide useful estimates of cell survival parameters for chondrocytes. However, in addition to the caveats that have been discussed previously some further comments should be made. Cell survival values by themselves are useful for comparisons with other tissues but they can in no sense be regarded as providing a sufficient description of the effects of radiation on the tissue. The natural history of the post-irradiation effects requires to be followed in some detail from irradiation to recovery so that effects of damage to blood vessels and other supporting tissues can be assessed at each dose level.

The system underlines the warning that care should be expressed in extrapolating from rodent to man. It is clear that the architecture of the human growth plate is dissimilar in important features from that of the rodent (Kember & Sissons, 1976), the main difference being the presence in man of the wide inert zone above the proliferation zone. This zone separates the cell columns from the epiphyseal blood supply. It is likely, therefore, that the growth plate chondrocytes in man may exist at a much lower oxygen tension than in the rat. Estimates of cell kinetics of the growth plate also show that the cell division rate is slower by a factor of 5 to 10 in man by comparison with the rat.

The cartilage system has the advantage over other cell survival systems that comparative experiments in the same animal are possible. This has been used in the experiments that showed lack of recovery under hypoxic conditions between split doses. The system's value would be enhanced if data from in vitro survival experiments with chondrocytes were available.

## THE CHONDROSARCOMA

The moderate success of the in situ clone system in growth cartilage promoted the search for a parallel system in the tumour of cartilage — the chondrosarcoma. Recovery clones have been observed in the DC II chondrosarcoma of mice by Swarm et al, (1964). Using the same transplantable tumour in Balb/c mice, attempts were made to repeat the observation and to quantitate the system. The tumours were transplanted subcutaneously into the thigh and were allowed to grow to a palpable size (about 1 $cm^3$) before irradiation but under these conditions few clones were observed. The study was continued therefore, following closely the protocol of Swarm et al (1964). The tumours were irradiated at 28 days after transplantation, that is, before they were palpable, and then removed after a further 28 days when the animals were killed.

A series of tumours were collected at 28 days after transplantation to estimate the numbers of cells at risk in irradiated tumours. These 28 day tumours varied in size from 2–200 mg and estimates of cell numbers varied from $2 \times 10^5$ to $3 \times 10^6$ per tumour. This variation introduces a basic imprecision to the later stages of the quantitation.

Clones were seen in many of the irradiated tumours (Fig. 13.4) and they were counted by a method similar

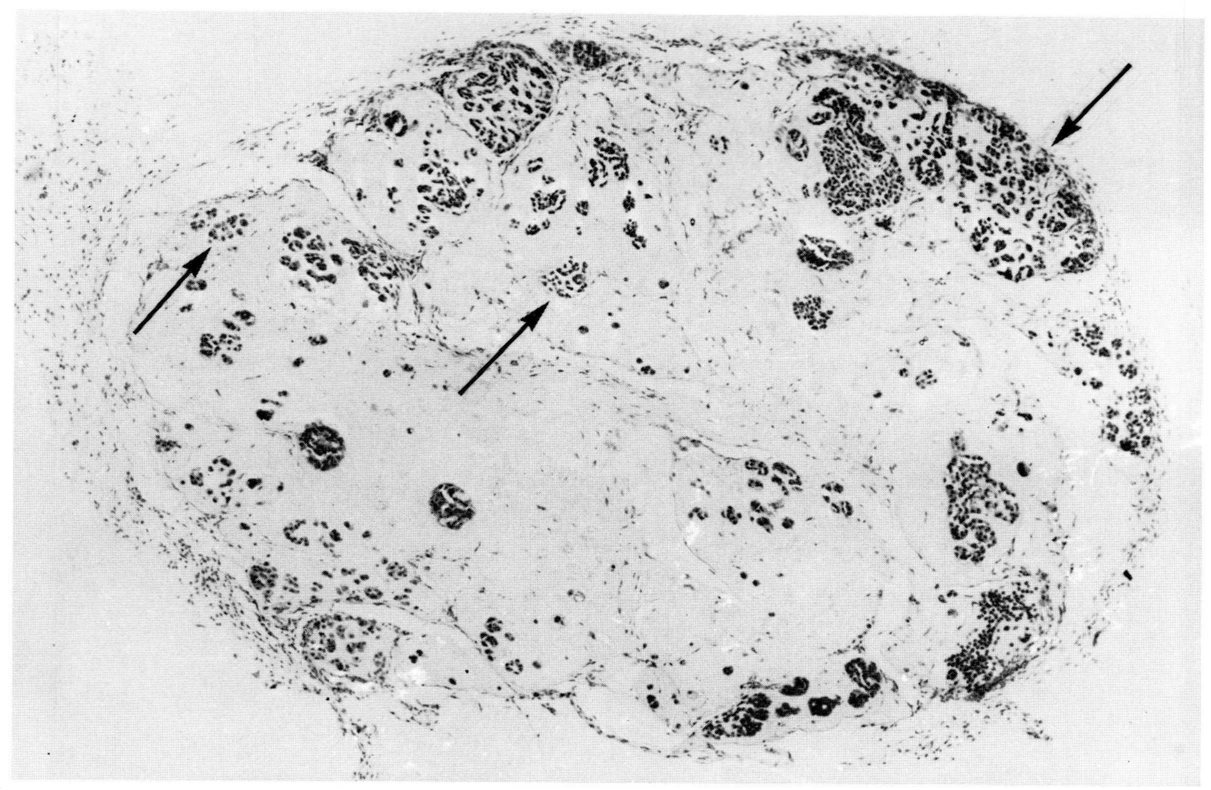

**Fig. 13.4** Recovery clones in irradiated chondrosarcoma. The arrows show clones of different sizes and appearances. Section through transplanted tumour at 28 days after irradiation. Stained H & E × 80.

to that used for the cartilage plate. The clones within any one tumour showed a range of sizes and might be confluent at one pole of the tumour and completely absent at the other. The unirradiated tumour is comprised of small clusters of cells and the clones also contain 'micro-clones' within the larger groupings. This poses the question 'What is a clone?' The assessment and counting of clones is even more subjective for the chondrosarcoma than for the cartilage plate system.

If the uncertainties in the system are accepted then survival fractions can be estimated and may be plotted against dose. A $D_o$ of 1.5 to 2 Gy was estimated based on an assumed extrapolation number of 6. It was disappointing to arrive at such poor data from a system that apparently has such potential. Perhaps the system is worthy of further development starting with a more repeatable transplantation technique using, possibly, a pellet containing a known number of tumour cells since the small 1 mm cubes of tumour that were implanted in our experiments could have originated from 'active' or 'inactive' regions of the original tumour. The DC II tumour also proved unreliable in radiobiological experiments based on measurements of tumour size during regrowth after irradiation (Marsden et al, 1980).

## ACKNOWLEDGEMENTS

These various studies were carried out with the support of the Medical Research Council and the Cancer Research Campaign.

## REFERENCES

Dahl B 1936 De l'effet des rayons X sur les os longs en développement et sur la formation de cal. Étude radiobiologique et anatomique chez le rat. Skrifter norske videnskaps Band 1, p 90

Gall E A, Lingley J A, Hilcken J A 1940 Comparative experimental study of 200 kilovolt and 1000 kilovolt Roentgen rays. American Journal of Pathology 16:605

Kember N F 1965 An in vivo survival system based on the recovery of rat growth cartilage from radiation injury. Nature 207:501

Kember N F 1967a Cell survival and radiation damage in growth cartilage. British Journal of Radiology 40:496
Kember N F 1967b Hypoxia and recovery in growth cartilage in vivo. International Journal of Radiation Biology 13:387
Kember N F 1969a Radiobiological investigations with fast neutrons using the cartilage clone system. British Journal of Radiology 42:595
Kember N F 1969b Growing bones on the computer — Some pitfalls of a computer simulation of the effects of radiation on bone growth. Cell and Tissue Kinetics 2:11
Kember N F 1972 Hydroxyurea and the differentiation of growth cartilage cells in the rat. Cell and Tissue Kinetics 5:199
Kember N F and Sissons H A 1976 Quantitative histology of the human growth plate. The Journal of Bone and Joint Surgery 58-B:426
Marsden J J, Kember N F and Shaw J E H 1980 Irregular Radiation response of a chondrosarcoma. British Journal of Cancer 41:88
Swarm R L, Correa J N, Andrews J R and Miller E 1964 Morphological demonstration of recurrent tumour following irradiation. Journal of National Cancer Institute 33:657
Withers H R 1967 The dose-survival relationship for irradiation of epithelial cells of mouse skin. British Journal of Radiology 40:187

# Clonal regeneration techniques in vitro applied to the thyroid

## HISTORICAL BACKGROUND

The thyroid gland has two main endocrine functions; first the secretion of the thyroid hormones triiodothyronine ($T_3$) and thyroxine ($T_4$) following the incorporation of iodine into the prohormone, thyroglobulin, and second the secretion of calcitonin by the C cells. The differentiated characteristics of the gland can thus be summarised as (1) the ability to take up iodine; (2) the ability to form thyroglobulin and iodinate; (3) the ability to produce $T_3$ and $T_4$; (4) the ability to produce calcitonin. In addition, the gland has a characteristic follicular appearance. The shape of the follicular cells varies with the state of activity of the gland, being flattened if the follicles are filled with colloid material and the gland is resting, becoming columnar when the gland is active and the follicles contain little colloid.

The epithelial cells lining the follicles selectively absorb iodide ions from the blood in the rich capillary plexuses surrounding the follicles. By the use of radioactive iodine it has been shown that ingested iodine is rapidly taken up by the thyroid gland and is concentrated to a level up to 50 times that of the plasma content. Autoradiographic studies on histological sections of thyroid glands of rats, which had been given radioactive iodine and amino acids, showed that the amino acids including tyrosine are used to synthesize large protein molecules in the rough endoplasmic reticulum of the basal and perinuclear cytoplasm. To these molecules carbohydrate groups are added in the cisternae of the endoplasmic reticulum and in the saccules of the Golgi complex, so forming a complicated molecule of thyroglobulin — a glycoprotein. Membrane-bound vesicles transport this glycoprotein to the apical surface of the cell, where it is released into the colloid. Iodine is added to the tyrosyl radicals of this molecule at or near the cell surface adjacent to the colloid.

Clonal techniques for studying survival or transformation of thyroid cells cannot be meaningful unless some of the characteristics of differentiated thyroid function are retained. Commonly used techniques for the study of thyroid response in vitro involve the use of thyroid slices (Dumont et al, 1980) or various tissue culture methods (e.g. Dickson et al, 1981; Ambesi-Impionbato, 1980; Miller et al, 1983; and the technique discussed in this chapter, O'Connor et al, 1980a and Mothersill et al, 1981, 1984ab). In addition an assay has been developed (Clifton et al, 1978) where treatments are carried out on the intact gland in vivo and this is then dissected, dissociated and injected into the fat pad of the rat where the cells grow (see Ch. 15).

Of the true in vitro methods, loss of differentiation and the need for frequent subculture procedures mean that the results obtained from clonal assays have uncertain relevance to the intact tissue. Furthermore, the study of long-term effects of for example drugs or radiation, is difficult.

The technique to be described here provides a substantial improvement on previously used methods, as it allows clonal assays to be performed on primary cultures of thyroid cells which continue to trap iodine, produce $T_4$ and retain a follicular morphology for up to three months in continuous culture without replacement of the initial culture medium. It thus provides a unique method for longterm studies. The method for primary culture of thyroid cells was described first by O'Connor et al (1980a). The minor modifications currently in use are incorporated in the method described below.

## PRIMARY CULTURE OF THYROID GLANDS

### Preparation of the cell suspension

Sheep thyroid glands are obtained from a nearby abbatoir, human glands are obtained during surgery. They are placed in Earle's $Ca^{2+}$ & $Mg^{2+}$ free balanced salt solution containing fungizone (4 $\mu$g/ml), gentamicin sulphate (150 $\mu$g/ml), penicillin (200 IU/ml) and streptomycin (200 $\mu$g/ml). All processing of the tissue is performed in a laminar air flow cabinet (Microflow); the glands are trimmed of excess fat, connective tissue and muscle, and diced to a final size of ~2 $mm^3$ using autoclaved surgical instruments. The chopped tissue is then placed in a sterile 100 ml container to which is added 40 ml 0.5 per cent w:v trypsin solution (Gibco Biocult Scotland) and 30 mg collagenase (Sigma type 4). The container is then placed on a heated magnetic stirrer, set at 37°C, for 30 minutes. The supernatant is poured off and filtered first through a standard blood filter and then through several layers of fine surgical gauze. An equal volume of medium containing 20 per cent serum is added to the filtrate to neutralize the trypsin activity. The cells are collected by gentle centrifugation at 200 × G for 15 min and stored at 4°C. The remaining tissue is incubated for three further 30 min periods as before. The cells from the four incubations are pooled and then plated into 5 ml of medium in small closed flasks (Nunc) with a growth area of 25 $mm^2$ at a density of (1–2) × $10^6$ and maintained in an air incubator at 37°C.

### Composition of growth medium

The growth medium normally used is basal medium (Eagle's) (BME) with the additives listed in Table 14.1 (see also Table 14.2). Other basic media successfully used include the following: Dulbecco's modified medium, Nutrient mixture F-12 (Ham's), Modified essential medium (Eagle's), McCoy's medium and Leibowitz medium. These were all found to give cultures similar to those obtained with BME, provided the additives remained constant (Murphy et al, 1983). Variations in additives caused considerable changes in the life span and degree of differentiation of the culture (Table 14.3). Lamb or horse serum were equally good but neither calf, fetal calf nor porcine serum were satisfactory. The hormone additives were all required but thyroid stimulating hormone (TSH) was the most essential, and differentiation was severely reduced in the absence of this substance. The antibiotic level is high and this point is discussed in detail later.

### Development of the primary culture

In order to maintain differentiated thyroid cells in culture it is essential to avoid medium changes at all times during the entire life span of the culture. Using this regimen the cells divide about twice during the first few days after plating and no differentiation is apparent. By day 5 the flasks are confluent and small follicles can be seen, particularly in areas of high cell density. Morphological differentiation continues to improve and reaches a constant state by day 14–16. By day 20–25 the cultures die unless treated as below.

**Table 14.1** Composition of growth medium used for primary culture of human or sheep thyroid cells (IU = international units)

| | |
|---|---|
| Basal medium (Eagle's) containing | |
| Lamb serum (for both human and sheep thyroid) | 20% v:v |
| Hydrocortisone | 0.1 μg/ml |
| Insulin | 10 mIU/ml |
| Potassium iodide | 1 μM |
| Penicillin | 20 IU/ml |
| Streptomycin | 20 μg/ml |
| Gentamicin Sulphate | 4 IU/ml |
| Fungizone | 1 μg/ml |
| Thyrotropin (TSH) | 17 mIU/ml |

**Table 14.2** Culture medium used by Miller et al (1983) for the growth of human thyroid cells

| | |
|---|---|
| MEM + Nutrient Mixture F12 | 1:1 |
| Glucose | 2 gm/l |
| Fetal calf serum | 2.5% |
| Insulin | 10 μg/ml |
| Hydrocortisone | 20 ng/ml |
| Transferrin | 5 μg/ml |
| Glycl-l-histidyl-L-Lysine acetate | 10 ng/ml |
| TSH | 10 mIU/ml |
| Somatostatin | 10 ng/ml |
| Epidermal growth factor | 10 ng/ml |

### The glucose metabolism of primary thyroid cells

During the culture period the cells have a unique metabolism which is not yet fully understood (Mothersill et al, 1980, 1981). For the first five days the glucose in the medium is rapidly used which results in the production of stochiometric amounts of lactate. From day 5 until the cells die the lactate is used and the exhaustion of the lactate levels in the medium invariably coincides with cell death (Fig. 14.1). Non-thyroid cells (CHO-KI or V79) do not use lactate (Mothersill et al, 1980). Differentiation is highly dependent on the metabolism of lactate and if the use of lactate is prevented either by changing the culture medium or by adding concentrated glucose solutions to the medium, dedifferentiation occurs.

### Extension of the life span of the primary culture

Differentiated thyroid cultures can be maintained for up to 100 days without medium changes by adding concentrated glucose solutions (5 mg/0.1 ml) to the medium one to two days before the lactate is depleted. Earlier addition of glucose results in loss of some differentiation (Mothersill et al, 1981, 1984a). Attempts to continue the life of the culture beyond 100 days were unsuccessful since the cells failed to use lactate after this time and cell death resulted following the use of the added glucose.

Addition of medium which has been previously 'conditioned' by thyroid cultures for 10 days can be used to extend the life span. However, this technique requires considerable planning where longterm experiments are envisaged and can lead to difficulties, for example with the longterm effects of chemicals which may modify the medium constituents.

**Table 14.3** Effect of changes in medium constituents on morphological differentiation of primary sheep thyroid cells at 15 days

| Treatment | Presence of follicles | Radio-iodine uptake (counts/$10^6$ cells/min $\times$ $10^{-3}$) |
|---|---|---|
| All constituents | **** | 12 ± 1.5 |
| No hydrocortisone | *** | 10.6 ± 0.6 |
| No insulin | *** | 9.9 ± 0.8 |
| No TSH | * | 2.5 ± 0.16 |
| All hormones + calf serum | ** | 5.6 ± 0.3 |
| All hormones + fetal calf serum | ** | 7.8 ± 0.8 |
| All hormones + porcine serum | * | 3.2 ± 0.1 |
| All hormones + horse serum | **** | 12.9 ± 0.8 |
| All constituents in MEM medium | **** | 13.3 ± 0.7 |
| All constituents in Dulbecco's medium | **** | 10.3 ± 1.9 |
| All constituents in Ham's F12 medium | **** | 14.3 ± 1.8 |
| All constituents in McCoy's medium | **** | 11.3 ± 1.8 |
| All constituents in Leibowitz medium | **** | 10.3 ± 0.8 |
| All constituents in medium 199 | **** | 12.4 ± 1.3 |

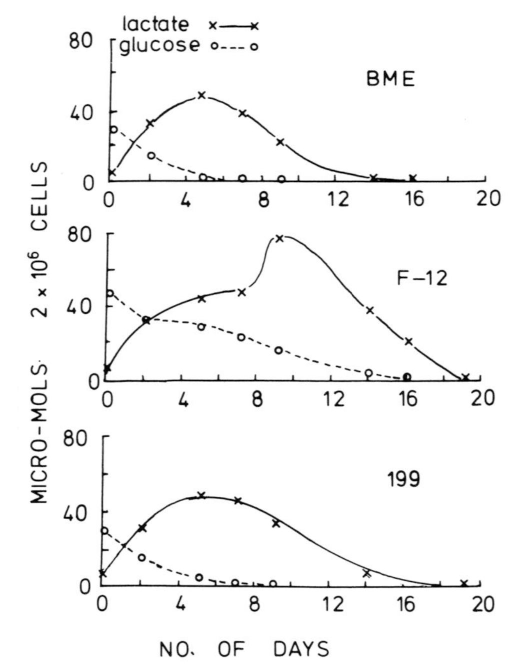


**Fig. 14.1** The levels of lactate and glucose detected in medium samples from primary thyroid cells cultured in three different media (BME, F-12 and 199). The initial glucose level in BME & 199 is 1 mg/ml. F-12 has an initial glucose level of 1.8 mg/ml.

## Evidence of thyroid function in primary thyroid cultures

Differentiated characteristics which can be observed in the primary culture include the appearance of follicles which contain PAS positive material (Fig. 14.2). Acid phosphatase activity around the follicles is also apparent indicating secretory activity. When they are metabolising lactate the cells take up radioactive iodine at a rate of up to 14 000 counts/$10^6$ cells/min which is 10–15 times the rate observed in non-thyroid cells. If glucose is added to the medium the iodine-trapping ability falls off (Fig. 14.3). Assay of $T_4$ in the medium of differentiated thyroid cultures shows a gradually increasing level in both human and sheep thyroid cells from trace levels to a maximum of 13 $\mu$g/100 ml, indicating that the cells are producing $T_4$.

## Problems associated with the primary culture

*Poor differentiation and contamination with fibroblasts*
This is frequently seen if the initial cell inoculum is less than 1 $\times$ $10^6$; the optimum inoculum is approximately 3 $\times$ $10^6$. The high inoculum allows rapid breakdown of glucose to lactate while preventing escessive growth of fibroblasts.

A second cause of poor differentiation is the use of small thyroids which are fibrous in texture.

*Longterm contaminants*
Cultures maintained for three months frequently develop slow-growing contaminants in the latter stages. These are usually yeast or moulds and are probably latent in the initial stages. Attempts at control are hampered as medium changes must be avoided. High levels of antibiotics and antimycotics in the initial culture are moderately successful but a particularly important control step was found to be the use of low centrifugation speeds during collection of the cells in the initial preparation of the tissue culture. Contamination is less of a

through a 10 ml standard pipette 15 times (note: in all previous steps pipetting should be kept to a minimum to avoid cell damage). The final cell suspension is filtered through nitex cloth (Tetko Inc.) with a 53 $\mu$m pore diameter to remove cell clumps. All cells from similar rats are pooled, pelleted and resuspended in 2–10 ml of medium.

**Mammary cell counting, dilution and transplantation**
A typical design for a mammary clonogen assay is summarized in Table 15.1.

The final enzymatically dispersed, washed suspension of mammary cells is serially diluted, 1:1, in 12 tubes. (Note: pipettes are always changed between dilutions.) A haemocytometer (phase contrast type) is filled with a sample from an appropriate tube, e.g. one with $5 \times 10^5$ cells/ml. The cells are counted with a phase microscope at 400 × magnification and the cell concentration is calculated for each dilution (cells/ml). The cell concentrations appropriate to the experimental design are chosen such that the actual number of cells per ml will be as close as possible to the desired number of cells per ml in the dilutions to be used. The final step in preparation for transplantation is the addition of an equal volume of 50 per cent brain homogenate to each suspension (see below).

We use a 0.10 ml syringe with a fixed 22 gauge 1½ inch needle (Hamilton — special order) to inject 0.03 to 0.06 ml aliquots of the cell suspensions into the graft sites. The white fat pads chosen for grafting are surgically exposed; the syringe needle is inserted into the pad just beneath its exposed surface. The fat pad is pinched around the needle with straight serated forceps to prevent leakage, and the suspension is injected slowly. The needle is removed by drawing it out through the forceps which are held in place for 10 seconds. The skin incision is then closed with wound clips.

The mammary cells are transplanted at one or two sites in the interscapular (i.s.) white fat pad. When one site is grafted, the cells are injected in the center of the pad; when two sites are to be employed, they are injected in the right and left wings of the pad.

**Scoring and calculation of clonogen number and survival**
Under appropriate host conditions and grafted cell numbers the grafts give rise to mammary structures which are normal in respect to both morphology and function (Gould et al, 1977). One such mammary structure may develop from a single cell (Gould & Clifton, 1977), i.e. the evidence indicates that this is 'probably' a clonal assay. Three weeks after grafting in MtT-bearing recipients, one can readily identify mammary structures in stained whole mounts of the graft sites. In preparation for scoring, the mammary cell-inoculated fats pads are removed at autopsy, spread on either cover slips or microscope slides and fixed in Bouin's fixative overnight. They are washed in running tap water for one hour, in acetone for one hour, and are stained with haematoxylin (Meyers) overnight followed by destaining in acid alcohol for one hour. They are then dehydrated in an alcohol series, passed into xylene and finally into mineral oil. An alternative whole mount staining method is presented in Chapter 16.

The stained whole mounts can be stored in mineral oil. For scoring, the stained fat pad is placed in a petri dish containing mineral oil and examined with a stereo dissection microscope for the presence of mammary structures. Three weeks after grafting, the growing mammary tissue usually consists of one or more spheres of cells surrounding a milk filled lumen (termed an alveolar unit — AU) (Fig. 15.1) or more frequently later on, the beginning of a branching ductal mammary gland (Fig. 15.2). If mammary structures are not initially

**Table 15.1** Design of a typical mammary clonogen assay

| Group | Desired number of cells per graft site in 0.03 ml* | Desired number of cells per ml | Actual number of cells per ml (tube number);† | Actual number of cells per site | Number of rats‡ | Number of sites |
|---|---|---|---|---|---|---|
| A | 16 000 | $5.2 \times 10^5$ | $4.0 \times 10^5$ (4) | 12 000 | 6 | 12 |
| B | 8000 | $2.6 \times 10^5$ | $2.0 \times 10^5$ (5) | 6000 | 6 | 12 |
| C | 4000 | $1.3 \times 10^5$ | $1.0 \times 10^5$ (6) | 3000 | 6 | 12 |
| D | 2000 | $6.6 \times 10^4$ | $5.0 \times 10^4$ (7) | 1500 | 6 | 12 |
| E | 1000 | $3.3 \times 10^4$ | $2.5 \times 10^4$ (8) | 750 | 6 | 12 |
| F | 500 | $1.7 \times 10^4$ | $1.2 \times 10^4$ (9) | 375 | 6 | 12 |
| G | 250 | $8.3 \times 10^3$ | $6 \times 10^3$ (10) | 188 | 6 | 12 |
| H | 125 | $4.1 \times 10^3$ | $3 \times 10^3$ (11) | 94 | 6 | 12 |

* 8 donor rats, 7-week-old female F-344 virgins
† haemocytometer cell count for dilution tube number 4 = $4.0 \times 10^5$ cell/ml
‡ 48 recipient rats, 7-week-old female F-344 virgins grafted 3 weeks previously with MtT

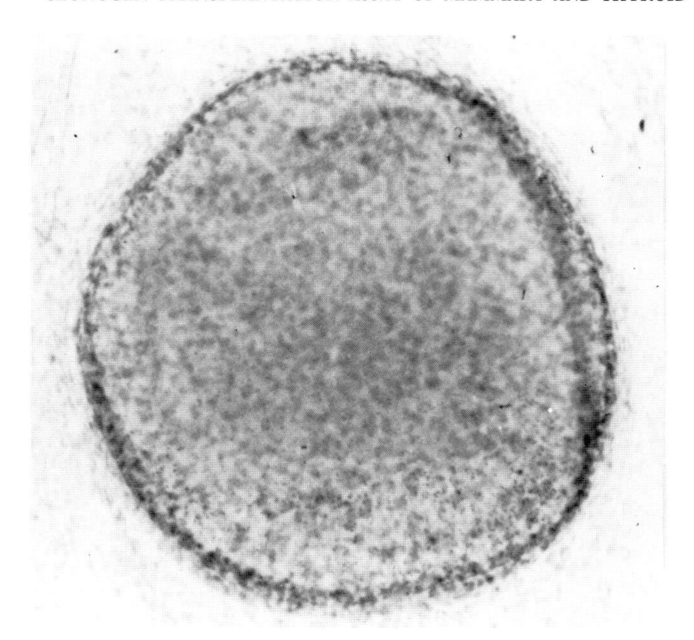

**Fig. 15.1** Milk-filled spherical alveolar unit (AU) stained with haematoxylin and viewed in whole mount with dissecting microscope; x1045.

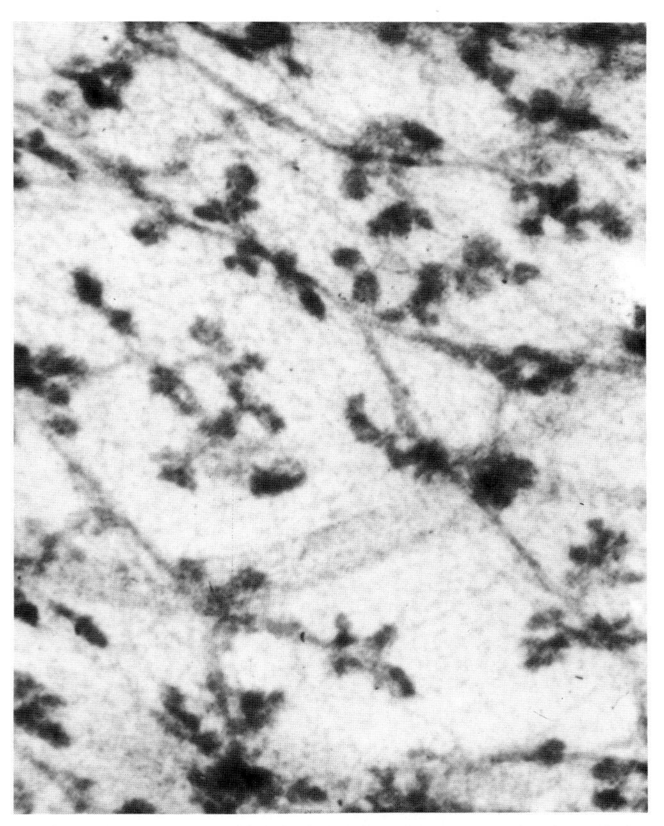

**Fig. 15.2** Branching mammary glandular structure which had developed 3 months after inoculation of monodispersed mammary cells in an MtT grafted adrenalectomized recipient. Haematoxylin-stained whole mount; x380.

found on first examination, the pad is dissected with watchmakers forceps in search of hidden structures. With experience, 99 per cent of AU can be identified by whole mount staining only. If there is a question as to the identity of the located structure, the area can be embedded in paraffin or epoxyresin, sectioned and stained for histologic examination. We recommend examination of many of these structures in both whole mounts and sections when these techniques are first being adopted.

Once each injection site has been scored as either positive or negative for mammary structures, the fraction of sites with successful transplants is related to the number of cells injected according to the model of Porter et al (1973):

$$\log M = \log K + S \log Z$$

where M is the average number of clonogens; Z is the average number of morphologically intact cells inoculated per graft site; S is a measure of slope; and K is the clonogenic fraction. A computerized iterative method which utilizes a maximum likelihood procedure is used to estimate both S and K and their 95 per cent confidence intervals. Log K and S are then used to calculate an AD50 (alveolar dose-50%) value for mammary cells, i.e. the average number of inoculated cells required to give rise to one or more AU in 50 per cent of the transplant sites (Gould & Clifton, 1977). Alternatively, this value may be estimated from the fitted curve (Fig. 15.3). For example, if S = 1, then the AD50 is that cell number which on the average contains 0.693 clonogens. In a typical experiment with untreated cells from young adult virgin F-344 rats, the AD50 ranges between 1500 and 2200 cells. The fraction of clonogens surviving may then be calculated by dividing the control AD50 by the AD50 of the irradiated or otherwise treated cells.

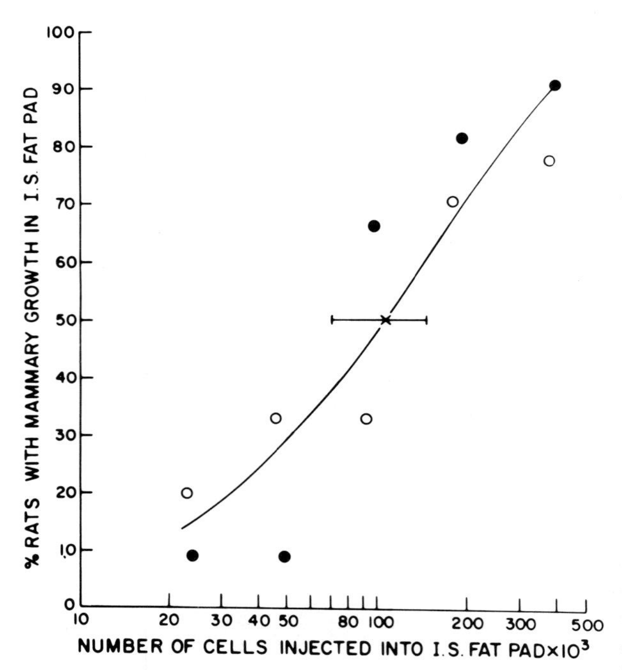


**Fig. 15.3** Cell dose/alveolar unit response relationship. Different symbols are replicate experiments. Line is calculated as indicated in text. The AD50 (injected number of cells giving alveolar units in 50% of the sites) is shown with 95% confidence limits.

**Thyroid clonogen assay**

The thyroid clonogenic assay is very similar to the mammary assay (Clifton et al, 1978; De Mott et al, 1979). The major differences are in the methods used for preparing monodispersed cells. The thyroid cells like the mammary cells may be irradiated or otherwise treated in vivo in the donor, in vitro before grafting, or in vivo in the recipient. In order to prepare thyroid cells, donor rats are killed by ether overdose, and the ventral neck area is shaved and washed as described above. They are then pinned on their backs and the thyroid is exposed by dissection and gently teased away from the surrounding tissue. Both lobes are removed, and no special effort is made to remove the isthmus area. The glands are scissor-minced or minced with opposed scalpel blades to yield pieces of less than 1 $mm^3$. The minced tissue is then incubated in collagenase solution (2 mg/ml); the crude collagenase employed is Cooper Biomedical Type II. The tissue is digested for two hours at 37°C with shaking. The cells are centrifuged out of the collagenase and resuspended in a cold (4°C) sterile pronase solution (1.25 per cent; Calbiochem). The suspension is shaken in an ice bath (4°C) for 90 minutes, after which the cells are pelleted and washed twice in room temperature serum-free medium; DNase (1 ml of 0.05 per cent per 40 ml) is added to the first wash. After the second was the cells are pooled and filtered as described above.

The method of counting, diluting and injecting thyroid cells is identical to that employed for mammary cells. It should be noted, however, that it is possible to use an additional four injection sites per rat and thus reduced the total number of rats needed. The four additional sites are located in the right and left inguinal mammary fat pad. On each side one injection is made in the area of the pad around the lymph nodes; the other is made higher on the hip area. When these four sites are used, we usually only make one injection into the i.s. area, i.e. five sites per rat. The experimental protocol is similar to that used for the mammary assay. An example is presented in Table 15.2.

The autopsy, fat pad staining, reading, scoring of follicular units (FU) and calculation of FD50 (50% follicular dose) values and of clonogen survival are performed

**Table 15.2** A sample thyroid clonogen assay

| Group | Desired number of cells per injection (0.03 ml)* | Desired number of cells per ml | Actual number of cells per ml (tube number)† | Actual number of cells per inj. | Number of rats‡ | Number of sites |
|---|---|---|---|---|---|---|
| A | 1000 | $3.3 \times 10^5$ | $3.8 \times 10^5$ (4) | 1152 | 4 | 20 |
| B | 500 | $1.7 \times 10^5$ | $1.9 \times 10^5$ (5) | 576 | 4 | 20 |
| C | 250 | $8.3 \times 10^4$ | $9.5 \times 10^4$ (6) | 288 | 4 | 20 |
| D | 125 | $4.2 \times 10^4$ | $4.8 \times 10^4$ (7) | 144 | | 20 |
| E | 62 | $2.1 \times 10^4$ | $2.4 \times 10^4$ (8) | 73 | | 20 |
| F | 31 | $1.0 \times 10^4$ | $1.2 \times 10^4$ (9) | 36 | | 20 |
| G | 15 | $5.2 \times 10^3$ | $5.9 \times 10^3$ (10) | 18 | | 20 |

* 4 donor rats, 5-week-old F-344 males
† haemocytometer cell count for dilution tube number 4 = $3.8 \times 10^5$ cells/ml
‡ 28 recipient rats, 5-week-old F-344 males which were thyroidectomized one day prior to grafting

as described above for mammary cells (Fig. 15.4). In a typical experiment with thyroid cells from untreated 5-week-old F-344 males, the FD50, i.e. the number of grafted thyroid cells required to produce one or more follicular structures (FU) (Fig. 15.5) per transplant site, usually falls between 50 and 100 monodispersed cells.

## DETAILS AND MODIFICATIONS OF THE TECHNIQUE

### Enzymes

The collagenase used to dissociate mammary and thyroid cells is a crude preparation; it contains a specified amount of collagenase activity but also includes a number of other enzymes. The type and amount of contaminating enzymes are determined by the way the collagenase is prepared, Cooper Biomedical, (Malvern, PA, USA) prepares crude collagenase by four methods, each resulting in collagenase mixed with a different spectrum of contaminating enzymes. These are termed CLS, CLS II, CLS III and CLS IV. Different tissues respond best to different mixtures. We have found CLS and CLS III best suited for mammary dissociation. CLS III has lower proteolytic activity than CLS. CLS II has been found best suited for thyroid dissociation. We have found it best consistently to use 2 mg/ml and not to adjust for variations in collagenase activity from batch to batch. Furthermore, the Cooper collagenase types are different from those prepared by other suppliers. If collagenase from another source is to be used, it should be tested. Finally, we find it desirable to test new lots of Cooper collagenase before purchase to avoid major variations in enzyme potency.

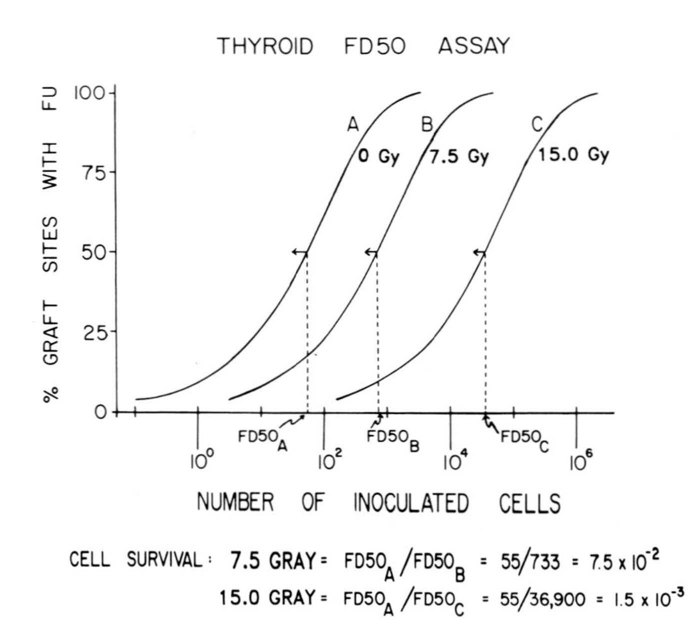


**Fig. 15.4** Effect of radiation on the cell dose-follicle unit response relationship and calculation of cell survival from values of FD50 (injected number of cells giving follicles in 50% of the sites).

cipient categories was not significantly different. Thus, the 8-fold difference in the LND50 values cannot be explained by a difference in the number of initially injected cells which ultimately remain at the site of injection.

The rate of DNA synthesis in the transplanted hepatocytes was studied by autoradiographic analysis of the transplanted nodules, At both 1 and 3 days after transplantation, a greater labelling index was seen in hepatocytes transplanted into hepatectomized hosts (15.8% and 7.6%) compared to the control recipients (2% and 3.7%). However, the labelling index of the transplanted hepatocytes between days 5 and 20 after transplantation was found to be similar in both categories of recipient hosts, and equal to 3.7 per cent. This is, however, 6 times larger than that observed in the normal liver; thus demonstrating that accelerated proliferative behaviour of the transplanted hepatocytes continues even after liver regeneration in situ is complete. This implies that stimulatory factors may always be present in control animals and/or the growth of hepatocytes transplanted into an ectopic site is regulated in a manner which is different from hepatocytes in situ. The results of these investigations demonstrate that the higher probability of nodule formation in hepatocytes transplanted into hepatectomized hosts is due to the increased stimulation of cellular replication in the hepatocytes during the early course of events after partial hepatectomy. Conversely, they indicate that the formation of the liver nodules that we observe 20 days after transplantation is not primarily due to cell aggregation since the likelihood of this occuring would not be expected to be affected by hepatectomy of the recipient animal.

The effect which the time between hepatectomy and cellular injection had on the clonability of the transplanted hepatocytes was also investigated. The LND50 value is reduced by 38 per cent, 71 per cent and 87 per cent of the control value when a 2/3 hepatectomy of the recipient animals preceeds cellular injection by 24, 12 and 6 h, respectively. When a 2/3 hepatectomy of the recipient hosts was performed 48 h prior to hepatocyte transplantation the LND50 value was found to be 16 000 cells, a value which is insignificantly different from that observed for control animals. These data demonstrate that the stimulatory effect of a partial hepatectomy maximizes shortly after surgery, and has essentially disappeared by 40 h. As a consequence, for reproducibility and maximum stimulatory effect, hepatocytes should always be uniformly injected 1–2 h after hepatectomy.

## Suspension media for hepatocytes

Peters & Hewitt (1974) observed that brain homogenate, rich in thromboplastin activity, enhanced coagulation and decreased the number of injected malignant cells required for tumour formation. With transplanted mammary (Gould & Clifton, 1977) and thyroid (Mulcahy et al, 1980) epithelial cells, it was also found that the presence of brain homogenate improved the clonogenicity of the cells. When hepatocytes are suspended in L15 and injected into 2/3 hepatectomized recipients, the LND50 value is 11 400 cells. However, mixing the hepatocytes with a 50 per cent homogenate suspension improves their clonability by a factor of 5. Thus, brain homogenate similary improves the clonogenicity of transplanted hepatocytes, and was originally used in all of our cell survival and liver regeneration studies.

A crude homogenate of brain tissue would not only contain substances which improve the clonogenicity of hepatocytes, but also factors such as proteolytic enzymes which could decrease the survival of the transplanted cells. We thus wished to find a suspending medium which would improve the clonability of cells, and would be more precisely defined than a crude suspension of brain homogenate. Hepatocytes maintained in vitro on floating collagen membranes have been shown to maintain higher levels of hepatospecific functions than those plated on a plastic substratum (Michalopoulos & Pitot, 1975). With this in mind we tested whether the collagen gel material used in vitro would further improve the clonogenicity of the transplanted liver cells. To test this, hepatocytes were mixed with the collagen gel material described above. The mixture of collagen and hepatocytes (2:1), which is liquid at 4°C, solidifies at 37°C thus forming a collagen gel matrix after injection into the fat pad. The LND50 value for hepatocytes injected with the collagen into 2/3 hepatectomized recipients is 240 cells (95% confidence interval, 190–290 cells) or approximately a factor of 10 less than that when brain homogenate is used as the cellular suspension material. Because of this substantial improvement in the clonability of hepatocytes, collagen gel is presently used in lieu of the brain homogenate.

### Age of the donor and recipient animal

The importance of these two variables on the clonability of hepatocytes has not been investigated extensively. However, preliminary data indicate that hepatocytes from young animals (28 days) are 2 to 3 times more clonable when injected into young recipients than into animals approximately 1 year old. Similarly, the cells from young animals are 2 to 3 times more clonogenic than those from old animals when transplanted into young recipient animals. Thus, the age of both the donor and recipient animal seem to be important in determining the reproductive capacity of transplanted hepatocytes, and to maximize the cloning efficiency of the transplanted hepatocytes, animals 80–100 g in weight are uniformly used as both the donor and recipient.

### Sex, strain and species of animal

Originally, all experiments were performed with female Fischer 344 rats. As previously stated the LND50 value for female hepatocytes transplanted with brain homogenate into hepatectomized female rats is 2100 cells. The LND50 value for male hepatocytes transplanted into male rats is 2700 cells (95% confidence interval, 970–4300 cells). The difference berween these two values is insignificant ($p>0.1$). Therefore, the assay can be equally utilized with either male or female Fischer 344 rats. Similar studies have also been performed with isogeneic Wistar/Furth (W/Fu) rats, and preliminary results demonstrate that W/Fu hepatocytes transplant equally well. We have also demonstrated the feasibility of transplanting cells from one sex into recipients of the opposite sex, and also of transplanting cells from either the Fischer 344 or W/Fu strain into (W/FuXF344) $F_1$ hybrid animals. Finally, we have successfully transplanted isolated human hepatocytes into athymic nude mice (Fig. 16.2; Strom et al, 1982), and preliminary results indicate the cells proliferate in response to a hepatectomy of the recipient mouse. The flexibility of this transplantation system is therefore quite extensive.

## INFORMATION WHICH CAN BE DEDUCED USING THIS TECHNIQUE

Thus far, we have used this transplantation system for further elucidating the mechanisms involved in liver regeneration (Jirtle & Michalopoulos, 1982), and the results of these studies have been summarized above. We have also focused our attention on estimating the genotoxic effects of ionizing radiation on parenchymal hepatocytes. Prior to the development of this assay system no information was available concerning the reproductive survival of parenchymal hepatocytes exposed to ionizing radiation. Our studies have shown that the $D_o$ value for parenchymal hepatocytes is 2.7 Gy (Fig. 16.6; Jirtle et al, 1981) which is large when compared to that reported for other normal tissues (Jirtle et al, 1982). However, hepatocytes were found to be rather inefficient in repairing radiation induced damage. This conclusion results from the fact that the maximum $D_q$ value observed for liver cells is 2.0 Gy, and of the normal tissues studied, only bone marrow stem cells have a smaller shoulder on the survival curve (Jirtle et al, 1982). The implication of this result is that, even with complete repair of damage between fractionated doses of low LET radiation, parenchymal hepatocytes would be relatively sensitive to conventional radiotherapy treatment.This provides a plausible explanation of why the liver is a radiosensitive organ (Ingold et al, 1965).

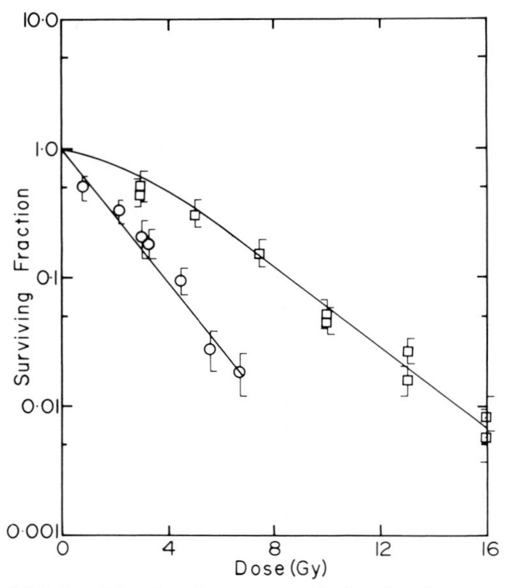


**Fig. 16.6** Surviving fraction versus radiation dose for parenchymal hepatocytes exposed to either $^{60}$Co radiation (□) or 14.3 MeV neutrons (○). The reproductive survival was assayed 24 h after radiation exposure. Error bars represent the standard error of the mean.

Exposure of hepatocytes in situ to 14.3 MeV neutrons reduces the $D_o$ value to 1.7 Gy and the n value to unity (Fig. 16.6). Also, it was found that hepatocytes exposed to neutrons were not able to repair potentially lethal damage (PLD), whereas there is significant PLD repair after exposure to low LET radiation (Jirtle et al, 1982). By comparing the survival curve for low LET radiation with that for neutrons, RBE values can be estimated as a function of neutron dose, and the values ranged from 4.2 at 0.5 Gy to 2.1 at 6.0 Gy.

We have also determined the radioprotective effect of the aminothiol WR-2721. This drug is of interest because it has been reputed to protect normal tissues to a greater extent than tumours, and also it has been shown to concentrate in liver tissue (Utley et al, 1976). A comparison of the survival curves for hepatocytes with (400 mg/kg) and without WR-2721 shows that at this drug dose the radiation dose modifying factor is 2.2. Of additional importance is the fact that this degree of protection is maintained at drug doses as low as 200 mg/kg; thus permitting its effective use during fractionated radiotherapy treatment where drug dose must be reduced because of toxicity. These data suggest that WR-2721 may be a useful adjuvant to the radiation treatment of metastatic lesions in the liver, and this hypothesis is currently being further investigated.

Thus far we have primarily investigated the genotoxic effects of ionizing radiation. However, this assay system can also be used to measure the genotoxic effects of chemical carcinogens and chemotherapeutic agents. Also since there exists a positive enzyme marker for putative preneoplastic hepatocytes (i.e. gamma glutamyltranspeptidase), this system also lends itself for use in systematically investigating the biology of the cellular alterations which occur during the early stages in hepatocarcinogenesis. Additionally, the ability to transplant hepatocytes enables one to investigate what has been referred to as 'seed or soil' problems. For example, is the probability of malignant transformation of hepatocytes dependent upon the cellular age and/or the age of the environment within with the cell resides? This type of question can only be unambiguously answered by the transplantation of cells. Finally, the ability to disperse enzymatically and transplant human hepatocytes into athymic nude mice (Fig. 16.2; Strom et al, 1982) provides a heterotransplantation system which should prove to be of considerable value in evaluating the comparative toxic effects of a variety of xenobiotic chemicals without the logistic difficulties in using the intact human liver. In conclusion, the in vivo assay system described in this chapter provides a quantitative method for addressing a variety of questions in the area of hepatopathology which, heretofore, was not possible.

## ACKNOWLEDGEMENTS

This work was supported by NIH research grant CA25951 and CA30241 from the National Cancer Institute.

## REFERENCES

Berry M N, Friend D S 1969 High yield preparation of isolated rat liver parenchymal cells. Journal of Cell Biology 43:506

Christensen B G, Jacobsen E 1949 Studies on liver regeneration. Acta Medica Scandinavica, Suppl. 234:103

Deschenes J, Valet J P, Marceau N 1980 Hepatocytes from newborn and weanling rats in monolayer culture: isolation by perfusion, fibronectin-mediated adhesion, spreading and functional activities. In Vitro 16:722

Ehmann U K, Hagasawa H, Petersen D F, Lett J T 1974 Symptoms of X-ray damage to radiosensitive mouse leukemia cells: asynchronous populations. Radiation Research 60:453

Ehrmann R L, Gey G O 1956 The growth of cells on a transparent gel of reconstituted rat-tail collagen. Journal of the National Cancer Institute 16:1375

Elkind M M, Whitmore G F 1967 The radiobiology of cultured mammalian cells. Gordon & Breach, New York, p 7

Elsdale T, Bard J 1972 Collagen substrata for studies on cell behaviour. Journal of Cell Biology 54:626

Emerman J T, Pitelka D R 1977 Maintenence and induction of morphological differentiation in dissociated mammary epithelium on floating collagen membranes. In Vitro 13:316

Fiala S, Fiala E S 1973 Activation by chemical carcinogens of γ-glutamyltranspeptidase in rat and mouse liver. Journal of the National Cancer Institute 51: 151

Finney D J 1964 Statistical method in biological assay, 2nd ed. Charles Griffin, London, p 570

Gould M N, Clifton K H 1977 The survival of mammary cells following irradiation in vivo: a directly generated single-dose-survival curve. Radiation Research 72:343

Hewitt H B, Wilson C A 1959 A survival curve for mammalian leukaemia cells in vivo. British Journal of Cancer 13:69

Higgins G M, Anderson R M 1931 Experimental pathology of the liver I. restoration of the liver in the white rat following partial surgical removal. Archives of Pathology 12:186

Hunt J M, Buckley M T, Onnink P A, Rolfe P B, Laishes B A 1982 Liver cell membrane alloantigens as cellular markers in genotypic mosaic rat livers undergoing chemically induced hepatocarcinogenesis. Cancer Research 42:227

Ingold J A, Reed G B, Kaplan H S, Bagshaw M A 1965 Radiation hepatitis. American Journal of Roentgenology 93:200

Jirtle R L, Michalopoulos G 1982 Effects of partial hepatectomy on transplanted hepatocytes. Cancer Research 42:3000

Jirtle R L, Biles C, Michalopoulos G 1980 Morphologic and histochemical analysis of hepatocytes transplanted into syngeneic hosts. American Journal of Pathology 101:115

Jirtle R L, McLain J R, Strom S C, Michalopoulos G 1982 Repair of radiation damage in noncycling parenchymal hepatocytes. British Journal of Radiology 55:847

Jirtle R L, Michalopoulos G, McLain J R, Crowley J 1981 Transplantation system for determining the clonogenic survival of parenchymal hepatocytes exposed to ionizing radiation. Cancer Research 41:3512

Jolanko H, Ruoslahti E 1979 Differential expression of α-fetoprotein and γ-glutamyltranspeptidase in chemical and spontaneous hepatocarcinogenesis. Cancer Research 39:3495

Laishes B A, Farber E 1978 Transfer of viable putative preneoplastic hepatocytes to livers of syngeneic host rats. Journal of the National Cancer Institute 61:507

Lee G, Makowka L, Kaku T, Tatematsu M, Finkelstein S, Medline A 1982 Proceedings of the American Association for Cancer Research 23:96

Leong G F, Grisham J W, Hole B V, Albright M L 1964 Effect of partial hepatectomy on DNA synthesis and mitosis in heterotropic partial autografts of rat liver. Cancer Research 24:1496

McGowan J A, Strain A J, Bucher N L R 1981 DNA synthesis in primary cultures of adult rat hepatocytes in a defined medium: effects of epidermal growth factor, insulin, glucagon and cyclic-AMP. Journal of Cellular Physiology 108:353

Michalopoulos G, Pitot H C 1975 Primary culture of parenchymal liver cells on collagen membranes. Experimental Cell Research 94:70

Michalopoulos G, Russell F, Biles C 1979 Primary cultures of hepatocytes on human fibroblasts. In Vitro 15:796

Michalopoulos G, Sattler C A, Sattler G L, Pitot H C 1976

Cytochrome P-450 induction by phenobarbital and 3-methylcholanthrene in primary cultures of hepatocytes. Science 193:907

Michalopoulos G, Cianciulli H D, Novotny A R, Kligerman A D, Strom S C, Jirtle R L 1982 Liver regeneration studies with rat hepatocytes in primary culture. Cancer Research 42:4673

Mito M, Ebatu H, Kusano M, Onishi T, Saito T, Sakamoto S 1979 Morphology and function of isolated hepatocytes transplanted into rat spleen. Transplantation 28:499

Miyazaki M 1978 Primary culture of adult rat liver cells: II. cytological and biochemical properties of primary cultured cells. Acta Medica Okayama 32:11

Mulcahy R T, Gould M N, Clifton K H 1980 The survival of thyroid cells: in vivo irradiation and in situ repair. Radiation Research 84:523

Peters L J, Hewitt H B 1974 The influence of fibrin formation on the transplantability of murine tumor cells: implications for the mechanism of the Revesz effect. British Journal of Cancer 29:279

Porter E H, Hewitt H B, Blake E R 1973 The transplantation kinetics of tumour cells. British Journal of Cancer 27:55

Potter V R, Walker P R, Goodman J I 1972 Survey of current studies on oncogeny as blocked ontogeny: Isoenzyme changes in livers of rats fed 3'methy-4-dimethylaminoazobenzene with colateral studies on DNA stability. Gann. Monog. Cancer Research 13:121

Rosenberg M R, Strom S C, Michalopoulos G 1982 Effect of hydrocortisone and nicotinamide on gamma glytamyltransferase in primary cultures of rat hepatocytes. In Vitro 18:775

Sattler C A, Michalopoulos G, Sattker G L, Pitot H C 1978 Ultrastructure of adult rat hepatocytes cultured on floating collagen membranes. Cancer Research 38:1539

Sirica A E, Richards W, Tsukada Y, Sattler C A, Pitot H C 1979 Proceedings of the National Acadamy of Science (USA) 76:283

Solt D, Farber E 1976 New principle for the analysis of chemical carcinogenesis. Nature 263:701

Strom S C, Jirtle R L, Jones R S, Novicki D L, Rosenberg M R, Novotny A, Irons G, McLain J R, Michalopoulos G 1982 Isolation, culture, and transplantation of human hepatocytes. Journal of the National Cancer Institute 68:771

Utley J F, Marlowe C, Waddel W J 1976 Distribution of $^{35}S$-labelled WR-2721 in normal and malignant tissues of the mouse. Radiation Research 68:284

# 17

*K.A. Mason H.R. Withers*

# Kidney tubules

## HISTORICAL BACKGROUND

The history of the development of a colony assay to study radiation-induced nephritis has been a long and at times confusing one. The primary problem has arisen from confusion over defining the target cell responsible for the late radiation damage observed in the kidney. A review of the literature from 1904, when Linser & Baermann first studied the response of the kidney to radiation, until the present time reveals that kidney tubules were the most frequently affected structure whether or not the investigator believed this to be primary radiation damage or secondary to vascular damage. There is ample evidence in the literature to support the idea that the target for late kidney damage is the epithelium of the renal tubule and that histological evidence of vascular changes appear later. Proponents of parenchymal cell depletion (Withers, 1976; Withers et al, 1980) and vascular damage (Casarett, 1976; Hopewell, 1980) as primary causes of tissue failure have previously debated this question.

Even though early investigators undoubtedly encountered difficulties in radiation dosimetry, some remarkable studies were carried out. In 1905 Schultz & Hoffman were the first to report damage to the renal tubules. Observing the response of the rabbit kidney from 1 h to 48 days after irradiation, they noted tubular destruction at the later times.

Many more investigations were carried out in the 1920s (Hartman et al, 1926; Hartman et al, 1927; Gabriel, 1926; Tsuzuki, 1926; Domagk, 1927; Willis & Bachem, 1927) with essentially common results, namely: (1) the changes in the kidney were progressive, (2) the tubules were the most severely damaged structure, and (3) the glomeruli were relatively well preserved, at least until the kidney became irreversibly atrophic.

In 1930 Bolliger & Laidley published a detailed histological description of radiation induced nephritis at times varying from hours to months after treatment. This was a landmark investigation for several reasons: (1) they described in detail five distinct histologic periods following irradiation of the kidney, (2) they used loss of kidney weight as a quantitative endpoint, (3) they were the first to correlate animal lethality with destruction of the kidney tubules, (4) they described both the process of degeneration of the tubules and the process of recovery, and (5) they were the first to describe what we now recognize as regenerating tubules at long times after irradiation. Had they correlated tubule depletion with dose, the present article would be describing a very old colony assay rather than a new one. The histological picture described by Bolliger & Laidley (1930) in dogs is essentially the same as that observed by the present authors using a mouse system, although the time course for development of the changes is somewhat different. The five histologic periods were described as follows:

1. 0–48 hours — period of acute congestion which rapidly resolves, leaving the kidney normal in appearance
2. 1–8 days — period of latency when no change from normal can be detected
3. 5–32 days — period of tubular change with degeneration of tubule cells ending in fibrous invasion and fatty degeneration of the convoluted tubules with animal lethality beginning late in this period
4. 21–60 days — period of development of fibrous tissue with progressive replacement of the degenerated parenchyma, normal appearing glomeruli and occasional islands of regenerating tubules
5. 60–230 days — final period of tubular depletion with concomitant weight loss and contracture of the kidney resulting in an increased density of glomeruli and vascular tortuosity.

During the fifth period (60–230 days), distinct islands of tubular regeneration were frequently observed. Bolliger & Laidley (1930) described radiation nephritis as a slowly-developing sclerosis in which fibrous tissue replaced the destroyed tubules and became hyalinized. The fibrous tissue surrounding the relatively undamaged glomeruli caused them to atrophy and undergo hyalinization, and the larger vessels which then served no useful function underwent proliferative endarteritis. In addition to the histological description of radiation nephritis, they made other important observations. First, even in animals which died of uremia on about day 40, discrete islands of tubular repair were evident and extensive repair of tubular damage was present in animals surviving six months or longer. Secondly, they noted that if nephrectomized animals had enough functional reserve to tide them over the period of maximal tubular destruction, far reaching anatomical repair of the kidney was possible.

In 1946, Lacassagne confirmed the above findings in mice, showing destruction of uriniferous tubules with limited influence on glomeruli at 300–616 days after roentgen therapy.

During the 1950s, Mendelsohn & Caceres (1953) and Huang et al (1954), using renal function tests, concluded that radiation caused a dose dependent decrease in tubular function which slowly recovered towards normal.

Phillips & Ross (1973) made the first report of a quantitative assessment of the radiation response of the kidney. Death from renal failure was correlated with the histological observation of radiation induced tubule depletion.

Glatstein (1973) measured the extraction of $^{86}$Rb by mouse kidneys as a means of determining blood flow through the organ after irradiation. Although the 40–50 per cent decrease in $^{86}$Rb extraction per gram of kidney after 19 Gy and the 25 per cent decrease after 15 Gy was interpreted as radiation induced reduction in blood flow, this reduction in $^{86}$Rb extraction could equally well be explained as the selective loss of proximal tubule cells available to extract the $^{86}$Rb.

In 1978, Jordan et al developed a six point qualitative histological grading system (Table 17.1) for nephritis based on their observation that tubular changes were the most significant and consistently-scorable damage. Vascular sclerosis occurred in only 3.8 per cent of the irradiated kidneys and was thought to be secondary to parenchymal cell loss. A consistent correlation between histologic tubular damage and mortality was demonstrated at three time intervals after irradiation: 5–7 months, 8–16 months, and 18–24 months. The X-ray dose necessary to produce a given level of effect in 50 per cent of the animals (Grade 3 or greater) was inversely related to the time after irradiation.

**Table 17.1** Histopathologic grading of renal tubular changes

| | |
|---|---|
| 0 | No significant tubular abnormality |
| 1 | Enlarged tubular epithelial nuclei |
| 2 | Tubular epithelial atrophy and cystic dilation |
| 3 | Single tubular collapse or marked epithelial thinning associated with enlarged tubular epithelial nuclei |
| 4 | Focal tubular collapse (<3 mm) |
| 5 | Extensive tubular collapse (<3 mm) |

In 1984, Withers & Mason reported on a technique to quantitate clonal regeneration of tubules in the mouse kidney after irradiation. To our knowledge, this is the first in situ clonal assay developed for a 'late effects' tissue and is analogous to stem cell assays in 'acute effects' tissues such as stomach, jejunum, colon, skin and testis except for the slow rate of expression of injury. The variation in rate of development of injury does not imply a change in the pathobiology of the radiation effect from parenchymal cell depletion to vascular compromise, but rather reflects merely differences in the turnover kinetics of the 'target' cells and their progeny.

## DESCRIPTION OF TECHNIQUE AND METHODOLOGY

The renal tubule assay (Withers & Mason, 1984) has only recently been developed. Consequently, the technique and methodology are based solely on this initial work. When this 'late effects' assay is further developed in both our own laboratory and others, it is to be expected that many technical modifications and refinements will be made, as has been the case for stem cell assays in 'acute effects' tissues. Therefore the following description of the methodology for the renal tubule assay should not be considered the only way to perform the assay or even necessarily the best, but as a technique in its infancy with further development yet to come.

The left kidneys of 10–12 week old $C_3$Hf/Kam female mice were irradiated with $^{137}$Cs gamma rays at a dose rate of 9.4 Gy per minute. The unirradiated right kidney was left in situ. It was deemed advisable to use the smallest field size possible to avoid irradiating a large volume of abdomen which might cause the animal to die from intestinal obstruction due to fibrosis before the completion of the renal tubule assay. A round 2 cm diameter field was the smallest that could be used and still ensure that the whole kidney was within the irradiated volume. The mice were anaesthetized with 55 mg/kg Nembutal (sodium pentobarbital) and taped to an irradiation jig, lying on their backs, with only the left kidney within the irradiation field. The kidney was localized within the field by palpation. The mice were irradiated with a special $^{137}$Cs small animal irradiator (Hranitzky et al, 1973) with parallel opposed beams directed toward the ventral and dorsal surfaces of the supine mouse.

In preliminary experiments it was determined that the number of surviving tubules decreased progressively, in a dose dependent manner, between about 30–52 weeks post-irradiation with no significant further change in the ratio of intact to depleted tubules between about 56–70 weeks. For all subsequent experiments with single doses, animals were sacrificed between 62–68 weeks post-irradiation. Although surviving tubules can be scored at about 14 months, a 16 month incubation period produced colonies with a more easily scorable size and appearance.

At the time of sacrifice, the irradiated kidney was removed and fixed in neutral buffered formalin. Histological preparations, cut at a thickness of 4 $\mu$m, were made from coronal cross sections at the level of the renal pelvis and stained with haematoxylin and eosin. All histological sections were scored using bright field microscopy with a 16x flat field objective.

Due to the large number of tubules in a kidney cross section a sampling technique was employed whereby only those tubules in the cortex in contact with the renal capsule were scored for viability (Hewitt, 1956). The number of tubules touching the renal capsule in coronal cross sections of normal intact kidneys of our mice at 60–80 weeks of age was 370 ± 5 (se). In irradiated animals, the number of regenerated tubules in contact with the capsule was counted in *complete* transverse sections and scored as a fraction of the number in similar transverse sections of unirradiated controls. Torn or incomplete histological sections were excluded from the analysis of results.

The uriniferous tubules in the renal cortex of a mouse are lined with a single layer of cuboidal (or columnar) epithelium (Fig. 17.1). These cells have a single spherical nucleus in an eosinophilic cytoplasm with a large volumetric ratio of cytoplasm to nucleus. In cross sections of normal proximal convoluted tubules 3 or 4 nuclei per tubule can be seen. In irradiated animals, tubules were scored as viable by the following criteria: (1) the tubule touched the renal capsule, (2) the tubule contained at least 2 epithelial cells each with a single spherical nucleus, (3) the cuboidal or columnar epithelial cells had a large amount of eosinophilic cytoplasm.

The microscopic appearance of an irradiated kidney 62 weeks after a dose of 14 Gy gamma rays is shown in Figure 17.2. Slightly hypertrophic proximal renal tubules are distributed among the ghosts of tubules completely depopulated of epithelium, with glomeruli showing little significant change. Other histological observations included frequent cysts and lymphocyte invasion in some areas of tubular degeneration but these changes were not dose dependent. Less frequent effects

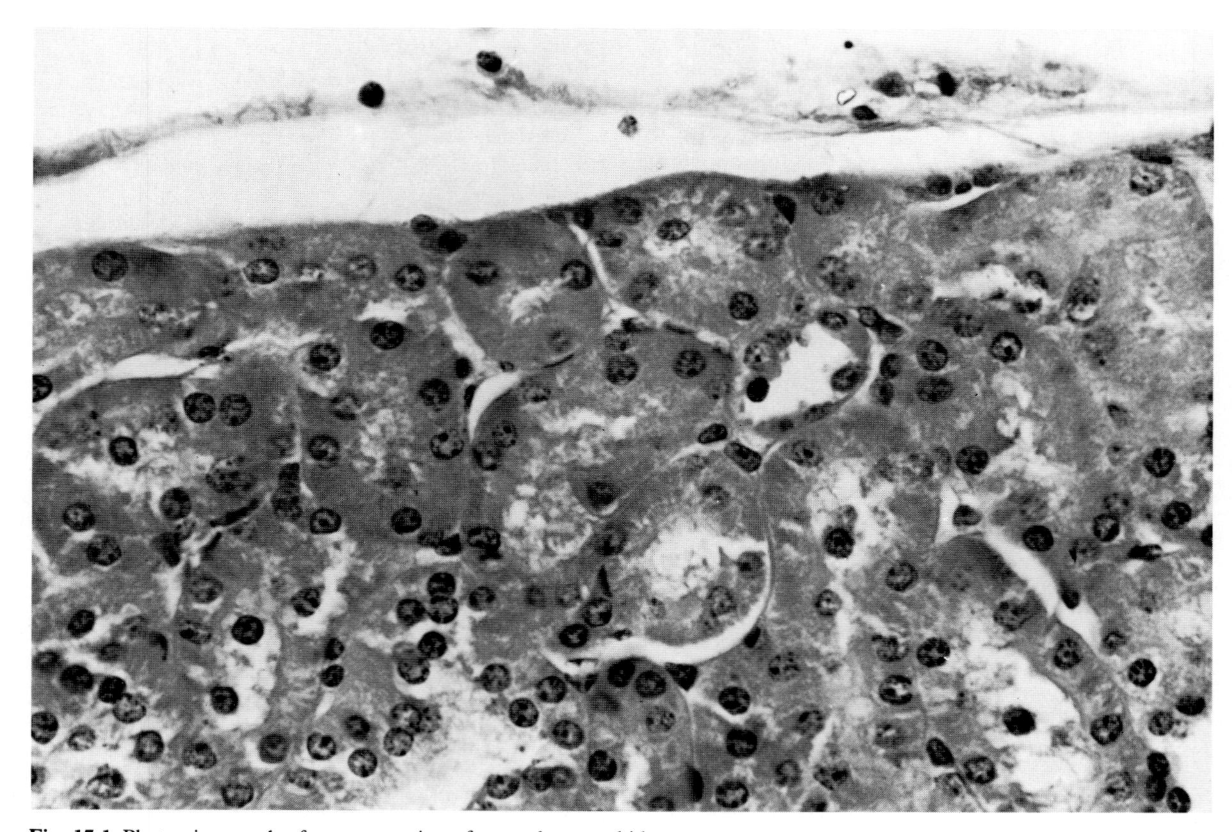

**Fig. 17.1** Photomicrograph of a cross section of normal mouse kidney.

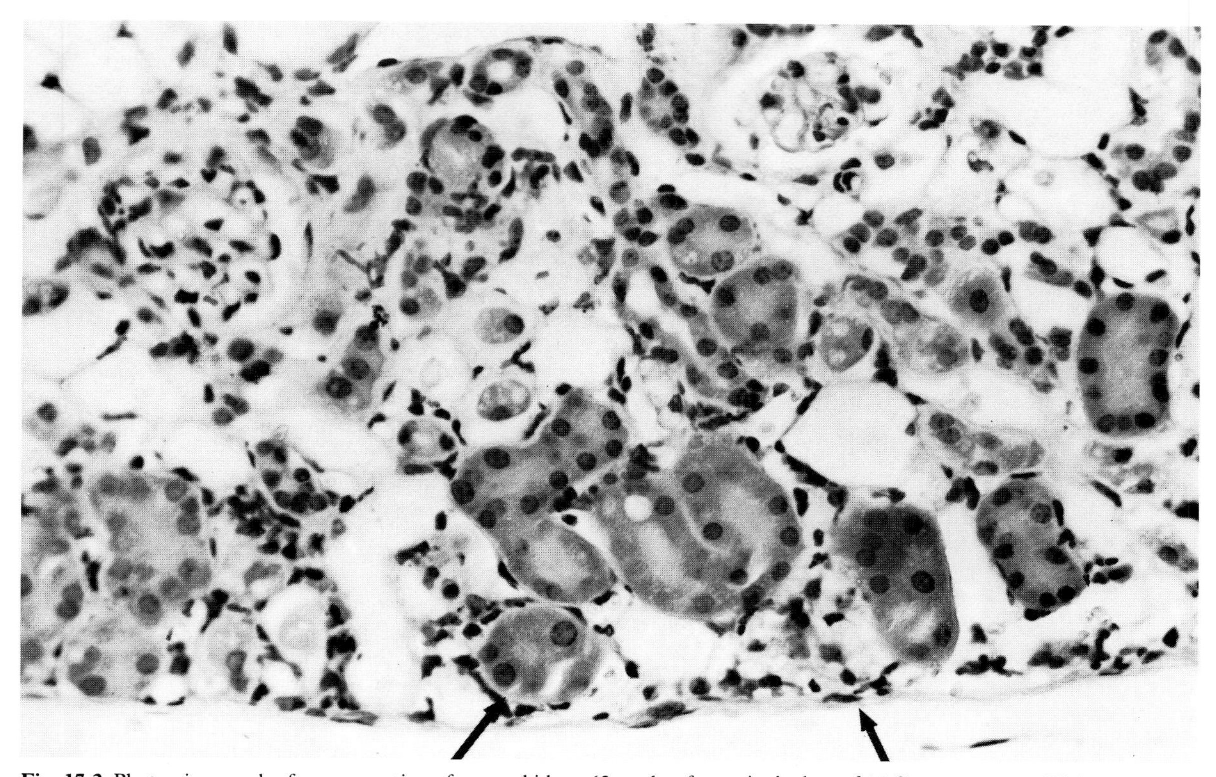

**Fig. 17.2** Photomicrograph of a cross section of mouse kidney 62 weeks after a single dose of 14 Gy gamma rays. Right arrow, the kidney capsule. Left arrow, example of surviving tubule adjacent to capsule.

included ectopic bone formation in 3–4 per cent of the kidneys and endarteritis in 2–3 per cent of mice irradiated with doses between 10.5–30 Gy.

A considerable proportion of the mice did not survive the time allowed for development of easily-counted colonies. At 16 months post-irradiation, an $LD_{50}$ of 13.95 Gy (13.44–14.49 Gy, 95 per cent CI) was associated with the survival of about 40 tubules touching the renal capsule. Renal lethality from doses of 10.5–15 Gy occurred at about 14 months post-irradiation while at doses from 15.5–18 Gy, death occurred most frequently at about 7 months. The reason for these unexpected deaths is being investigated in recent experiments. They were probably not all due solely to renal insufficiency, but may also have resulted from radiation-induced complications such as intestinal fibrosis or hypertension.

## CONSTITUTION OF A COLONY AND PROBLEMS ASSOCIATED WITH ITS MEASUREMENT

Our definition of a renal tubule colony in a histological cross section, as a tubule touching the renal capsule, with at least two cuboidal epithelial cells composed of a large amount of eosinophilic cytoplasm and a single spherical nucleus, is one of convenience for scoring purposes. Using these criteria, tubule survival does not change significantly between about 56 to 70 weeks post-irradiation, in our mice.

The number of regenerated tubules in contact with the capsule (for ease of counting) was scored in complete coronal histological sections and expressed as a fraction of the number in similar transverse sections of unirradiated controls. The number of regenerated tubules decreased logarithmically as a function of dose. The slope of the curve relating the logarithm of tubule survival to dose is similar to that for mammalian cells in conditions where it is known that the assay measures the independent survival of single cells (Elkind & Whitmore, 1967). Based on the assumption that tubules can regenerate from single surviving epithelial clonogens (see below), a correction was made, as it has been for other in situ clonogenic assays, to account for the increasing multiplicity of surviving cells per tubule at lower doses. Assuming random survival of renal tubule epithelial cells, the average number of surviving cells per tubule, from Poisson distribution statistics, is $-\ln f$, where f is the proportion of tubules in which no cells survived. A survival curve plotting total number of surviving cells

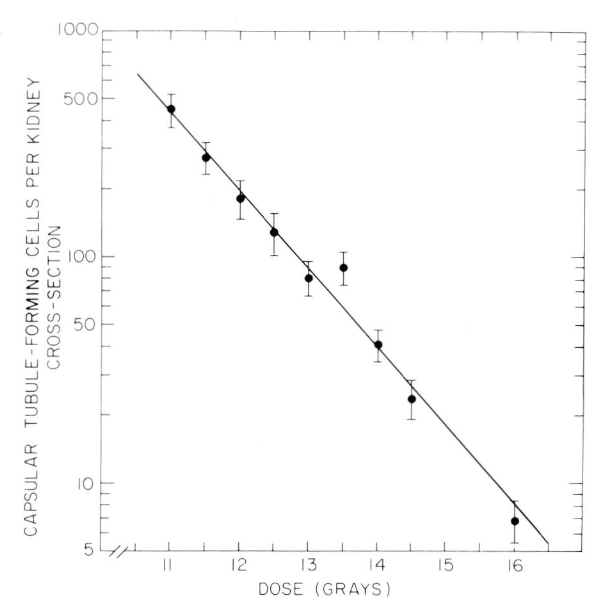


**Fig. 17.3** Single-dose survival curve of mouse renal tubule cells plotted on semi-logarithmic coordinates. Tubules were scored at 16 months post-irradiation. $D_o$ = 1.25 Gy (1.13–1.41 Gy, 95 per cent CI). Error bars represent the mean ± SE. Counts of tubules were converted to tubule-forming units by applying a Poisson correction.

in tubules that contact the renal capsule as a function of dose is shown in Figure 17.3. The mean lethal dose, that is, the dose that results in the random reduction of cell survival to $e^{-1}$, was 1.25 Gy (1.13–1.41 Gy, 95% CI).

It seems likely that renal tubules respond to radiation as independent units with a capacity to regenerate themselves from single cells, but without the capacity to repopulate adjacent nephrons. However, it should be noted that if the tubule required the integral survival of n tubule cells *as a group*, the survival probability for the whole group would be $p^n$ (where p is the probability of survival of a single cell within the group), and the experimentally-determined slope of the survival curve would be 1/nth the slope characteristic of single cells within the group (assuming all cells within the group were of equal radiosensitivity). Since the experimentally-measured slope is similar to that for a wide variety of mammalian cells measured in conditions where single cells are known to have survived independently, the value for n for renal tubule cells is unlikely to be other than 1. Although the slope of the curve in Figure 17.3 indicates that the radiosensitivity of the cells that regenerate the nephron is similar to that of other mammalian cells such as the epithelium of the gastrointestinal tract (Withers & Elkind, 1969; Chen & Withers, 1972; Withers & Mason, 1974), spermatogenic tubules (Withers et al, 1974) and skin (Withers, 1967), the rates of response of these various tissues to insults such as radiation differ markedly from one another. Thus, in the mouse, the time required for depletion and partial regeneration of the epithelium after a single dose of radiation is about 3 days for jejunum, 5 days for colon, 10 days for stomach, 30 days for testis and more than 300 days for renal tubules. It is generally agreed that the mechanism of radiation injury for 'early-responding' tissues such as jejunum, colon, stomach, testis and skin is depletion of parenchymal cells, and it is reasonable to assume that the same mechanism applies for injury to the renal tubule epithelium.

The survival of renal tubule cells is a logarithmic function of radiation dose, with a dose survival relationship similar to that of mammalian cells in vitro, which implies that killing of these cells is random, independent, and, therefore, a direct effect of radiation, not an indirect effect from changes in vasculature.

## PRESENTATION OF DATA

Since the renal tubule assay is limited to a single study (Withers & Mason, 1984), the data derived from this technique have thus far been presented in only one way. It should be noted, however, that renal tubule data are amenable to presentation in other ways, such as those discussed in Chapter 5 on microcolonies in the gastrointestinal epithelium.

Figure 17.3 shows the response of renal tubules exposed to a single dose of radiation, plotting the number of capsular tubule-forming cells per kidney cross section as a function of radiation dose on semi-logarithmic coordinates. Because it is known that irradiated cells survive randomly, and since the number of intact tubules per cross section decreased logarithmically with dose, the counts of viable tubules were converted to numbers of capsular tubule-forming cells assuming, as with other in situ assays, that the number of such units followed Poisson distribution statistics (Ch. 1). This correction accounts for the greater average multiplicity of cells surviving in the tubules at lower doses and makes the survival curve slightly steeper than it would be if no correction were made (i.e. if it was assumed that all the regenerated tubules had originated from a single surviving cell rather than from a Poisson distribution of 1, 2, 3, etc. cells). The correction is based on the frequency of tubules that contained 0 cells, i.e. 370 minus the number of regenerated tubules.

## MODIFICATIONS TO THE MATERIALS AND THEIR CONSEQUENCES

### Animals

Thus far, renal tubule depletion has been quantitated

only in a mouse system (Withers & Mason, 1984). However, similar radiation induced histopathological changes in the kidney have been documented for mice (Phillips & Ross, 1973; Jordan et al, 1978), rats (Rich et al, 1961), rabbits (Schulz & Hoffman, 1905; Redd, 1960), dogs (Bolliger & Laidley, 1930; Hoopes, personal communication), Rhesus monkey (Raulston et al, 1978), and man (Luxton, 1962). There is no a priori reason to expect that the pathobiology or radiobiology of the kidney would differ significantly among different mammalian species although there may be differences in the time course of renal tubule depletion. Confirmation of the applicability of the renal tubule assay to diverse species awaits further investigation.

### Sex

The renal tubule assay was developed using only female mice. There is at least one known sex related renal histological difference between male and female mice. The lining of Bowman's capsule in most glomeruli consists of cuboidal cells in the male but flattened epithelioid cells in the female (Gude et al, 1982). Also males and females differ in their susceptibility to tubular necrosis from inhalation of chloroform (Hewitt, 1956). To what extent subtle sex-related or other differences in renal histopathology or physiology might affect renal function or cellular response to radiation is unknown at this time.

### Histology sections

Although coronal cross sections of kidney were used for the renal tubule assay, there is no reason that longitudinal histological sections of the kidney cannot also be used. The number of tubules touching the capsule in normal kidneys in our mice was about 370 in coronal cross sections and about 570 in longitudinal sections. In both cases, there are approximately 20 tubules touching the capsule per 1 mm of length. Only complete, untorn histological sections were used for analysis of results. If high quality histological sections are not available, torn or incomplete sections could be used by expressing all data as the number of renal tubules touching an arbitrary length (e.g. 1 cm) of renal capsule. However, it should be borne in mind that (a) shrinkage of the kidney is probably dose-dependent and (b) some areas of a damaged kidney (perhaps areas of extensive tubular degeneration) may be more susceptible to tearing or fragmentation during the preparation of histological sections. If some areas of the kidney are preferentially torn away or lost, then expressing the data as number of tubules per fixed length in these sections may be invalid. It therefore seems preferable to use *complete* sections rather than an arbitrary length of capsule to express data in terms of surviving tubules.

### Diet

Recent data demonstrates that rats maintained on diets with very low (4 per cent) or very high (50 per cent) dietary protein levels may exhibit different histological, functional and time related renal changes in response to radiation (Mahler et al, 1982). Animals used for the renal tubule assay should be maintained on standard laboratory diets with known and consistent levels of protein so that dietary effects on the assay can either be standarized for intra- and inter-laboratory comparisons of data or, alternatively, dietary protein levels can be altered at will to test their effect on the radioresponse of renal tubules. Mice used in the development of the renal tubule assay were maintained on a sterile commercial mouse pellet (Wayne Sterilizable Lab-Blox) containing a minimum crude protein level of 24 per cent.

## MODIFICATIONS TO METHODS AND THEIR CONSEQUENCES

### Nephrectomy

Redd (1960) has suggested that the extent to which a kidney may recover from radiation injury is largely dependent on the presence of another normal kidney. If there is another kidney present, the damaged kidney may recover less, or at a slower rate than if the damaged kidney itself was called upon to maintain vital functions. If this is true, details of the renal tubule assay would vary with the presence or absence of an unirradiated kidney. Redd (1960) has demonstrated that the presence of a normal kidney may protect against radiation-induced hypertension.

From the above observations it seems possible that unilaterally irradiated animals may survive longer than either bilaterally irradiated or nephrectomized animals but their regenerative response may be weaker in the absence of a need to maintain critical functions. A comparison of $LD_{50}$ data at 16 months post-irradiation suggests that this may not be a highly significant factor. Phillips & Ross (1973) reported an $LD_{50}$ of about 13 Gy for mice nephrectomized prior to irradiation of the remaining kidney while we have obtained an $LD_{50}$ value of 13.95 Gy for unilaterally irradiated mice whose normal kidney was left in situ.

If animals used for the renal tubule assay are nephrectomized prior to irradiation, it is important to allow sufficient time between the surgical procedure and irradiation for complete compensatory hypertrophy and hyperplasia of the remaining kidney. Sagerman (1964) reported on a patient treated for Wilms tumour who exhibited unexpectedly severe renal sensitivity when irradiated 48 hours after unilateral nephrectomy. Rollason (1949) has shown that the post-nephrectomy regenerative response reaches a plateau after about 3 weeks in the rat although other investigators have reported values of 10–14 days (Caldwell et al, 1970).

**Oxygenation**

The kidney has a high oxygen requirement (Caldwell & Wittenberg, 1974). Therefore changes in technique may easily alter the oxygen status of the kidney at the time of irradiation.

Steckel et al (1974) reported on the protection against renal damage observed in 10 out of 18 dogs and 3 out of 5 patients by inducing vasoconstriction with epinephrine. More recently, Baker et al (1982) have reported the first known instance of normal tissue protection by hyperthermia. Kidneys of mice heated to 42.5°C one hour before irradiation excreted less protein into the urine than kidneys irradiated without prior heating. They proposed that this unexpected kidney protection by prior hyperthermia might have resulted from less blood flow after the heat treatment.

## INFORMATION THAT CAN BE DEDUCED USING THE ASSAY

The renal tubule assay has not yet been used extensively. It is the first assay in situ for survival of the target cells for a 'late' radiation injury. Not only can it yield useful information about the kidney itself but it may provide data that can be used to model the responses of other slowly-responding tissues. Useful future studies include the measurement of the half time for repair, time of onset, rate and magnitude of repopulation, dose rate and fractionation effects, and number of clonogenic cells per tubule. Whether the whole tubule, including both proximal and distal segments, can be regenerated from one surviving cell is another question that will probably be answered only by microdissection of kidneys at various times after irradiation.

In addition to measuring radiation responses, it should be possible to use this assay to study the response of slowly-proliferating cells to other insults such as hyperthermia and chemotherapy and combinations of modalities. Data on late responses to radiation and other cytotoxic agents are scarce relative to those for acute effects. Clonogenic assays such as this, which permit study of the cellular basis for late injury, should prove useful in reducing the gap in knowledge between acute and late radiation sequelae.

## ACKNOWLEDGEMENTS

This investigation was supported in part by PHS Grant numbers USPHS CA-29644 and USPHS CA-31612, awarded by the National Cancer Institute.

## REFERENCES

Baker D G, Sager H T, Elkin D, Constable W, Rinehart L, Wills M, Savory J, Lacher D 1982 The response of kidney to ionizing radiation combined with hyperthermia induced by ultrasound. Radiology 145:515

Bolliger A, Laidley J W S 1930 Experimental renal disease produced by X-rays: Histological changes in the kidney exposed to a measured amount of unfiltered rays of medium wave length. The Medical Journal of Australia 1:136

Caldwell P R B, Wittenberg B A 1974 The oxygen dependency of mammalian tissues. The American Journal of Medicine 57:447

Caldwell W L, Hattori H, Rhamy R K 1970 Effect of irradiation on renal enlargement following uninephrectomy in the rabbit. Journal of Urology 103:399

Casarett G W 1976 Basic mechanisms of permanent and delayed radiation pathology. Cancer 37:1002

Chen K Y, Withers H R 1972 Survival characteristics of stem cells of gastric mucosa exposed to localized gamma irradiation in $C_3H$ mice. International Journal of Radiation Biology 21:521

Domagk G 1927 Die rontgenstrahlenwirkung auf das gewebe im besonderen betrachtet an den nieren. Morphologische und funktionelle veranderungen. Beitraege zur Pathologischen Anatomie and zur Allgemeinen Pathologie 77:525

Elkind M M, Whitmore G F 1967 The radiobiology of cultured mammalian cells. Gordon & Breach, New York.

Gabriel G 1926 Die beeinflussung von tierorganen durch rontgenbestrahlung. Strahlentherapie 22:107

Glatstein E 1973 Alterations in rubidium-86 extraction in normal mouse tissues after irradiation. An estimation of long-term blood flow changes in kidney, lung, liver, skin and muscle. Radiation Research 53:88

Gude W D, Cosgrove G E, Hirsch G P 1982 Histological atlas of the laboratory mouse, Plenum Publishing Corporation, New York, p 21

Hartman F W, Bolliger A, Doub H P 1926 Experimental nephritis produced by irradiation. American Journal of the Medical Sciences 172:487

Hartman F W, Bolliger A, Doub H P 1927 Functional studies throughout the course of roentgen-ray nephritis in dogs. The Journal of the American Medical Association 88:139

Hewitt H B 1956 Renal necrosis in mice after accidental exposure to chloroform. British Journal of Experimental Pathology 37:32

Hopewell J W 1980 The importance of vascular damage in the development of late radiation effects in normal tissues. In: Meyn R E, Withers H R (eds) Radiation biology in cancer research. Raven Press, New York, p 449

Hranitzky E B, Almond P R, Suit H D, Moore B S 1973 A cesium-137 irradiator for small laboratory animals. Radiology 107:641

Huang K C, Almand J R, Hargan L A 1954 The effect of total body x-irradiation on hepatic and renal function in albino rats. Radiation Research 1:426

Jordan S W, Key C R, Gomez L S, Agnew J, Barton S L 1978 Late effects of radiation on the mouse kidney. Experimental and Molecular Pathology 29:115

Lacassagne A 1946 Influence of wave-length on certain lesions produced by irradiation of mice. Proceedings of the Royal Society of Medicine 34:605

Linser C, Baermann H 1904 Uber die lokale und allgemeine wirkung der rontgenstrahlen. Fortschritte auf dem Gebiete der Roentgenstrahlen 7:996

Luxton R W 1962 The clinical and pathological effects of renal irradiation. Progress in Radiation Therapy 2:15

Mahler P A, Oberley T D, Yatvin M B 1982 Histologic examination of the influence of dietary protein on rat radiation nephropathy. Radiation Research 89:546

Mendelsohn M L, Caceres E 1953 Effect of x-ray to kidney on renal function of dog. American Journal of Physiology 173:351

Phillips T L, Ross G 1973 A quantitative technique for measuring renal damage after irradiation. Radiology 109:457

Raulston G L, Gray K N, Gleiser C A, Jardine J H, Flow B L, Huchton J I, Bennett K R, Hussey D H 1978 A comparison of the effects of 50 $MeV_{d \rightarrow Be}$ neutron and cobalt-60 irradiation on the kidneys of Rhesus monkeys. Radiology 128:245

Redd B L Jr 1960 Radiation nephritis. Review, case report, and animal study. American Journal of Roentgenology, Radium Therapy and Nuclear Medicine 83:88

Rich J G, Glagov S, Larsen K, Spargo B 1961 Histological studies of rat kidney after abdominal x-irradiation. Archives of Pathology 72:388

Rollason H D 1949 Compensatory hyperplasia of the kidney of the young rat with special emphasis on the role of cellular hyperplasia. Anatomical Record 104:263

Sagerman R H 1964 Radiation nephritis. Journal of Urology 91:332

Schulz A, Hoffman B 1905 Zur wirkungsweise der rontgenstrahlen. Deutsche Zeitschrift fuer Chirurgie 79:350

Steckel R J, Collins J D, Snow H D, Lagasse L D, Barenfus M, Anderson D P, Weisenburger T 1974 Radiation protection of the normal kidney by selective arterial infusions. Cancer 34:1046

Tsuzuki M 1926 Experimental studies on biological action of hard roentgen rays. American Journal of Roentgenology and Radium Therapy 16:134

Willis D A, Bachem A 1927 Effects of roentgen rays upon kidney. American Journal of Roentgenology and Radium Therapy 18:334

Withers H R 1967 The dose-survival relationship for irradiation of epithelial cells of mouse skin. British Journal of Radiology 40:187

Withers H R 1976 Late tissue damage: Radiobiological considerations. Cancer 37:1002

Withers H R, Elkind M M 1969 Radiosensitivity and fractionation response of crypt cells of mouse jejunum. Radiation Research 38:598

Withers H R, Mason K A 1974 The kinetics of recovery in irradiated colonic mucosa of the mouse. Cancer 34:896

Withers H R, Mason K A 1984 Late radiation response of kidney assayed by tubule cell survival. In preparation

Withers H R, Hunter N, Barkley H T Jr, Reid B O 1974 Radiation survival and regeneration characteristics of spermatogenic stem cells of mouse testis. Radiation Research 57:88

Withers H R, Peters L J, Kogelnik H D 1980 The pathobiology of late effects of irradiation. In: Meyn R E, Withers H R (eds) Radiation biology in cancer research. Raven Press, New York, p 439

# 18

*H.S. Reinhold J.W. Hopewell G.H. Buisman*

# Colony regeneration techniques in vascular endothelium

## HISTORICAL BACKGROUND

In investigations into the effect of irradiation on the wound repair of the skin of rats, Takahashi (1930) concluded that newly developed capillaries were very radiosensitive. This investigation was based on histological observations without quantitative analysis. It was not until the fifties that van den Brenk (1955a, b), quantified vascular proliferation, using the Sandison-Clark observation chamber implanted into rabbits' ears. Using this system, van den Brenk (1955b) concluded that irradiation with doses between 15 and 20 Gy (i.e. below the tolerance dose) did not inhibit vascular growth. Moreover, even after a dose as high as 40 Gy, which precluded normal budding, the vessels had not lost their intrinsic capacity for growth. Subsequently, van den Brenk (1959) determined a dose-effect relationship for the rate of vascular regeneration in the rabbit ear chamber and concluded, amongst other things, that regenerating vascular endothelium possesses great recovery potential after radiation damage. It should be noted that the time period in these studies for vascular proliferation and progression was of the order of one to two months. While the proliferation of endothelial cells probably played a role, many other developments such as the rearrangement of the microcirculation may take place during this protracted period of time.

In the early 70s, determinations of the radiosensitivity of endothelial cells were made in vitro (Nias, 1974; DeGowin, 1976). However, data in vivo are of utmost importance and it was on this basis that four groups of workers set out independently to develop methods for evaluating the radiosensitivity of capillary endothelium in situ, using reproductive cell survival as the endpoint. This has led to the development of several interesting methods which will be dealt with in more detail in the following paragraphs.

## REQUIREMENTS FOR MEASURING COLONY REGENERATION IN CAPILLARY ENDOTHELIUM

There is a remarkable similarity in the way the various groups of investigators developed their methods. All authors working in this field have recognized that it is essential that the system must be one in which capillary endothelial cells can be irradiated with various doses and that following irradiation the capillary cells have to be stimulated to proliferate into some kind of colony or its equivalent. This process must be amenable to quantification. The following paragraphs will deal with various aspects of the assay methods.

The requirements of a tissue structure, for the quantitative assay of capillary proliferation, can be summarized as follows:

1. The tissue has to be depleted of blood vessels, but has to be capable of supporting any newly formed blood vessels. The latter must be supplied from supporting vessels at its periphery.
2. The structure has to be suitable for applying a substance that stimulates capillary proliferation.
3. The structure must be such that capillary colonies can be seen and in some way or another quantified.

The stimuli that cause proliferation of endothelial cells, fall into one of the following groups:

1. stimuli released by growing tissues
2. stimuli involved in the replacement of worn-out endothelial cells.
3. stimuli released by wounds
4. stimuli involved in changes in vessel diameter
5. abnormal stimuli, inducing for example, tumour vessels or teleangiectasis.

In order to quantify the end result, the colonies, or their equivalent have to be counted. This is a difficult problem as it is impossible to recognize all the endothelial cells in a tissue. Therefore, all groups working in this field have used systems in which the presence of blood vessles (containing blood) in one form or another were used, rather than endothelial cells. This is probably justified in view of the fact that there will obviously be a close relationship between the presence of blood vessels and the number of endothelial cells present. However, there is a problem in such approaches in the interpretation of cell survival in capillary endothelium in the sense of 'single cells giving rise to single colonies'. It should be realized that endothelial cells in vivo never develop into discrete colonies, as the nature of these cells is to form an inter-connecting system of vessels. Blood vessel growth into a depleted or avascular part of a tissue is complex. Growth is organized and other cell types, like fibroblasts, proliferate in addition to endothelial cells. Moreover, endothelial cells migrate towards the growing tips of capillary buds. These characteristics of regenerating capillary sprouts have been recognized by all investigators in this field and none has claimed to have developed a scoring system for single cell colonies. However, it must be assumed that a fractional survival of endothelial cells will express itself as a proportional decrease in the number of regenerating units, i.e. capillary sprouts. The same probably holds true for any other tissue constituent from which a fractional survival is determined (van den Brenk, 1972).

The various endothelial assays in vivo have another series of common characteristics. In the first place, they all use as a vascular 'bed' a kind of two-dimensional, flat, tissue structure. This may be external or internal. Examples of the external surface are pig skin (Hopewell & Patterson, 1972) and the dog cornea (Gillette et al, 1975; Fike & Gillette, 1978). For the internal surface an artificial pouch in the rat skin was used by Van den Brenk (1972) amd Reinhold & Buisman (1973). Interestingly, the reason for choosing rat skin was different for each author. Van den Brenk based his system on the inflammatory-assay system of Selye, while Reinhold & Buisman were partly guided by the observation of Hewitt (1956) that formic acid applied to a subcutaneous air pouch in mice induced vasoproliferation, and also by a system developed by Tobin et al (1962) who used the rat subcutaneous air pouch for gas exchange studies. The latter reported that the vascularity in the air pockets increased with time. The rat subcutis has also been used by Folkman et al (1971) to assay tumour angiogenic factor (TAF), a factor which stimulates endothelial proliferation.

With regard to the angiogenic stimulus, all assay systems apply some kind of stimulus to cause vascular proliferation in a vascular or circulation-depleted tissue. For this purpose Hopewell & Patterson (1972) used reimplanted skin grafts, Gillette et al (1975) and Fike & Gillette (1978) a wounded cornea, while Reinhold & Buisman (1973) caused a depletion of the vasculature of the rat subcutis by freezing. In the transplanted pig skin and the wounded cornea, metabolic factors within the tissue probably served as the stimuli for angiogenesis. However, the angiogenic stimuli in the rat skin were artificial ones. Van den Brenk (1959) used croton oil in the subcutaneous pouch and Reinhold & Buisman (1973) used a filter pad which maintained a local acidic milieu.

The time allotted for endothelial proliferation before quantitative evaluation varied from two days in pig skin to seven days for the dog's cornea. For the rat subcutis a period of 12 (Reinhold & Buisman, 1973) to 13 days (Van den Brenk, 1972) appears to be optimal for the surviving endothelial cells to proliferate sufficiently to form 'colonies' that are large enough to count.

The method of determination is, as mentioned before, generally based on the presence of blood vessels, either unstained or stained with intravenously injected dyes. The presence of blood vessels can be assessed on an 'all-or-nothing' basis, resulting in an $ED_{50}$ value for the presence of regenerating blood vessels in the dog's cornea (Gillette et al, 1975). The presence of circulating capillary loops, visualized after an intravenous injection of dye has been used with success (Hopewell & Patterson, 1972). The number of isolated 'macro-colonies' of blood vessels, measuring about one to five mm in diameter in the irradiated rat skin, was the endpoint used by Van den Brenk (1972, 1974). Finally, volumetric-like assays were performed by Fike & Gillette (1978) on the vascularization in wounded cornea of dogs using histological sections and in the rat subcutis using quantitative angiography (Reinhold & Buisman, 1973, 1975). Again, all authors recognized that endothelial cells are an integral part of the lining of an interconnecting system of vessels. The fractional decrease in the selected endpoint had therefore to be used as an index of the probable reproductive survival of individual cells. It should be mentioned that endothelial cells have a capability of migration in order to accomplish the maintenance of an uninterrupted endothelial lining, even in spite of the loss of cells due to radiation.

## ASSAY SYSTEMS

### The capillary loop assay in the pig dermis

In the normal tissue of adult mammals, vascular endothelial cells proliferate extremely slowly. However, after tissue injury proliferation of these cells may be enhanced. An increased proliferation of endothelial cells is known to occur in the process of wound healing in skin, and in split-thickness skin grafts a good vascular supply in the graft would appear to be well established within five days (Rubin et al, 1960; Young & Hopewell, in preparation). However, there remains some dispute as to whether a graft is completely revascularised from the underlying bed, or if vessel sprouts from the bed simply join up with pre-existing channels in the skin graft (Conway et al, 1952).

The growth of a capillary sprout from the vascular bed could represent proliferation by a single endothelial cell. Such proliferation could provide a method for estimating cell survival in irradiated tissue, provided the capillary sprouts could be identified and counted (Hopewell & Patterson, 1972).

In preliminary studies, grafts were prepared in unirradiated pig skin in a fashion similar to that of clinical split-thickness skin grafts. The thickness of the graft was varied (230–1600 μm). After removing a graft it was turned through 180° and placed back on its original donor site. When an intravital dye, Disulphine blue, was injected intravenously after 24 hours, no dye entered any of the skin grafts. However, when the dye was injected after 48 hours, dye entered all the grafts and was seen as points of blue. Each point of blue dye was taken to represent an individual capillary loop entering the skin graft. This is indicated diagramatically in Figure 18.1. In thin (230–600 μm) grafts the points of blue were so numerous as to colour the graft almost completely; with thicker grafts (750–1600 μm) capillary loops were less numerous. This finding is consistent with the known vascular architecture of the dermis, the superficial papillary and subpapillary layers are highly vascularised, while the deeper reticular dermis is less so. The vessel density in the reticular dermis is 10 per cent of that in the papillary dermis (Young & Hopewell, 1980). A thick graft taken at the level of the reticular dermis would consequently have fewer cut vessels from which revascularisation could occur. This suggested that a graft thickness of 750 μm might be optimal for the evaluation of endothelial cell radio-sensitivity.

Female large white pigs used for the study of endothelial cell survival were admitted to the animal house when 10–12 weeks of age and weighing 20–25 kg. Two weeks later the hair to the right of the mid-dorsal line was shaved and four 4 × 4 cm fields were tattooed on the skin. The median edge of the fields was a standard 4 cm from the midline and the fields were separated by a 4 cm gap. While under anaesthesia (Berry et al, 1974) skin fields on a total of six pigs were irradiated with a range of doses of X-rays. At an interval of three weeks after irradiation split-thickness skin grafts were taken using a Zimmer electric dermatome using the following standard procedures.

On the irradiated flank a 4 cm strip of skin was removed which included both irradiated tissue and the normal skin between the treated areas. A strip of normal skin of similar width and length was taken from the opposite flank. The areas of grafted skin were transposed so that a normal graft was placed on an irradiated 'bed', while an irradiated graft was placed on a normal vascular bed on the opposite flank. The grafts were pressed firmly into place and were found to adhere within a few

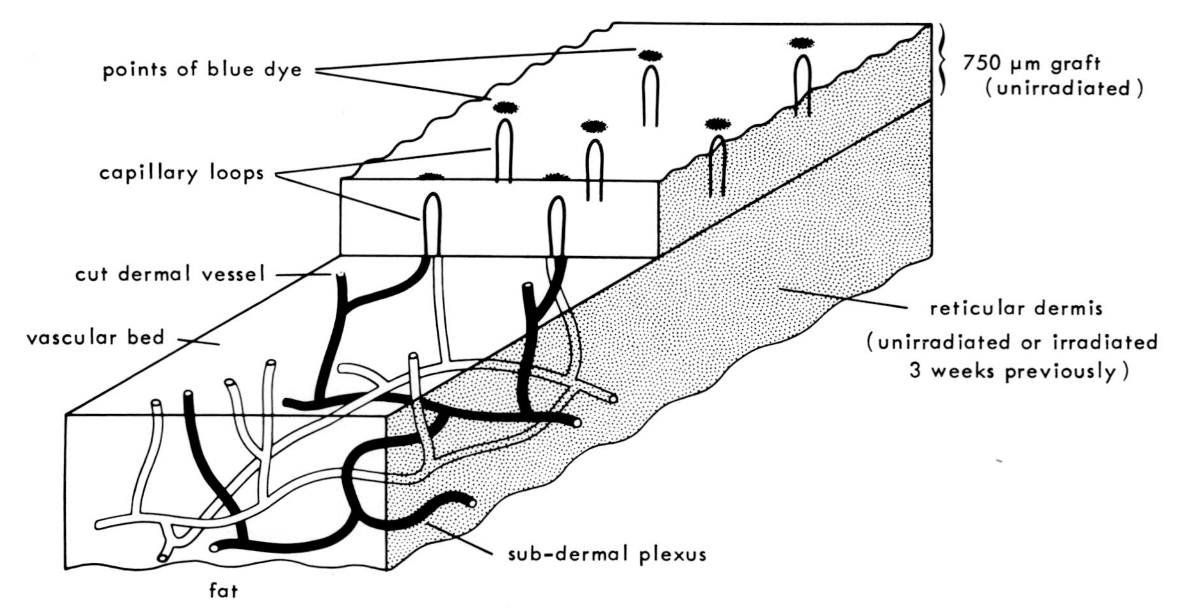


**Fig. 18.1** Block diagram showing the cut vessels in the vascular bed which form the origin of the capillary loops in a free skin graft. These capillary loops are visualized as 'points of blue dye' by intravenous injection of the intravital dye Disulphine blue 48 hours after skin grafting (transposition).

minutes; sutures at the edge of the graft were also used to help to keep the graft in place. No dressings were used.

In the first two animals 230 μm thick skin grafts were used. Injection of dye 48 hours after surgery showed a well developed system of capillary loops with a uniform distribution in normal grafts on a normal vascular bed. However, in grafts on an area of 'bed' irradiated three weeks previously with doses of 10–20 Gy, a poor and irregular pattern of neo-vascularisation had developed. The number of capillary loops, particularly those in normal 230 μm thick grafts placed on a normal vascular bed, were too numerous to allow density counts to be made accurately. The disparity between the vascularity of thin grafts on irradiated and unirradiated vascular 'beds' was more evident on day five, and by day eight grafts placed on areas irradiated with 15 and 20 Gy had sloughed. The grafts on regions irradiated with 10 Gy were partly lost and only grafts placed on areas treated with 5 Gy 'took'.

In all subsequent studies thicker, 750 μm, grafts were used. Animals that received such grafts three weeks after irradiation showed fewer points of blue dye after 48 hours and these could be resolved and counted.

When disulphine blue was given after five days all the grafted areas turned blue at the same time and with the same intensity as normal skin, suggesting that a completely adequate vascular supply had been established in the skin graft. All skin grafts including those on areas of vascular bed irradiated with 10–20 Gy 'took' and remained viable.

The relative dose-related changes in the density of points of blue dye counted 48 hours after surgery in pig skin grafts have been plotted in Figure 18.2. After an initial shoulder region the capillary density was found to decline exponentially with dose. The slope of the exponential portion of the curve suggests a $D_o$ value of approximately 10.3 Gy for endothelial cells in pig dermis. This $D_o$ value obtained for endothelial cells in pig skin was based on counting capillary loops 48 hours after skin grafting. This result was similar to those obtained in a rat subcutis model (Rinhold, 1972), although in this instance sprouts were counted five days after endothelial cell stimulation; after this time in the pig the anastomosis of vessels made such counts impractical.

The observations on the behaviour of grafted skin on an irradiated vascular bed would also suggest a different pattern of re-vascularisation in thick and thin skin grafts. Thin grafts must depend on the complete re-vascularisation of the graft by capillary loops from the bed. Failure of capillary loops to develop because of the radiation sterilisation of endothelial cells resulted in the partial or total loss of the graft. With thick grafts, the graft must survive by the union of capillary loops from the bed with the existing anastomosing channels in the

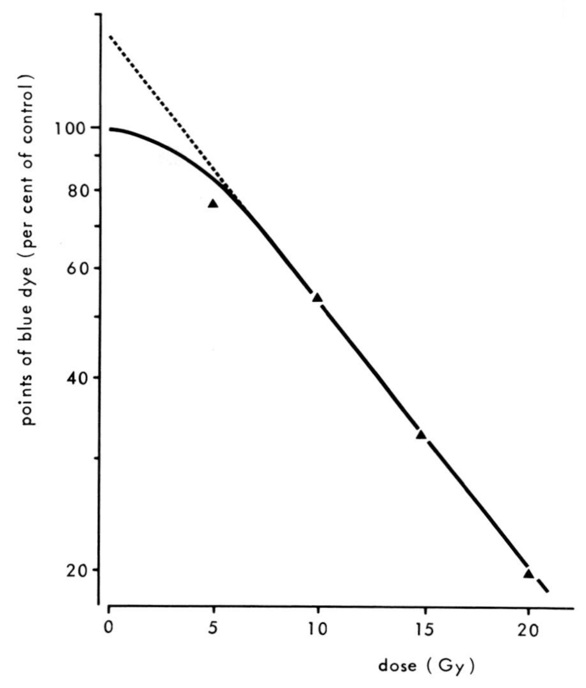


**Fig. 18.2** Dose-related changes in the relative number of points of blue dye in skin grafts of the pig 48 hours after surgery. Skin grafts were prepared 3 weeks after irradiation. The $D_o$ is 10.3 Gy (Hopewell & Patterson, 1972).

graft. Thus a graft on an irradiated bed can survive even though only a very limited number of loops are able to develop.

## The volumetric assay on dog cornea

This method was developed on the basis of Gillette's vascular $ED_{50}$ assay of dog cornea (Gillette et al, 1975). The method, as worked out by Fike and Gillette (1978) and Fike et al, (1979) can be summarized as follows: a portion of the dog's cornea, 8 mm in diameter, was excised from the central lamella with a calibrated corneal trephine set at a depth of 0.3 mm. The 8 mm lesion left an intact area, 4 mm wide between the lesion and the limbus. In order to maintain a continuing angiogenic stimulus for a period of one week, the epithelial covering was peeled from the wound each day. After seven days, a representative section of the cornea was removed. Histological sections were made and the degree of neovascularization was determined from these sections by means of a modified Chalkley counting technique. This involved the use of a 36-point grid, and 'hits' were counted when a blood vessel intersected a point on the scoring grid. A total of 2000 hits (skipping any open spaces in the histological sections) was counted per specimen. The percentage of capillary volume was derived from the

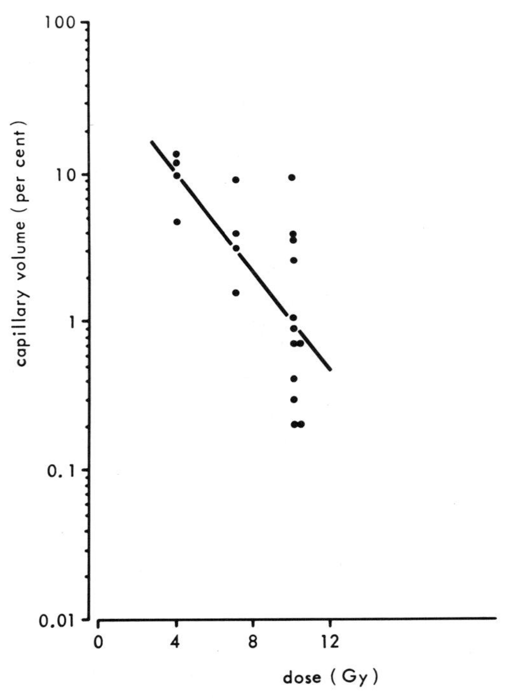


**Fig. 18.3** Dose response curve for slowly proliferating capillary endothelial cells after $^{60}$Co gamma-irradiation of the dog's cornea. Individual data points are shown. The $D_o$ is 2.65 Gy (Fike & Gillette, 1978, with permission).

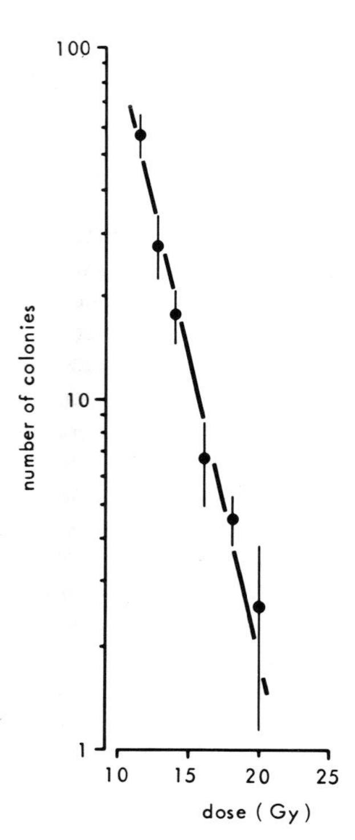


**Fig. 18.4** Incidence of macrovascular colonies in subcutaneous Selye pouches in the rat. The rats were irradiated 5 min after the pouches were produced. The $D_o$ is 2.4 Gy (van den Brenk, 1972, with permission).

number of capillary hits relative to the total number of hits. The survival curve obtained from this data is shown in Figure 18.3. The $D_0$ value was 2.65 Gy.

### The macro-colony assay

Van den Brenk designed this assay system on the basis of the subcutaneous granuloma pouch 'Selye pouch' of the rat (van den Brenk, 1972; van den Brenk et al, 1974). Following the preparation of a 30 ml subcutaneous air sac, the granuloma pouch develops as an inflammatory-type reaction to the injection of 1 ml of 0.5 per cent croton oil in mazola oil. In 'normal' conditions the pouch is covered on the inside with a confluent layer of vessel-rich granulomatous tissue. However, after irradiation, a dose-dependent decrease in the development of granuloma tissue occurs, and after doses exceeding 10 Gy the granuloma surface has thinned to the extent that individual capillary 'macro-colonies' can be distinguished. The 'macro-colonies' consisted of vascular foci, 1–5 mm in diameter. This aspect of the method is similar to that of the techniques for skin and gut developed by Withers, details of which can be found elsewhere in this volume. In the hands of van den Brenk this system proved to be a reliable way to measure the radiation sensitivity of the cells responsible for vascular 'macro-colonies' for a range of single doses between 10 and 20 Gy. Lower doses yielded confluent colonies, which could not be counted, whereas higher doses produced too few colonies per pouch to be useful for quantitative analysis. The angiogenic stimulus in this system, as discussed previously, was the reaction to croton oil and the time between the application of the stimulus and the assay was 13 days (van den Brenk, 1972) or 14–16 days (van den Brenk et al, 1974). With regard to the radiobiological information obtained with this system, a $D_o$ of about 2.4 Gy was derived (Fig. 18.4), and the $D_2$-$D_1$ recovery was approximately 1.8 Gy. A larger repair capacity was observed if the time between irradiation and the application of the angiogenic stimulus was delayed for 2–3 weeks (van den Brenk, 1974).

### The angiographic assay

The assay as used by two of the present authors (Reinhold & Buisman, 1973) was based on a quantitative evaluation of the impaired efficiency of the re-vascularization of a vascular-depleted area. As in van den Brenk's study (1972) the subcutis of the rat was selected as a useful tissue for the assay. In order to obtain an essentially two-dimensional plane the capillaries were allowed to grow only in a thin sheet of subcutis. To achieve this the subcutis on the back of the animal was distended by the subcutaneous injection of 30 ml of air (Hewitt, 1956; Tobin et al, 1962). Such a distended balloon-shaped subcutis was also a very convenient site to irradiate. The area in which capillaries were forced to proliferate was prepared as follows.

A superficial medial incision of about 2 cm long was made in the uppermost part of the epilated skin of the air pouch. The incision was made in such a way that only the superficial skin was cut; the underlying subcutis was left intact. Next, the skin was gently separated from the subcutis by means of a small spatulum to a diameter of about 15 mm. In this way, a semi-isolated sheet of subcutis was obtained. This was accessible through the incision in the overlying skin. The blood vessels in the central part of this sheet of subcutis were destroyed by means of freezing. To do this a stainless steel cylinder with an internal diameter of 9 mm was introduced through the skin so that it rested on the thin layer of subcutis (Fig. 18.5). Next, a small Teflon plunger, pre-cooled to -196°C by immersion in liquid nitrogen, was lowered through the stainless steel cylinder and left in contact with the subcutis for 30 seconds. In this way, the (endothelial) cells in an area of the subcutis, 9 mm in diameter, were destroyed while surrounding structures were shielded from freezing through their contact with the stainless steel cylinder. After removal of the shield and plunger, one drop of a suspension with angiogenic properties was dripped onto the frozen subcutaneous area. This suspension consisted of a mixture of uric acid crystals (0.02 $g.ml^{-1}$) in a seven per cent bovine albumin solution to which Ampicilline (2.5 $mg.ml^{-1}$) was added. Before closing the wound with an autoclip (Clay Adams) a pad was placed between the cutis and the subcutis. The pad measured 14 mm in diameter and consisted of several layers; a sheet of Nucleopore membrane (General Electric Co., California, USA) faced the subcutis; attached to this were three layers of Micropore surgical tape (3M Corporation) in which a mixture of uric acid crystals and lithium lactate was embedded (3 mg and 0.03 $mg.cm^{-2}$, respectively); the pad was covered on the side of the overlying skin with a layer of self-adhesive Teflon tape (no. 549, 3M Corporation). The animals were then returned to their cages and — for the 'survival' assays — a 12 day period

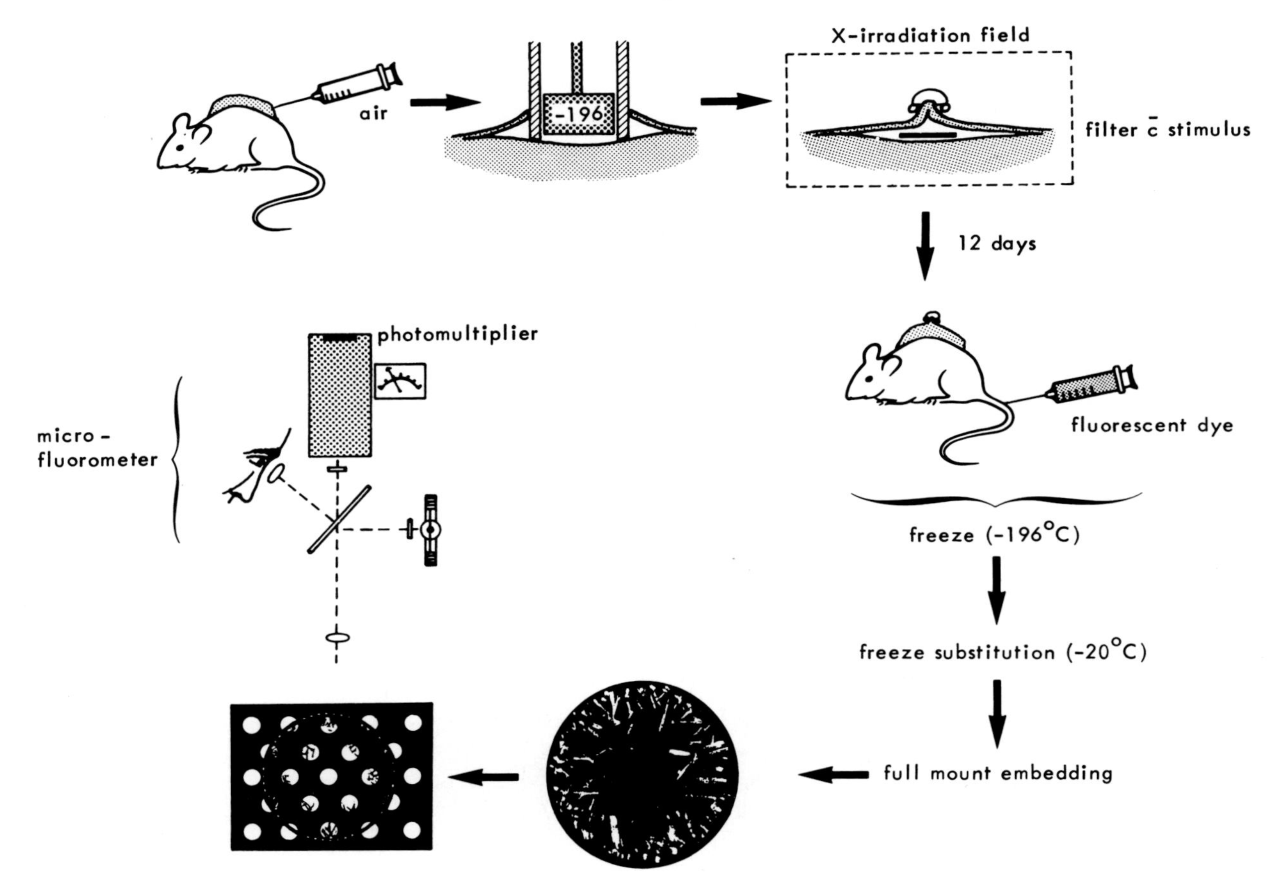


**Fig. 18.5** Schematic representation of the various steps in the technique used for assaying the proliferative capacity of endothelium (Reinhold & Buisman, 1973).

of capillary growth was allotted. Evaluation of the proliferative capacity of the endothelium, as mentioned previously, was performed by angiographic means. Under light ether anaesthesia the animals were injected intravenously with a 1 ml volume of a two per cent solution of a fluorescent dye; 'Blankophor G' (I. G. Farben, Germany). 60 seconds later the animals were sacrificed in liquid nitrogen. The isolated subcutis, including the filter pad and a 1 cm margin of skin, was then excised while still frozen and subsequently freeze-substituted for some days in tetrahydrofuran. The subcutaneous sheet bearing the proliferating capillaries was then detached and embedded in a nonfluorescent plastic (lamellon 2760; Scado-Archer-Daniels). After curing, this plastic has a refractive index of approximately 1.54, so that the embedded tissue became transparent.*

Quantification was carried out by means of a microfluorometer, i.e. a fluorescent microscope with a photomultiplier attachment. The idea behind this was that (a) the highest contrast between the dye-filled blood vessels and its supporting layer of subcutis was achieved with angiography with fluorescent dyes rather than light-absorbing dyes; (b) the UV light (365 nm) was efficiently absorbed by the dye in the vessels, so that a quantitative measure was obtained which represented mainly the surface (or silhouette) of the vascular system. Assuming the density of the endothelial cell covering of the vessels would be approximately constant, the intensity of the fluorescence measured would reflect the numbers of endothelial cells in the preparation.

The circular area from which the fluorescence was evaluated, in each preparation, was 9 mm in diameter. However, limitations in the optics of the fluorescence microscope required multiple sampling of areas 1.6 mm in diameter. The sampling sites were indicated by an overlay with holes drilled close together in an isometric pattern. The margin of the frozen area (and thus the area to be measured) was indicated by a ring, drawn with indian ink. Five 'background' and 28 measured points per preparation were assessed. 'Background' consisted of the composite autofluorescence of the coverglass, the tissue and the plastic used for embedding. The 'fluorescence value' of a preparation was defined as the total amount of fluorescence, corrected for the background of those measured points that had values differing significantly (+1 SD) from the mean background values of that preparation. If the 'survival' after high doses was very low, the 'fluorescence value' could be zero. This can create problems with the mathematical-statistical analyses as one has to use a logarithmic transformation. This situation was dealt with by the assumption that the distribution of the values was log-normal and that the few points at, or below the threshold (background) value could be accounted for by extrapolation. Such a correction can of course only be made when, for example, two of the 20 determined points are below the threshold value. The correction provided a means to avoid an unjustified shift in the mean values which occasionally would have occurred at very low survival levels. Preliminary experiments indicated a uniform growth rate of the capillaries, as demonstrated by fluorescence values between seven and 15 days after stimulation (Fig. 18.6). The twelfth day appeared to be a suitable time for assaying dose-response relationships (Fig. 18.7).

The dose-effect relationship, derived in this way, is shown in Figure 18.8. The $D_o$ value was about 1.7 Gy, the extrapolation number (n) approximately 7 and the $D_q$ value, 3.4 Gy. The value of $D_2$-$D_1$ obtained in other studies (Reinhold & Buisman, 1975) was about 2.9 Gy.

*Recently Epotek (Epoxy Technology Inc., PO Box 567, Billerica, Ma. 01821, USA) has become available and is more convenient for this purpose

## DISCUSSION

The four different assays in vivo with capillary endothelium are all based on recording the presence or absence of capillaries, rather than on counting endothelial cells. Moreover, they are all based on assaying growing capillaries. Normal capillary endothelium has a very low turnover rate (Tannock & Hayashi, 1972) and therefore, in contrast to many other 'in vivo' colony assays, the capillary endothelium has to be stimulated to proliferate. The latter is achieved in two of the assays (pig skin, dog cornea) by wounding, and in the others by the application of some kind of stimulus that either evokes an inflammatory reaction, or mimics that of anaerobic metabolism. An advantage of the application of such an imposed proliferative stimulus is that the moment of its application can be predetermined. This allows the investigation of repair phenomena that may operate in cells that are damaged by irradiation or other agents, but that do not proliferate actively at that time. This repair phenomenon has been investigated by van den Brenk et al, (1974) and Reinhold & Buisman (1975), with strikingly similar results. From both investigations it may be concluded that capillary endothelium is capable of repairing a sizeable amount of radiation-induced damage, if a period of two or more weeks has elapsed before the cells are stimulated to proliferate. This phenomenon also expresses itself in a change in the shape of the 'survival' curve.

The range of survival levels that can be investigated with a method that is strongly dependent on the integrity, as well as on the vascular function, of a tissue, depends very much on which of the systems is employed. The system based on capillary loops in pig skin is limited to doses that cover the first decade in survival. The

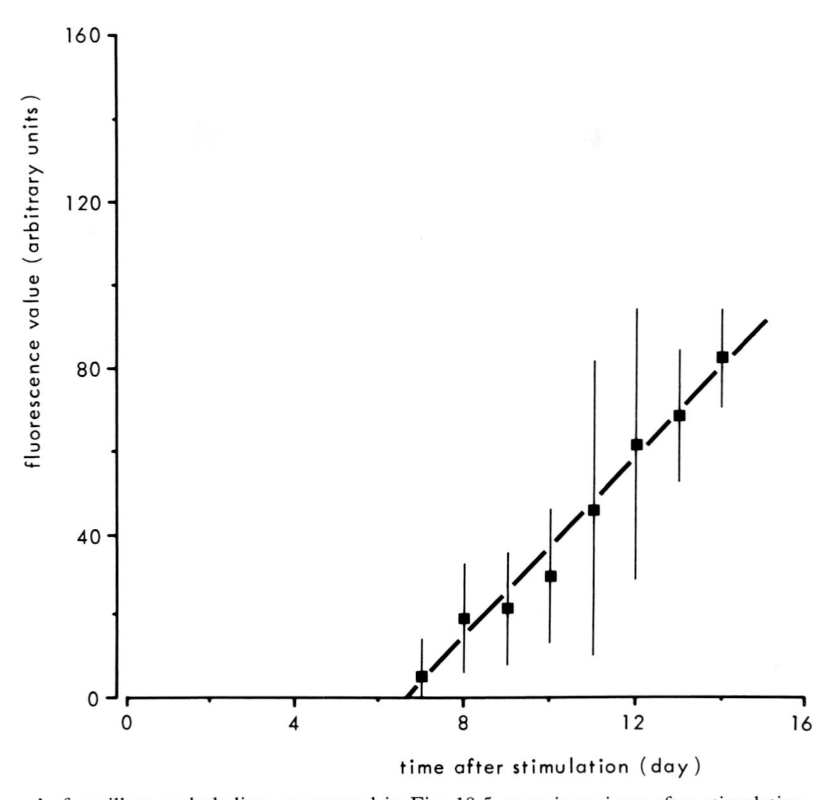


**Fig. 18.6** 'Growth rate' of capillary endothelium as assayed in Fig. 18.5 at various times after stimulation.

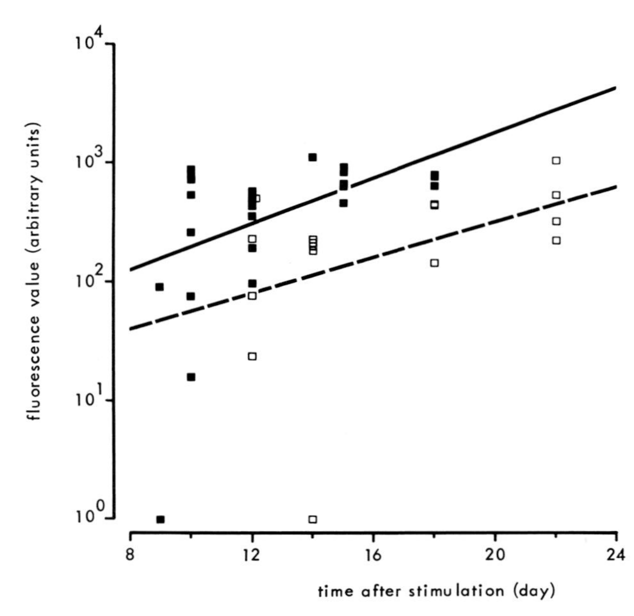


**Fig. 18.7** Comparison of growth rates for irradiated (5 Gy) (dashed — versus non-irradiated) rat skin. There is no significant difference in slope and 12 days seems to be an appropriate day for assay.

systems using the cornea in dogs and the rat subcutis cover two decades in survival, and the results using the rat granuloma pouch is probably most useful between the 2nd and 3rd decades. In some assay systems unfortunately the 100 per cent level (at zero dose) cannot be determined, due to the confluency of the units to be determined because their density is too high. The density of capillaries in most tissues is unknown, and is often a point of controversy. It can roughly be estimated to range from 25 to 2500 capillaries per mm² for tissues like fat, muscle and brain. Even the lowest conceivable density of 25 capillaries per mm² cannot be evaluated by any 'colony-counting' method, even with magnification. This means that with for example the rat granuloma pouch technique, the highest survival level that can be measured is probably between $10^{-2}$ and $10^{-3}$.

A problem that cannot be resolved with the present techniques is how the results are to be interpreted in terms of 'single cell survival' in the conventional manner. There is hardly any information available on the number of endothelial cells required to cover a unit length of blood vessel. Moreover, it should be realised that the endothelium cannot be separated into single cells. Even if an individual cell could form a single

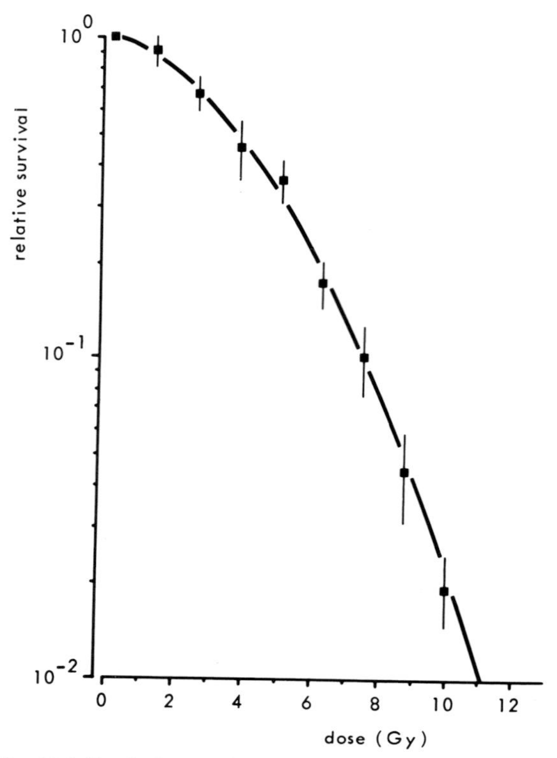


**Fig. 18.8** Survival curve for capillary endothelium, determined with the system shown in Fig. 18.5. The $D_o$ is 1.7 Gy, n is seven (Reinhold & Buisman, 1973). Recalculation via the L-Q model yielded values for alpha of 0.08 $Gy^{-1}$, for beta 0.03 $Gy^{-2}$ and for alpha/beta of 2.64Gy.

sprout-like capillary, this joins with other sprouts after growing to a length of a few hundred microns. Only if a blood circulation is established in such loops, does further sprouting occur. This factor limits the possibility of counting the number of capillary sprouts as a measure of cell survival. The assay of endothelial cell colonies, a short period of time after the application of a stimulus, will only allow cells to divide a few times and thus it could be argued that these assays are not a true measure of reproductive viability; that is the capacity of cells to undergo 'unlimited proliferation'. There may be contributory factors like endothelial cell stretching and/or the performance of one or two non-lethal cell divisions by otherwise 'clonogenic non-surviving' cells may interfere. Therefore, cell survival may be overestimated in studies with short proliferation times and this would explain some of the large $D_0$ values obtained (Hopewell & Patterson, 1972; Reinhold, 1972).

While, at first sight, the macro-colony assay (van den Brenk, 1972) seems to provide the strictest criterion for measuring division potential and as a consequence would give the highest sensitivity, some of the other methods must be considered equally strict. When one realizes that capillary endothelium by nature will never form *isolated* colonies of more than 50 cells, any assay system is bound to measure some form of network. Whether this consists of macro-colonies, or some type of brush-border sprouting is rather arbitrary. Most important is that sufficient time elapses for a substantial number of divisions to take place. It is, therefore, encouraging that the results of the assays with assay times of one week or longer (van den Brenk, 1972; Reinhold & Buisman, 1973 Fike & Gillette, 1978) agree remarkably well ($D_o$, 1.7–2.65 Gy). This is in contrast to the aforementioned studies with short times allowed for proliferation, giving high $D_o$ value (Hopewell & Patterson 1972; Reinhold, 1972). However, in slowly dividing cells, such as in the endothelium, the concept of unlimited cell proliferation may be misplaced. For the long-term integrity of the tissue any individual cell may be required to undergo only a few divisions in its lifetime (Hopewell, 1983).

The vascular system, of which the endothelial cells are an important constituent, forms an intergral part of almost all tissues, including tumours. The radiation response of these tissues will depend on the interaction between vascular and parenchymal elements. This chapter has focused on the important endothelial response in tissues, irrespective of the response of other tissue components. It has been shown that measurements can be made and radiobiological parameters have been established.

## ACKNOWLEDGEMENTS

The studies of skin grafting in irradiated pig skin were supported by a grant from the Cancer Research Campaign.

The authors wish to acknowledge Drs H. A. S. van den Brenk, J. F. Fike and E. Gillette for advice on this manuscript.

## REFERENCES

Berry R J, Wiernik G, Patterson R J S 1974 Skin tolerance to fractionated X-irradiation in the pig — how good a predictor is the NSD formula? British Journal of Radiology 47:185

Conway H, Joslin D, Rees T D, Stark R B 1952 Observations on development of circulation in skin grafts. Plastic and Reconstructive Surgery 9:557

De Gowin R L, Lewis L J, Mason R E, Borke M K, Hoak

J C 1976 Radiation-induced inhibition of human endothelial cells replicating in culture. Radiation Research 68:244

Fike J R, Gillette E L 1978 $^{60}$Co gamma and negative pi-meson irradiation of microvasculature. International Journal of Radiation Oncology Biology and Physics 4:925

Fike J R, Gillette E L, Clow D J 1979 Repair of sublethal radiation damage by capillaries. International Journal of Radiation Oncology Biology and Physics 5:339

Folkman J, Merler E, Abernathy C, Williams G 1971 Isolation of a tumor factor responsible for angiogenesis. Journal of Experimental Medicine 133:275

Gillette E L, Maurer G D, Severin G A 1975 Endothelial repair of radiation damage following beta irradiation. Radiology 116:175

Hewitt H B 1956 The quantitative transplantation of sarcoma 37 into subcutaneous air pouches in mice. British Journal of Cancer 10:564

Hopewell J W 1983 Radiation effects on vascular tissue. In: Potten C S, Hendry J H (eds) Cytotoxic insult to tissue. Churchill Livingstone, Edinburgh, p 228

Hopewell J W, Patterson T J S 1972 The effects of previous X-irradiation on the re-vascularisation of free skin grafts in the pig (abstract). Biorheology 9:45

Nias A H W 1974 In: Friedman M (ed) Biological and clinical basis of radiosensitivity. Thomas, Springfield, p 156

Reinhold H S 1972 Radiation and the microcirculation In: Vaeth J M (ed) Frontiers of radiation therapy and oncology 6. Karger, Basel, p 44

Reinhold H S, Buisman G H 1973 Radiosensitivity of capillary endothelium. British Journal of Radiology 46:54

Reinhold H S, Buisman G H 1975 Repair of radiation damage to capillary endothelium. British Journal of Radiology 48:727

Rubin P, Casarett G, Grise J W 1960 The vascular pathophysiology of an irradiated graft. American Journal of Roentgenology 83:1097

Takahashi T 1930 The action of radium upon the formation of blood capillaries and connective tissue. British Journal of Radiology 3:439

Tannock I F, Hayashi S 1972 The proliferation of capillary endothelial cells. Cancer Research 32:77

Tobin C E, Liew H D van and Rahn H 1962 Reaction of the subcutaneous tissue of rats to injected air. Proceedings of the Society for Experimental Biology and Medicine 109:122

van den Brenk H A S 1959 The effect of ionizing radiations on capillary sprouting and vascular remodelling in the regenerating repair blastema observed in the rabbit ear chamber. American Journal of Roentgenology 81:859

van den Brenk H A S 1972 Macro-colony assay for measurement of reparative angiogenesis after X-irradiation. International Journal of Radiation Biology 21:513

van den Brenk H A S, Sharpington C, Orton C, Stone M 1974 Effects of X-radiation on growth and function of the repair blastema (granulation tissue) I. Wound contraction. International Journal of Radiation Biology 25:1

Young C M A, Hopewell J W 1980 The evaluation of an isotope clearance technique in the dermis of pig skin. A correlation of functional and morphological parameters. Microvascular Research 20:182

# 19

*R. Cox W.K. Masson*

# Quantitative clonogenic cell techniques in studies with human diploid fibroblasts

## INTRODUCTION

Cells having a fibroblastic morphology, i.e. of an elongated spindle shape, constitute a major cell population in many mammalian tissues. Such cells are known to produce collagen and elastin which have both supportive and mechanical functions in tissues. The main reason that clonogenic studies with cultured human diploid cells have focussed on fibroblasts is that such cells are relatively easy to grow in monolayer culture. Indeed, biopsy specimens from the majority of human tissues yield primary cultures with cells of fibroblastic morphology as the dominant population.

Unlike the vast majority of established human cell lines, i.e. those derived from tumour material or after treatment with oncogenic viruses, human fibroblast cell strains predominantly show a stable diploid complement of chromosomes which is almost invariably representative of the somatic cells of the human subject from which they were originally derived. It should be noted, however, that the majority of studies with human fibroblasts are performed not from a direct interest in such cells but because they provide the most convenient cellular model for genetic and physiological studies with normal human diploid cells. Clonogenic techniques play a major role in quantitative studies on cellular response to genotoxic agents such as ionising radiation, UV and chemical mutagens. The introduction of reliable clonogenic techniques for human fibroblasts has facilitated the study of normal cellular response to such agents and, through the study of human genetic diseases such as xeroderma pigmentosum (Arlett & Lehmann, 1978) and ataxia-telangiectasia (Bridges & Harnden, 1982), some understanding of the genetic factors which influence cellular response.

## HISTORICAL PERSPECTIVE

Clonogenic techniques for human fibroblasts were first described in the classical studies of Puck and co-workers in the late 1950s (Puck et al, 1957). However, whilst these studies had a major impact on cellular radiobiology, with few exceptions subsequent cell survival experiments during the 1960s used established cell lines rather than human fibroblasts. The early 1970s witnessed a renewed interest in studies with human fibroblasts; this interest has since grown with considerable emphasis being placed on studies with cells from donors with genetic disorders which may be associated with hypersensitivity to genotoxic agents.

## GENERAL EXPERIMENTAL CONSIDERATIONS

### Biopsy material

In principle, cultures of human fibroblasts may be initiated from almost any solid human tissue. Practical and ethical considerations determine that the most commonly used source is the skin but if fetal or autopsy material is available other tissues such as lung, peritoneum or muscle may be used. To reduce the possibility of culture contamination, biopsy material is best obtained under sterile or semi-sterile conditions and the excised tissue immediately transferred to fresh sterile culture medium (containing appropriate antibiotics). If necessary, excised tissue may be stored at 4°C and at a pH of 7.4 until initiation of the culture, which should be performed within 12 h of excision. Such conditions are not critical and in this laboratory we have frequently initiated vigorous cultures from skin biopsies that have travelled for 3 days at ambient temperatures in the UK postal system. Long term storage of biopsy material at liquid nitrogen temperature is feasible but should not be relied upon. It is worthwhile noting that only small quantities of tissue (skin biopsy of $<5$ mm diameter) are necessary for initiating a culture and, to minimise scarring, clinicians will usually only excise skin fragments of such a size.

### Initiation of primary cultures

A number of methods for initiation of monolayer fibroblast cultures are available. These and the general technique of cell culture are described in detail by Paul

(1975). Broadly, for initiation the options are (a) to dissociate the biopsy into small aggregates of cells with trypsin (0.1%) or trypsin plus collagenase (0.25%) in calcium and magnesium free isotonic saline at 37°C for 10–12 min and seed the cells directly into a suitable culture vessel or (b) surgically to divide the biopsy into small fragments (i.e. chop it up) and anchor each fragment to the growth surface of the vessel with an overlayed glass coverslip or a plasma-clot matrix. Method (a) does not work well for skin which is relatively resistant to enzyme dissociation but is a good method for soft tissues such as lung from which large numbers of primary culture cells may be obtained. Method (b) may be used for all tissues and the plasma-clot technique of Harnden (1960) is probably the most reliable method for initiation cultures from skin fragments. In this laboratory the following plasma-clot method is used: the biopsy specimen is suspended in ~2 ml of chick embryo extract and cut into ~1 mm cubes. Up to 6 single drops of chicken plasma (Difco) are placed in a 25 $cm^2$ plastic culture flask (Corning or Nunc). Single cubes of tissue are then transferred in single drops of embryo extract and mixed with each of the drops of plasma in the flask. When all 'spots' have been seeded, the flasks are gassed with 95 per cent air/5 per cent $C_{02}$ and clotting is usually complete within 1 hour. Culture medium is then added and the flasks incubated for culture.

A large number of different formulations of culture medium are used for human fibroblasts. In our experience Eagles Minimal Essential Medium is quite adequate for growth. The serum supplement is, however, more critical. Horse, human and newborn calf sera do not appear to adequately satisfy the growth requirements of human fibroblasts and in our experience much better growth rate in monolayers and a higher efficiency of colony formation is obtained with medium supplemented with fetal calf serum (usually at 10% v/v).

Using the plasma clot method, outgrowth of cells from a skin fragment is usually observed within 5 days of initiation. However, in some instances very slow initiation occurs and in such situations growth should be maximised by careful pH control and by changing the culture medium at weekly intervals. The initial outgrowth is often composed of epithelial cells but as growth proceeds cells of a fibroblastic morphology begin to dominate the culture. After 2–3 weeks of incubation at 37° the halo of dividing fibroblasts around each fragment has usually reached a diameter of ~10 mm. The primary culture may then be dissociated by a 5 min incubation at 37° with a solution of 0.1 per cent trypsin (Difco) plus 0.4 mg $ml^{-1}$ ethylenediamine tetra-acetic acid (EDTA), the cells are harvested by centrifugation (1000 rpm), then resuspended in culture medium, counted and finally are seeded into a suitable culture vessel at a density of ~$5.10^3$ cells $cm^{-2}$. This constitutes the 1st passage and this and subsequent cultures are termed 'secondary' and are given passage numbers accordingly.

**Maintenance of secondary cultures**

Since human diploid fibroblasts have a finite life-span (usually between 50 and 80 generations for normal cells) and the colony-forming capacity of cells in a population tends to decay with age, it is crucial to store large stocks of early passage cells at liquid nitrogen temperature. In this laboratory we use a 1 to 3 split for routine subculture and usually store five ampoules each of $10^6$ cells at liquid nitrogen temperature at each passage between numbers 1 and 4. By re-stocking each time an ampoule series has become exhausted, stocks may be maintained almost indefinitely, thus retaining a pool of early passage cells with high viability for colony studies. Such storage and re-use of cultures should be linked to biochemical and/or microscopic tests for mycoplasma contamination of cultures (Chen, 1977; Marcus et al, 1980).

**Growth of colonies and measurement of cell survival.**

Unlike the majority of established cell lines, the colony-forming efficiency (CFE) of single human fibroblast cells at low seeding density is usually less than 20 per cent. More importantly, there also appears to be a density-related stimulation of CFE in fibroblast cultures, which is probably associated with oxygen depletion and/or production of growth-factors by metabolising cells. The determination of cell survival after exposure to radiation or chemical agents usually requires that the seeding density of cells is increased with increasing dose, and in order to obtain qualitatively reliable dose-response data it is important to control for the effects of cell density on CFE. The use of 'feeder' cells irradiated with a supralethal dose of X-rays (~30 Gy) is the most convenient solution to this problem (Cox & Masson, 1974; Deschavanne et al, 1981); the metabolic activity of the reproductively inactive feeder cells being sufficient to sustain maximum CFE of viable 'experimental' cells. In this laboratory maximum CFE of most strains of human fibroblasts is observed at a total cell density (experimental cells plus feeder cells) of around $10^3$ cells $cm^{-2}$.

In the determination of dose-response data the number of irradiated (or chemically treated) cells is increased with dose so that at all doses an approximately equal number of surviving colonies may be scored. Accordingly, to maintain constant cell density the number of feeder cells added to dishes decreases with dose. In terms of the CFE of experimental cells, no advantage is gained by preseeding with feeder cells in order to condition the medium and generally it is more convenient that feeder and experimental cells are seeded together. We do not yet fully understand the complex interactions that occur between feeder and viable cells during the initiation of colony growth but empirical observations suggest that the most reliable dose-response data are obtained when feeder cells and experimental cells are isogenic and we routinely use feeder and experimental cells which are derived from the same initial culture.

In the determination of dose-response data the mode of irradiation (or chemical treatment) may be determined by experimental considerations. However, irradiation of monolayers and growth in situ of surviving cells does minimise possible interaction between radiation-induced cellular damage and insult from trypsin. For such irradiations of monolayers, experimental and feeder cells may be allowed an 18 h post-seeding incubation before irradiation; in general, after such a period cellular multiplicity is minimal and no correction of the survival data is necessary.

The number of experimental cells to be seeded in a given dose-response determination depends upon the expected viability of the untreated population. For normal human fibroblasts in early passage a CFE of between 30 and 70 per cent is usually observed. This figure for CFE may be considerably lower for cells in late passage and for cells from donors with some genetic disorders. It is important to establish an approximate figure for CFE before attempting full scale dose-response studies. Perhaps the major experimental variable in determining the CFE of human fibroblasts is the quality of the serum in the culture medium. Fetal calf serum (FCS) is considerably better in this respect than serum which may be obtained from other sources. In this laboratory we test a number of batches of FCS with a 'standard' human fibroblast strain and select batches that gives maximum CFE together with good growth.

In our experience, human fibroblasts are considerably more sensitive to standard cell culture procedures than the majority of established cell lines. The following list of procedural suggestions, whilst not exhaustive may be found useful.

1. If possible use early-passage cultures.
2. Maintain cultures in exponential growth; if plateau phase cultures are used, regular medium changes are necessary to maintain maximum viability.
3. Dissociate monolayer cultures with trypsin plus EDTA; trypsin alone is inefficient and over-treatment greatly reduces viability.
4. Ensure that trypsin is adequately washed from the cell suspension and the suspending medium is at the correct pH.
5. Avoid over-vigorous use of the pipette when resuspending the cells and minimise the period during which they are held in suspension.

For the majority of human fibroblast strains, growth adequate for scoring individual colonies will occur during a 14–18 day incubation period at 37°C in a gas phase of 95 per cent air plus 5 per cent $CO_2$ and without the need for any changes of culture medium. We would usually aim to obtain 50–100 colonies per 90 mm diameter dish and then score stained (e.g. Giemsa or Azur A) colonies with the aid of a low power (×100) binocular microscope using the criterion of more than 50 cells in a colony. Under the conditions specified here there is rarely any problem in distinguishing colonies from the background of giant non-viable feeder cells. The scoring problems that do arise are usually the result of slow or interrupted growth. Such conditions, which apply in the case of slowly-proliferating colonies of aging or genetically abnormal cells, or in some colonies of normal early passage cells surviving relatively high doses of radiation or chemicals, tend to produce diffuse colonies (through cell migration) which are inherently difficult to score. Extending the incubation period to 21–24 days and/or changing the culture medium weekly may reduce the problem. Unfortunately, such scoring problems are frequently insoluble and when they occur at a high frequency they adversely affect the quality of the data obtained.

## RADIATION DOSE-RESPONSE CURVES FOR CELL SURVIVAL IN CULTURED HUMAN FIBROBLASTS

An example of the X-ray dose-response curve for the inactivation of exponentially-growing early-passage human fibroblasts is given in Figure 19.1. The most striking feature of the form of the dose-response curve is the absence of a shoulder at low dose (Cox & Masson, 1974, 1975). Whilst some X-ray survival curves of cultured human fibroblasts show a small degree of curvature (Cox & Masson, 1975; Deschavanne et al, 1981) it is clear that the large shoulders seen in the X-ray dose-response relationships for the inactivation of established cell lines do not occur under conventional X- and γ-ray irradiation conditions with human fibroblasts. Detailed discussion of the biological and biophysical implications of these difference are beyond the scope of this paper and the reader is referred to the review article of Goodhead (1980) on the mechanistic aspects of radiation dose-response relationships. In general terms, however, studies with plateau-phase cultures of normal and putatively repair-deficient human fibroblast strains indicate a strong association between the presence of a shoulder on cell survival curves and the activity of post-irradiation recovery processes on radiation-induced potentially lethal cellular damage (Cox et al, 1981; Cox, 1982).

Cell survival data for cultured human fibroblasts are available for a wide range of radiation qualities (Cox et al, 1977; Cox & Masson, 1979). Although the form of

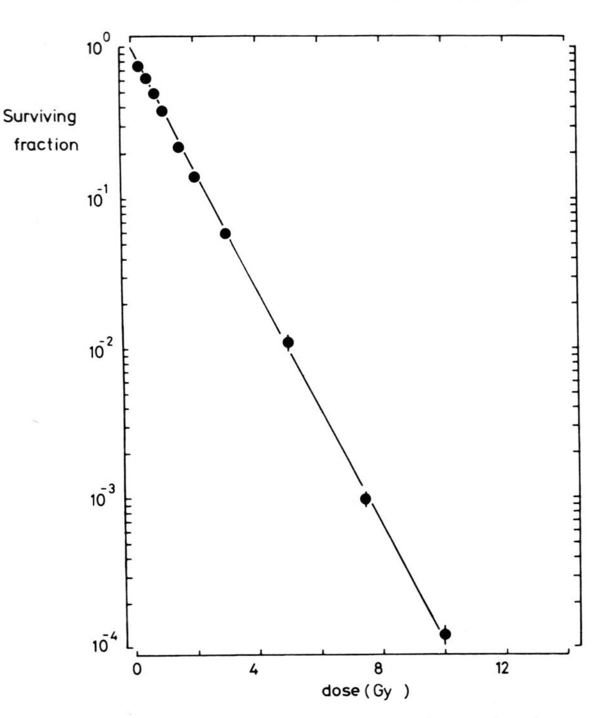


**Fig. 19.1** Survival of HF19 lung fibroblasts from a female human fetus after irradiation in monolayer with 250 kVp X-rays. Vertical bars indicate standard errors, where they are not shown they are smaller than the symbol. The data points weighted by the inverse of their variance were fitted by the method of least squares to the equation $S/S_0 = \exp(-D/D_o)$ where $S/S_0$ is the surviving fraction and D is the dose. From this, $D_o = 1.12$ Gy.

survival curves and absolute radiosensitivity are undoubtedly different in diploid human fibroblasts and established cell lines, the RBE/LET relationships for cell survival are very similar; these data may have important implications for radiobiological theory (Goodhead et al, 1980).

In terms of accuracy, intra-experimental variation in radiation dose-response relationships for human fibroblasts is comparable with that of established cell lines. However, considerable inter-experimental variation in dose-response data is frequently observed in human fibroblast cultures. Much of this variation is probably associated with selection during culture initiation and in subsequent growth. It is also important to recognise that the radiosensitivity measured in vitro is influenced by many genetic and physiological factors. It is therefore not surprising that there is a considerable inter-strain variation in 'normal' radiosensitivity ($D_o$ range 1.0–1.6 Gy; Cox & Masson, 1980). Variation in dose-response data in comparative experiments with a single cell strain may be minimised by controlling the obvious major experimental variables, i.e. serum quality and passage number. Comparative experiments with different cell strains (e.g. in studies on the genetic components of radiation response) are more problematical. In such experiments, data should be obtained with cells from as many donors as is practically possible, and relatively small differences ($<0.3$ Gy) in $D_o$ between cell strains should be interpreted with caution.

Similar considerations apply to dose-response determinations for the inactivation of human fibroblasts with toxic and mutagenic chemicals with the additional factor that, since cellular dosimetry of such agents is subject to more experimental variables than that of ionising radiation, greater variation in cellular response may be anticipated.

## CONCLUSIONS

Provided that the limitations imposed by the finite in vitro life span in vitro are fully considered, human diploid fibroblasts provide an excellent model in vitro for measuring the response of human cells to ionising radiation, UV and chemical agents. In addition to measurements of cell survival, human fibroblasts have been used for quantitative determination of gene mutation (Cox & Masson, 1976; Maher et al, 1976), cell transformation (Kakunaga, 1978; Borek, 1980), and recently for DNA-mediated gene transfer (Debenham et al, 1984). Of particular importance is the potential of the techniques with human fibroblasts for studies of some human genetic diseases. Such studies have already contributed greatly to our understanding of the genetic components of cellular sensitivity to genotoxic agents, an area of cell physiology which may have far reaching implications for human health (Bridges, 1982).

## ACKNOWLEDGEMENTS

The work of the authors is partly supported by CEC contract number BIO-E-412-81-UK.

## REFERENCES

Arlett C F, Lehmann A R 1978 Human disorders showing increased sensitivity to the induction of genetic damage. Annual Review of Genetics 12:95

Borek C 1980 X-ray induced neoplastic transformation of human diploid cells. Nature 283:776

Bridges B A 1982 Some DNA repair deficient human syndromes and their implications for human health. In: Sugimura T, Kondo S, Takebe H (eds) Environmental mutagens and carcinogens. University of Tokyo Press, Tokyo, p 47

Bridges B A, Harnden D G 1982 Ataxia-telangiectasia — a cellular and molecular link between cancer, neuropathology and immune deficiency. Wiley, Chichester, p 422

Chen T R 1977 In situ detection of mycoplasma contamination in cultures by fluorescent Hoechst 33258 stain. Experimental Cell Research 104:255

Cox R 1982 A cellular description of the repair defect in ataxia-telangiectasia. In: Bridges B A, Harnden D G (eds) Ataxia-telangiectasia. Wiley, Chichester, p 141

Cox R and Masson W K 1974 Changes in radiosensitivity during the in vitro growth of diploid human fibroblasts. International Journal of Radiation Biology 24:193

Cox R, Masson W K 1975 X-ray survival curves of cultured human diploid fibroblasts. In Alper T (ed) Proceedings of the 6th L H Gray Conference, the Institute of Physics. Wiley, Chichester, p 217

Cox R and Masson W K 1976 X-ray induced mutation to 6-thioguanine resistance in cultured human diploid fibroblasts. Mutation Research 37:125

Cox R, Masson W K 1979 Mutation and inactivation of cultured mammalian cells exposed to beams of accelerated heavy ions. III Human diploid fibroblasts. International Journal of Radiation Biology 36:149

Cox R, Masson W K 1980 Radiosensitivity in cultured human fibroblasts. International Journal of Radiation Biology 38:575

Cox R, Masson W K, Weichselbaum R R, Nove J, Little J B 1981 The repair of potentially lethal damage in X-irradiated cultures of normal and ataxia-telangiectasia human fibroblasts. International Journal of Radiation Biology 39:357

Cox R, Thacker J, Goodhead D T 1977 Inactivation and mutation of cultured mammalian cells by aluminium characteristic ultrasoft X-rays. II Dose-response of Chinese hamster and human diploid cells to aluminium X-rays and radiations of different LET. International Journal of Radiation Biology 31:561

Debenham P G, Webb M B T, Masson W K, Cox R 1984 DNA-mediated gene transfer into human diploid fibroblasts derived from normal and ataxia-telangiectasia donors: parameters for DNA transfer and properties of DNA transformants. International Journal of Radiation Biology 45:525

Deschavanne P J, Fertil B, Malaise E P and Lachet B 1981 Radiosensitivity and repair of radiation damage in human HF19 fibroblasts. International Journal of Radiation Biology 38:167

Goodhead D T 1980 Models of radiation inactivation and mutagenesis. In: Meyn R E, Withers H R (eds) Radiation biology in cancer research. Raven Press, New York, p 231

Goodhead D T, Munson R J, Thacker J, Cox R 1980 Inactivation and mutation of cultured mammalian cells by radiations of different LET. IV Biophysical interpretation. International Journal of Radiation Biology 37:135

Harnden D G 1960 A human skin culture technique used for cytological examinations. British Journal of Experimental Pathology 41:31

Kakunaga T 1978 Neoplastic transformation of human diploid fibroblasts by chemical carcinogens. Proceedings of the National Academy of Sciences (USA) 75:1334

Maher V M, Ouellette L M, Curren R D, McCormick J J 1976 Frequency of ultraviolet light-induced mutations is higher in xeroderma pigmentosum variant cells than in normal human cells. Nature 261:593

Marcus M, Lavi U, Nattenberg A, Rottem S, Markowitz O 1980 Selective killing of mycoplasmas from contaminated mammalian cells in cell cultures. Nature 285:659

Paul J 1975 Cell and tissue culture, 5th ed. Churchill Livingstone, Edinburgh, p 430

Puck T T, Morokovin D, Marcus P I, Cieciura S J 1957 Action of X-rays on mammalian cells. II Survival curves of cells from normal human tissues. Journal of Experimental Medicine 106:485

*E.P. Malaise M. Guichard P.J. Deschavanne*

# Primary cultures from lung and kidney

## Part 1 *LUNG*

### HISTORICAL BACKGROUND

Radiobiologists have tried several approaches to explain the radiosensitivity of lung tissue. First, they tried to evaluate the importance of radio-induced histopathological lesions in the lung of small rodents and to measure the lag-time before their appearance (Kurohara & Casarett, 1972; Maisin, 1970; Phillips & Margolis, 1972). Others have quantified the radio-induced fibrosis by biochemical methods (Law et al, 1976).

However, these studies have not furnished a specific answer to the following question: is the late damage primarily due to indirect factors, that is essentially to lesions to the connective tissue and/or to the capillary vessels, or is it due to a high intrinsic radiosensitivity of the lung cells? Nobody, to our knowledge, has been able to demonstrate the appearance of regenerating foci in the irradiated lung due to the proliferation of surviving cells, as has been observed for example in the skin (Withers, 1967) and in the small intestine (Withers & Elkind, 1968). This is probably due to the low turnover rate of the lung cells. Therefore we developed a technique for growing lung cell colonies in vitro that was sufficiently reliable to permit the direct study of the dose-effect relationship for various cytotoxic agents and, in particular, for ionizing radiations (Guichard et al 1980).

### DESCRIPTION OF THE TECHNIQUE AND METHODOLOGY

Our preliminary experiments, in which we used conventionally housed Balb/c mice mostly failed, as the cultures were often contaminated by bacteria in spite of the systematic use of antibiotics. We therefore used animals maintained in a sterile environment, including food, water, air and litter (Guichard et al, 1977). For this purpose we used heterozygous mice resulting from a cross between homozygous athymic nude male mice (nu/nu) and heterozygous Swiss female mice (nu/+).

**Preparation of cell suspensions and cell culture**

Three-to five-week-old animals were killed by cervical dislocation and immersed in an antiseptic solution for about 10 minutes (a 0.1% solution of cetrimonium bromide in ethanol). The lungs were separated from the trachea and from the main bronchial stem. They were then finely dissected with scissors in a watch glass and immersed in 40 ml of a calcium-free saline solution containing 0.25% crystallized trypsin. Preliminary experiments had shown that the addition of ethylenediaminetetra-acetic acid (EDTA) to this solution did not improve the results. The trypsinisation was allowed to continue for 25 minutes at 37°C with constant stirring. It was then stopped by the addition of 20 per cent calf serum. The cell suspension was centrifuged at 1000 rpm for five to seven minutes and the pellet was resuspended in a few ml of minimum essential medium (MEM with Earle's salts). The single cells were separated by filtration through a stainless steel filter with square pores, with the length of the sides being 18 $\mu$m. The cells were counted without staining. They were then seeded in plastic flasks containing feeder cells in MEM supplemented with antibiotics (15 mg gentamicin and 50 mg kanamycin per litre) and 20 per cent fetal calf serum. The culture medium was changed seven, and ten to 12 days after the beginning of the culture. Cultures containing colonies with at least 50 cells were fixed, stained and scored.

The irradiation was performed either in vivo or on cells which had been previously seeded in vitro. In the latter case there was a delay of 18 hours between the plating of single cells and the irradiation. Using mass cultures, we had previously checked that, during this 18 h time interval, the number of cultured cells had not significantly increased.

**Feeder cells**

The influence of two types of feeder cells was tested. A suspension of lung cells was prepared as described above. They were cultured in such proportions that the final number of cells per flask (feeder cells + cells whose clonogenicity had to be tested) was always $10^5$. 24 h after inoculation, the feeder cells received a single dose of 30 Gy. The medium was then changed before the viable lung clonogenic cells were added. In order to simplify a rather clumsy technique, we later prepared feeder cells from a primary culture of lung cells which had been allowed to grow until confluent. This allowed us to have at our disposal a great number of lung cells.

We have also prepared feeder cells from a nontransformed human fibroblast cell line: HF19. As in the case of mouse lung cells, we prepared a suspension of HF19 cells and inoculated each flask with $10^5$ cells, to which we administered 30 Gy 24 hours later. Another method of preparation of feeder cells was tested in parallel: the HF19 cells were first irradiated, then inoculated.

### Pulmonary alveolar macrophages (PAM)

In order to distinguish those colonies that might originate from macrophages, we prepared a PAM suspension according to the method of Soderland & Naum (1973). The trachea and the bronchi were washed several times under sterile conditions, using a catheter connected to a syringe. The PAMs were collected and seeded in MEM medium supplemented with 20 per cent fetal calf serum.

## PRESENTATION OF THE DATA

### Cell yield

The average number of isolated lung cells collected per mouse was 228 000 (95% confidence limits: 195 000–262 000). The average weight of the mouse lungs used in these experiments being 140 mg, the cell yield was thus $1.6 \times 10^6$ cells per gram of lung tissue.

### Plating efficiency (PE)

In the absence of feeder cells, the average PE was 2.3 per cent (Table 20.1).

In the presence of lung feeder cells, we usually observed a slight increase in PE from 2.3 to 3.4 per cent on the average. This small effect of lung feeder cells was found to be independent of their mode of preparation (primary culture, or cells subcultured once after having reached the plateau phase).

When the feeder cells were nontransformed HF19 fibroblasts, an appreciable increase in the number of control colonies was always observed, on the average by a factor of 3.4 (Table 20.1). The results were the same whether the HF19 cells were irradiated before or after inoculation.

Finally, the simultaneous use of two types of feeder cells (HF19 and lung cells) also improved appreciably the PE (which was multiplied on average by a factor of 4.7; Table 20.1).

For obvious practical reasons, we preferred to continue our experiments using HF19 feeder cells at a concentration of $10^5$ cells per 25 $cm^2$ culture flask. It was, indeed, easier to obtain HF19 cells in great numbers than lung cells from mice and the use of HF19 cells diminished the risk of contamination.

Some preliminary experiments were performed with eight to 14 months old mice. With these older mice it was usual to observe PEs lower than one per cent (in spite of the use of feeder cells). This is in contrast with the PEs found when mice a few weeks old were used. However, since with old mice the variations in PE were greater than with young mice (from 0.7% to 5.3%), one cannot state, solely on the evidence of preliminary experiments, that the PE of the cells of pulmonary origin decreases significantly with the age of the mice.

**Table 20.1** Influence of feeder cells on the plating efficiency of mouse lung cells

| Feeder cells | Plating efficiency |
|---|---|
| None | 2.3% |
| Murine lung cells | 3.4% |
| HF19 fibroblasts | 7.9% |
| Murine lung cells + HF19 fibroblasts | 10.7% |

### Characteristics of the colonies

The lung cell colonies present a rather variable morphology; however, two well-defined, easily recognizable types of colonies are mostly found. Firstly, colonies of cells which we have called fibroblast-like (Fig. 20.1A), and, secondly, colonies of epithelial-like cells (Fig. 20.1B). The first are composed of elongated, fusiform-type cells, sometimes criss-crossing each other; like the cells in colonies of human fibroblasts, these colonies have very irregular borders. The other epithelial-like colonies consists of small polyhedral cells, which are never found overlapping each other and which have a high nuclear-cytoplasmic ratio. These epithelial-like cell colonies are well delineated. In order to evaluate the stability of the morphological characteristics of these two cellular types, we sampled separately colonies of each type and used them as inocula for mass cultures. These cultures were regularly subcultured for almost a year. During this time the cells have always kept their initial morphological type.

In our experiments we always found colonies of both fibroblast- and epithelial-like cells. The proportions of these two types of colonies were of the order of 40 per cent and 60 per cent, respectively. We have not observed significant differences in these proportions either as a function of the culture conditions, or as a function

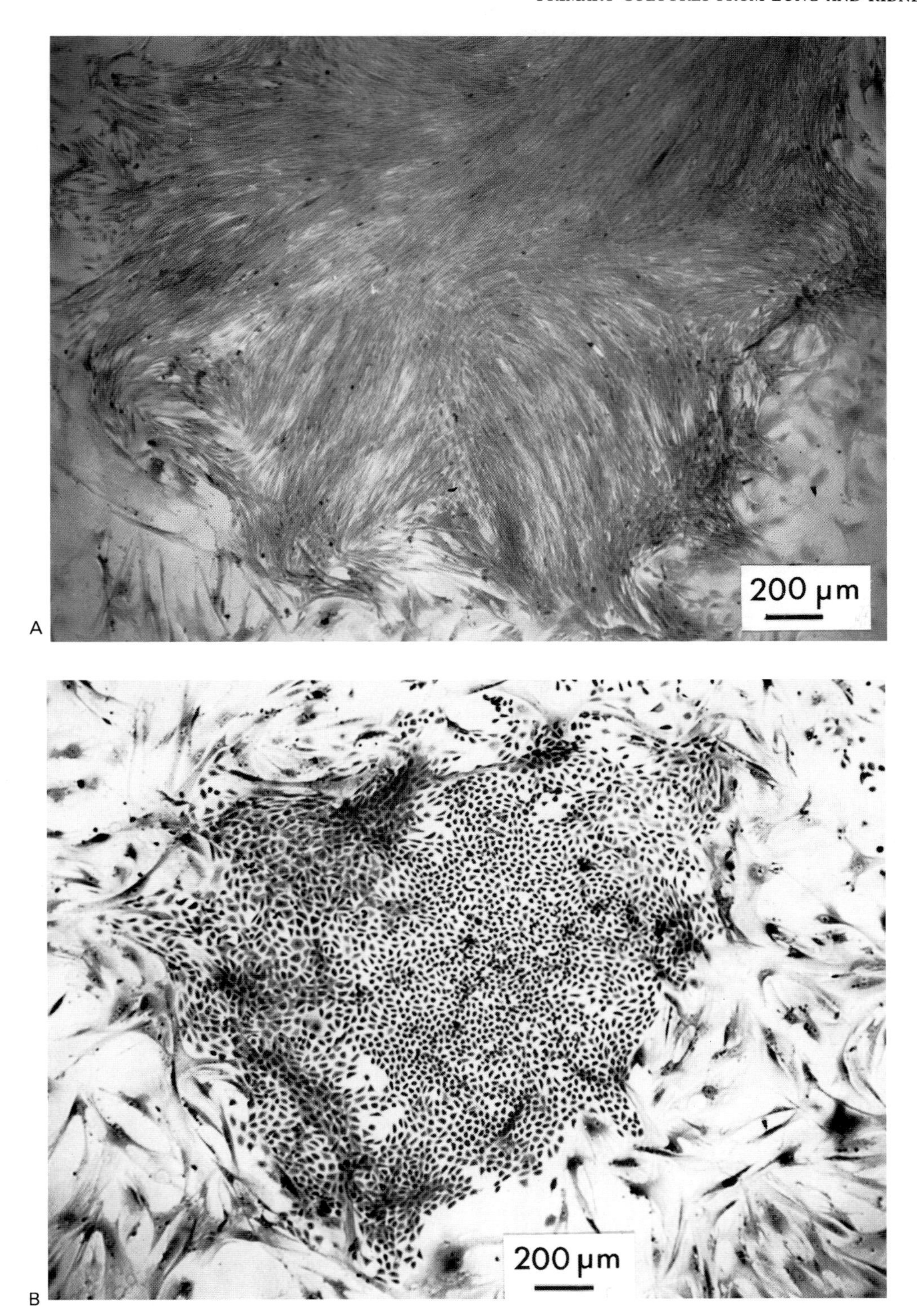


**Fig. 20.1A** Colony of fibroblast-like lung cells. This colony is surrounded by HF19 fibroblast feeder cells. Crystal-violet (magnification × 43).
**B** Colony of epithelial-like lung cells. This colony is surrounded by HF19 fibroblast feeder cells. Crystal-violet (magnification ×43).

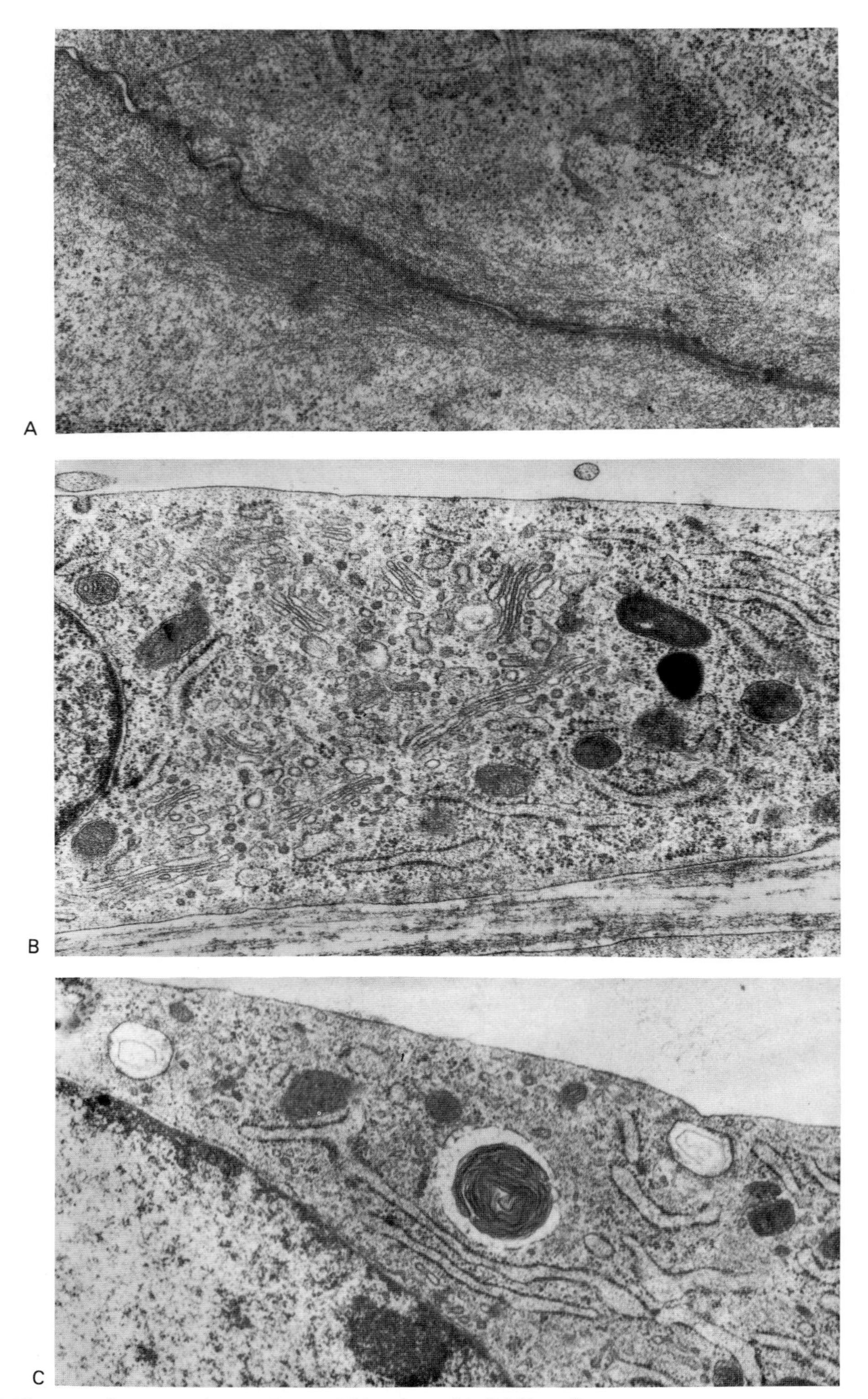

**Fig. 20.2A** Electron microscope picture of two neighbouring epithelial-like cells.
The cells are tightly linked by functional complexes surrounded by tonofilaments (magnification ×27 300).**B** Electron microscope picture of two neighbouring fibroblast-like cells. No intercellular structures are observed between these two cells (magnification ×27 300). **C** Myelinic whorl observed in an epithelial-like cell (electron microscope examination) (magnification ×21 700).

of the type of feeder cells. The examination, under the electron microscope, of the epithelial-like cell colonies showed cells tightly linked by functional complexes, surrounded by tonifilaments (Fig. 20.2A) (Malaise et al, 1982). This suggests that the epithelial-like cells are indeed of epithelial origin. On the other hand, no intercellular structure was observed in the fibroblast-like cell colonies (Fig. 20.2B). In addition, one could observe inside the epithelial-like cells some lamellar bodies which are observed also in vivo, in particular in type II pneumocytes (Fig. 20.2C).

We have been able to grow PAMs as colonies starting from macrophages by adding conditioned medium (medium having been in contact with PAMs grown as a mass culture until confluent) to the MEM medium supplemented with 20 per cent fetal calf serum. When placed in the presence of formalized yeast, all PAM colonies have always exhibited a marked phagocytosis. We have later used this method to detect the eventual presence of colonies of macrophage origin amongst those obtained from lung cells isolated by trypsinisation. None of these colonies has shown phagocytosis. This suggests, therefore, that none of the latter colonies is of macrophagic origin.

## INFORMATION THAT CAN BE DEDUCED USING THE TECHNIQUE

The culture method which we have developed has allowed us to obtain a survival curve of the cells of pulmonary origin virtually down to a survival of $10^{-3}$

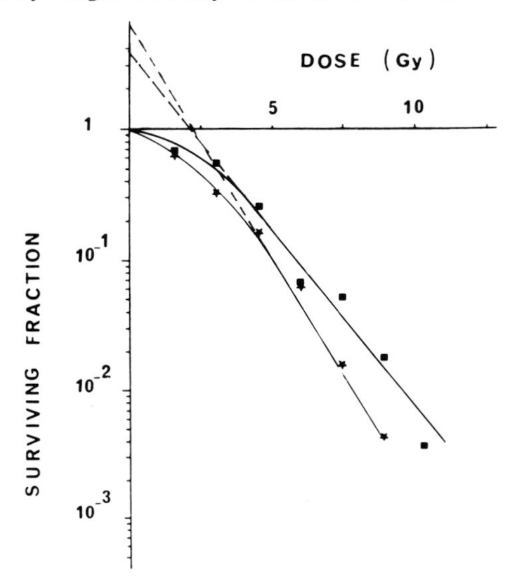


**Fig. 20.3** Survival curves of murine lung cells. Stars: cells were plated and then irradiated in vitro 18 hours after plating. Squares: cells were irradiated in vivo in air-breathing mice and plated immediately afterwards.

(Fig. 20.3) (Guichard et al, 1980). When irradiated in vitro 18 hours after inoculation, the murine lung cells present a survival curve whose parameters are as follows: n = 6.4 (5.7 – 7.1) and $D_o$ = 1.24 Gy (1.21–1.27). No significant difference in radiosensitivity could be detected between cells giving rise to fibroblast-like colonies and those giving rise to epithelial-like colonies.

When cells were irradiated in situ in air-breathing mice and seeded immediately afterwards, they demonstrated a survival curve with the following parameters: n = 3.9 (3.0–4.9) and $D_o$ = 1.58 Gy (1.52–1.64) (Fig. 20.3). If one considers the surviving fraction at each radiation dose, no significant difference could be detected between the two protocols (irradiation in vivo immediately followed by plating, or seeding followed by irradiation in vitro), although at equal doses the surviving fraction was always found to be slightly higher for cells irradiated in vivo.

Within the limits of the technique (that is down to a survival level of the order of $10^{-3}$) no subpopulation of radioresistant cells could be detected with irradiation in vivo (Fig. 20.3). When the animals were irradiated 10 minutes after having been asphyxiated by breathing nitrogen, the survival curve of the lung cells indicated the presence of a radioresistance corresponding to a dose modification by a factor of 2.5 (Guichard et al, 1980).

In conclusion, it is possible using colonies in vitro to study the inherent radiosensitivity of mouse lung cells. These do not seem to be particularly radiosensitive, whether the irradiation is performed in vivo or in vitro (Deschavanne et al, 1981). In addition, the cells undergo repair of sublethal and potentially lethal lesions. The high sensitivity of lung tissue for late lesions, e.g. emphysema and late sclerosis, cannot be explained by a particularly high inherent radiosensitivity of the lung cells. It seems likely, therefore, that the late lesions depend on other factors such as lesions of the connective and vascular tissue and/or a reduction below a critical level of the stem cells of the pulmonary parenchyma.

# Part 2 *KIDNEY*

## HISTORICAL BACKGROUND

Like the lung, the kidney is also a radiosensitive normal tissue. When a dose of 23 Gy is administered to the kidney of an adult, symptoms of an acute nephritis usually appear, followed by late renal sclerosis (Rubin & Casarett, 1972). For this reason, the kidney is considered a critical tissue in radiotherapy, in particular in the treatment of seminomas, lymphoblastomas, or neuroblastomas by extended abdominal fields (Maier, 1972).

Research has been undertaken to try to understand the causes of the radiosensitivity of renal tissue. Physiopathological studies, involving the incorporation of $^{86}$Rb by the kidney, were performed in mice after fractionated irradiation (Glatstein et al, 1975). Other studies were directed towards an evaluation of the importance of repair processes during fractionated radiation. During such studies, the recovery dose was measured, that is the additional dose of radiation necessary to produce a certain level of response when the radiation dose is delivered in two sessions instead of all at once (Hopewell & Berry, 1974; Phillips & Fu, 1975). However, the studies mentioned above did not use a direct approach, which might furnish information on the radiosensitivity of the kidney cells. This is why we developed, three years ago, a tissue culture technique applicable to mouse kidney cells (Deschavanne et al, 1980). This technique allowed us to study directly the survival curve of clonogenic cells of renal origin, as well as the effect of possible repair of sublethal and potentially lethal lesions (Guichard et al, 1982).

## DESCRIPTION OF THE TECHNIQUE AND METHODOLOGY

### Mice

As before, preliminary experiments, performed with kidneys from Balb/c mice living in an ordinary atmosphere gave poor results due to contamination of the cultures. It therefore seemed to us of utmost importance to use, as in the case of lung cells, specific pathogen-free animals. We used 3–5 week old heterozygous animals nu/+ obtained by a cross between Swiss nu/+ females and athymic nu/nu males.

### Preparation of kidney cells

The suspensions of kidney cells were prepared according to a method very similar to that described for the lung cell (Deschavanne et al 1980).

The mice were killed by cervival dislocation and immersed for about 10 minutes in a one per cent solution of cetrimonium bromide in ethanol. The two kidneys were removed under conditions of integral asepsis. They were then cut with scissors in a watch glass. The fragments were immersed in 35 ml of a modified (calcium-free) Dulbecco saline solution containing 0.25 per cent trypsin. The incubation was continued for 25 minutes at 37°C and the solution was constantly stirred. The trypsinisation was stopped by adding 20 per cent calf serum. The cell suspension was centrifuged at 1000 rpm for 7–10 minutes. The cell pellet was resuspended in MEM and the suspension was filtered through a stainless steel filter to recover only single cells. There were then inoculated in MEM (Earle's salts) supplemented with 20 per cent fetal calf serum and antibiotics (50 mg kanamycin and 15 mg gentamicin per litre). The medium was renewed at the end of a week and the colonies were scored after 12 days of culture and after fixation and staining.

### Feeder cells

We have tested the influence of two types of feeder cells: kidney cells having the same origin as those whose clonogenicity was to be tested and nontransformed human fibroblasts (HF19).

The renal cells were obtained from several kidneys treated together. Firstly, the feeder cell suspensions were treated in the same way as those whose clonogenicity was to be tested. The kidney feeder cells were inoculated into 25 cm$^2$ plastic flasks. Their number was adjusted so that the total kidney cell number per flask (feeder cells + cells to be tested) was always $10^5$. The kidney feeder cells were incubated for 18 hours (time necessary for the anchorage of most cells) before irradiation with a dose of 30 Gy. Immediately after irradiation, the medium was renewed and the kidney cells to be tested were inoculated.

The HF19 feeder cells were obtained by irradiating (with 30 Gy) exponentially growing cells, which were then inoculated into culture flasks ($10^5$ cells per flask). Finally, we have also tested the association of mouse kidney cells with HF19 cells as to their possible feeder role. In this case, the total number of cells per culture flask was always $2 \times 10^5$ ($10^5$ HF19 cells + $10^5$ renal cells, the latter consisting of feeder cells + cells to be tested).

### Peritoneal macrophages

A culture of these cells was prepared in order eventually to distinguish amongst the kidney cell colonies, those originating from macrophages. The peritoneal cavity was rinsed several times and the cells thus obtained were inoculated in the same medium used for the cells of renal origin.

## PRESENTATION OF THE DATA

### Cell yield

On the average, 303 000 cells were collected from the two kidneys of a single mouse (95% confidence limits: 238 000–368 000). Since the average weight of the two kidney was 250 mg, the cell yield was $1.2 \times 10^6$ cells per gram of kidney.

### Plating efficiency

In the absence of feeder cells, the PE was, on the av-

**Table 20.2** Influence of feeder cells on the plating efficiency of mouse kidney cells

| Feeder cells | Plating efficiency |
|---|---|
| None | 0.9% |
| Murine kidney cells | 3.2% |
| HF19 fibroblasts | 7.9% |
| Murine kidney cells + HF19 fibroblasts | 8.5% |

erage, lower than 1 per cent (Table 20.2).

The presence of feeder cells of renal origin increased appreciably the PE (by a factor of about 3.6) (Table 20.2). In the presence of HF19 feeder cells, the PE was clearly increased (by a factor of about 8.8) (Table 20.2). The simultaneous use of HF19 feeder cells and of feeder cells of renal origin improved the PE in proportions similar to those of HF19 cells alone (i.e. by a factor of about 9.4.) (Table 20.2).

These results encouraged us to continue our radiobiological experiments using only HF19 cells as feeder cells ($10^5$ cells per 25 $cm^2$ culture flask). Some preliminary experiments were performed with 8-month-old mice; we obtained an average PE of 0.7 per cent in spite of the presence of the HF19 feeder cells.This was an exceptionally low PE, which we have not encountered, even individually, in 3- to 5-week-old mice.

### Characteristics of the colonies

The renal cell colonies present essentially two types of morphology. There are, on the one hand, colonies of fusiform cells, which we have called fibroblast-like cells (Fig. 20.4A). These colonies have irregular borders. They are quite similar in appearance to the fibroblast-like cell colonies from mouse lung. On the other hand, there are colonies of polygonal cells, larger than the fibroblast-like cells, and which have a rather low nuclear-

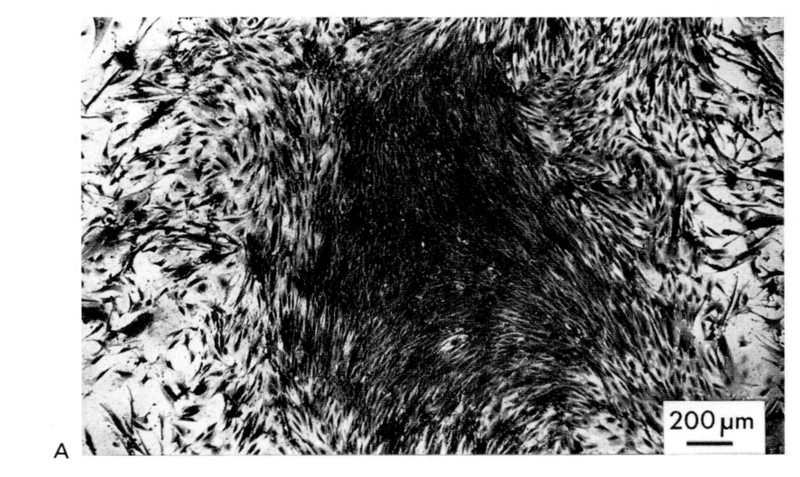


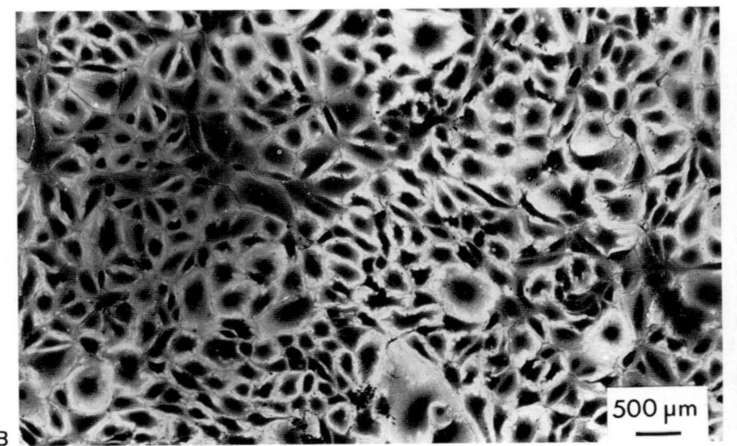

**Fig. 20.4A** Colony of fibroblast-like kidney cells. This colony is surrounded by HF19 fibroblast feeder cells (magnification ×30).**B** Colony of epithelial-like kidney cells (magnification ×76). **C** Colony of epithelial-like kidney cells. This colony is surrounded by HF19 fibroblast feeder cells (magnification ×30).

cyto plasmic ratio (Fig. 20.4B). These colonies, which we have called epithelial-like cell colonies, are better delineated that the fibroblast-like cell colonies (Fig. 20.4C).

In order to evaluate the stability of the morphological character of these two cell types, colonies were sampled separately and reinoculated in the presence of HF19 feeder cells. The fibroblast-like cells were regularly subcultured and maintained as mass cultures for a year. They always maintained the same appearance. In contrast, the epithelial-like cells always ceased to develop after the second subculture. In all our experiments we always observed both colonies of fibroblast-like cells and colonies of epithelial-like cells. The proportions of these two types of colonies are of the order of 10 to 20 per cent and 80 to 90 per cent respectively.

The peritoneal macrophages were phagocytic engulfing formalized yeasts. When applied to colonies developed from renal cells, this test showed that no cells in the colony were phagocytic. We therefore concluded that, under our culture conditions, the cell colonies of renal origin were not of macrophagic origin.

## INFORMATION THAT CAN BE DEDUCED USING THE TECHNIQUE

The culture method which we have described has allowed us to study the survival curve of cells of renal origin. Due mostly to the feeder effect of the HF19 cells, it was possible to administer radiation doses resulting in survival rates as low as $10^{-3}$. When irradiated in vitro 18 hours after inoculation, the kidney cells showed a survival curve with the following parameters: $n = 2.9$ and $D_o = 1.33$ Gy (Fig. 20.5).

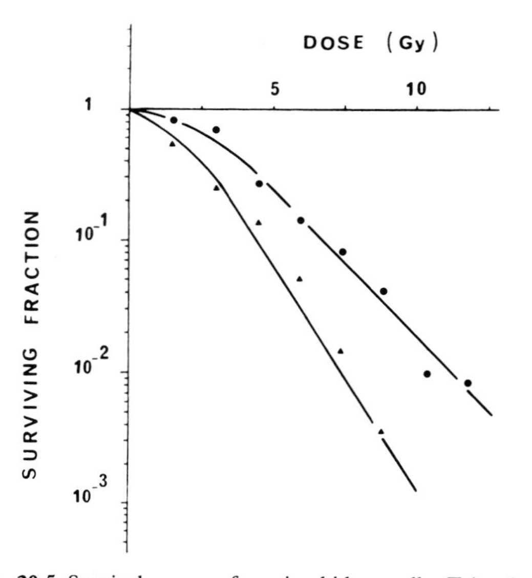


**Fig. 20.5** Survival curves of murine kidney cells. Triangles: cells were plated and then irradiated in vitro 18 hours after plating. Circles: cells were irradiated in vivo in air-breathing mice and plated immediately afterwards.

The survival curve for cells irradiated in vivo and seeded as soon as possible after irradiation was different from the preceding one (Fig. 20.5). We found $n = 3.05$ and $D_o = 1.99$ Gy. It would therefore seem that the radiosensitivity is lower when cells are irradiated in vivo than when they are irradiated in vitro. Although it is not possible to disprove the possible role of a redistribution in the cycle after the initiation of the culture, the most likely interpretation is that difference in radiosensitivity is due to an effect of contact-resistance which may be similar to that observed in certain cell lines grown as spheroids (Durand & Sutherland, 1973). Another possible interpretation is that a fast PLD repair might occur during the few minutes between the end of the irradiation and the beginning of the culture. The difference in radiosensitivity corresponds to a DMF of 1.5 (Fig. 20.5).

When the animals were irradiated 10 minutes after having been asphyxiated by breathing nitrogen, the kidney cells appeared more resistant, the o.e.r. measured at a survival level of $10^{-2}$ being equal to 2.5 (Deschavanne et al, 1980).

In conclusion, by using colonies in vitro it is possible to obtain a PE sufficiently high and stable to permit a study of the intrinsic radiosensitivity of mouse kidney cells. Since mouse kidneys show a radiosensitivity comparable to that of man (Deschavanne et al, 1980), it seems possible that the study of the radiosensitivity of mouse kidney cells might be useful in our understanding of the high radiosensitivity of the human kidney. The mouse kidney cells do not seem to be exceptionally radiosensitive. In addition, they present in vivo a phenomenon of radioresistance which might be interpreted as related to an intercellular contact. They are also the sites of repair of sublethal lesions (Guichard et al, 1982). The high radiosensitivity of the kidney cannot, therefore, be attributed to a high radiosensitivity of these kidney cells. It seems likely that late radiation-induced sclerosis depends on the reduction, below a critical level, in the kidney stem cells. The difficulty in maintaining a permanent culture of epithelial-like cells would seem to support this interpretation. One cannot, however, exclude the possible role of radiation-induced lesions in the connective tissue and in the vasculature.

## ACKNOWLEDGEMENTS

The authors would like to thank Miss Mireille Lahon for her skillful secretarial assistance. This work is supported by La Ligue Nationale Franąise Centre Le Cancer (Comité des-Hauts-dé Seine).

## REFERENCES

Deschavanne P J, Guichard M, Malaise E P 1980 Radiosensitivity of mouse kidney cells determined with an in vitro colony method. International Journal of Radiation Oncology, Biology and Physics 6:1551
Deschavanne P J, Guichard M, Malaise E P 1981 Repair of sublethal and potentially lethal damage in lung cells using an in vitro colony method. British Journal of Radiology 54:973
Durand R E, Sutherland R M 1973 Dependence of the radiation response of an in vitro tumor model on cell cycle effects. Cancer Research 33:213
Glatstein E, Brown R G, Zanelli G D, Fowler J F 1975 The uptake of rubidium-86 in mouse kidneys irradiated with fractionated doses of X-rays. Radiation Research 61:417
Guichard M, Gosse C, Malaise E P 1977 Survival curve of a human melanoma in nude mice. Journal of the National Cancer Institute 58:1665
Guichard M, Deschavanne P J, Malaise E P 1980 Radiosensitivity of mouse lung cells measured using an in vitro colony method. International Journal of Radiation Oncology, Biology and Physics 6:441
Guichard M, Deschavanne P J, Malaise E P 1982 Repair of sublethal and potentially lethal damage in mouse kidney cells in situ using an in vitro colony method. International Journal of Radiation Biology 41:105
Hopewell J W, Berry R J 1974 The predictive value of the NSD system for renal tolerance to fractionated X-irradiation in the pig. British Journal of Radiology 47:679
Kurohara S S, Casarett G W 1972 Effects of single thoracic X-ray exposure in rats. Radiation Research 52:263
Law M P, Hornsey S, Field S B 1976 Collagen content of lungs of mice after X-rays irradiation. Radiation Research 65:60
Maier J G 1972 Effects of radiations on kidney, bladder and prostate. In: Vaeth J M (ed) Radiation effects and tolerance, normal tissue. Basic concepts in radiation pathology. Frontiers of radiation therapeutic oncology. Karger, Basel, p 196
Maisin J R 1970 The ultrastructure of the lungs of mice exposed to supralethal dose of ionizing radiation on the thorax. Radiation Research 44:545
Malaise E P, Guichard M, Deschavanne P J, Noguès C 1982 Radiosensitivity of pulmonary cells measured with a colony assay. Radiation Research 91:308
Phillips T L, Margolis L 1972 Radiation pathology and the clinical response of lung and oesophagus. In: Vaeth J M (ed) Radiation effect and tolerance, normal tissue. Basic concepts in radiation pathology. Frontiers of radiation therapeutic oncology. Karger, Basel, p 254
Phillips T L, Fu K 1975 Derivation of time-dose factors for normal tissues using experimental endpoints in the mouse. In: Caldwell W L, Tolbert D D (eds) Proceedings of the conference on time-dose relationship in clinical therapy. Madison, Middleton p 42
Rubin P, Casarett G 1972 A direction for clinical pathology. The tolerance dose. In: Vaeth J M (ed) Radiation effect and tolerance, normal tissue. Basic concepts in radiation pathology. Frontiers of radiation therapeutic oncology. Karger, Basel, p 1
Soderland S C, Naum Y N 1973 Growth of pulmonary alveolar macrophages in vitro Nature 245:150
Withers H R 1967 The dose-survival relationship for irradiation of epithelial cells of mouse skin. British Journal of Radiology 40:187
Withers H R, Elkind M M 1968 Dose-survival characteristics of epithelial cells of mouse intestinal mucosa. Radiology 91:998

# Assay of colony-forming ability in established cell lines

## HISTORICAL BACKGROUND

Culture of animal cells in vitro has been possible since the early part of this century (Burrows, 1910; Carrel, 1912; 1913) and extensive studies of the nutritional requirements of cultured animal cells (mainly fibroblasts) were made in the early 1950s (Morgan et al, 1950) and by Eagle and his co-workers throughout the period 1950–1960. During this time the absolute requirement of cultured animal cells for at least 13 amino acids and 7 vitamins for continued proliferation in vitro was established. In 1955 Eagle published details of a synthetic medium containing 27 essential growth factors and defined the concentration of each required. However, for optimal growth even of high density cultures, this medium required supplementing by either horse serum (5–10% for mouse fibroblasts) or human serum (10% for human HeLa cells). During this period many investigators were also interested in the clonal growth of mammalian cells in vitro and were puzzled by the observation that conditions which apparently supported optimal growth of cells at high density were far from optimum for the survival, let alone the proliferation, of isolated single cells. A notable experiment which considerably increased our understanding of this phenomenon was performed by Sanford and co-workers in 1948 (Sanford et al, 1948). They demonstrated that the culture media used could support cell multiplication only after conditioning by pre-exposure to metabolically active cells. When only a few cells were placed in a large volume of medium they were unable to condition the medium sufficiently to support multiplication. However, when a single isolated cell was sealed in a capillary tube with a very small volume of medium it was able to condition the medium itself and begin multiplication. This experiment led to the establishment of the now widely used clone 929 of mouse L cells.

Sanford's technique was, however, not suitable for the quantitative analysis of clonal growth of large numbers of single cells which had been subjected to various treatments which was the goal of Dr T. T. Puck and his co-workers at the University of Colorado. Their solution to the problem of conditioning the medium was to use heavily irradiated cells as feeder layers (Fisher & Puck, 1956). Irradiated cells were known at that time to continue to metabolise protein and RNA but not to divide. They demonstrated that when a small number of viable cells was plated together with the irradiated cells, the viable cells were capable of continued proliferation and were able to form well-defined colonies. This improvement in methodology led to the publication of the first survival curve for X-irradiated mammalian cells (Puck & Marcus, 1955). A short time after the clonal growth was achieved with a feeder layer, it became evident that the principal need for the feeder layer was to compensate for the damage done to cells during trypsinisation and preparation of a single cell suspension. Modification of these procedures made it possible to establish clonal growth without feeder layers (Puck et al, 1956).

This early work was performed using 30 per cent whole serum in the medium and in addition the requirements for clonal growth of HeLa cells were known to be less stringent than those of other cells. It soon became obvious from the work of Ham and his collaborators throughout the late 1950s and early 1960s that additional factors were required for clonal growth of other mammalian cell lines. During this period many formulations for synthetic media suitable for clonal growth of other normal and malignant cells were published. Although it is presently still not possible to culture many animal cells in the complete absence of added serum it is now possible to obtain clonal growth of a wide variety of animal and human cell lines and this has led to an explosion in our knowledge of cell biology over the last thirty years.

## METHODOLOGY

### Introduction

Established cell lines, i.e. those which will proliferate indefinitely in vitro, have now been derived from a wide variety of animal and human sources. The listing and/or detailed description of such cell lines is outside the scope of this chapter and individual cell lines will only be referred to where a particular point requires illustration. The reader is referred to the catalogues of the several commercial companies (e.g. Flow Labs, Gibco) now supplying reagents for tissue culture, for available cell lines and the formulations of the many different media now available.

Amongst the established cell lines commonly used, two distinct growth habits prevail. Cells either grow attached to the substratum (glass or plastic) or are anchorage-independent and grow in non-stirred suspension culture. Typical of the former category are the widely used epithelial-like HeLa cell lines, derived from a carcinoma of the cervix, and fibroblastic type V79 and CHO cells derived from Chinese hamster, lung and ovary respectively. Anchorage-independent cell lines include a large number isolated from mouse leukaemias, P388, L1210, L5178Y, to name but a few, and more recently lymphoblastoid cell lines developed from human blood. The methodology for assay of colony survival differs somewhat for these two groups and will be dealt with separately. Some of the points apply to maintenance of stock cultures of both types of cell lines and this will be indicated.

### Cloning

A clone, by definition, is assumed to represent the progeny of a single cell as distinct from a colony which may arise from one or more cells.

Cell lines which attach to glass or plastic can be cloned by plating into petri dishes containing glass coverslips or by the use of cloning rings. In the first method a dilute cell suspension of about 5 cells/ml is prepared, and 5 ml aliquots are distributed into 60 mm diameter petri-dishes each containing several sterile glass coverslips. The dishes are incubated for 7–10 days and then inspected using an inverted phase-contrast microscope for coverslips on which colony growth has occurred. The appropriate coverslip can then be removed under sterile conditions and placed in a second petri dish to allow expansion of the population. Alternatively, cells are plated at low density as above and allowed to grow and form colonies. The medium is then removed from the dish and individual colonies can be isolated by surrounding them with a sterile cylinder (glass or metal) which is fixed to the plastic surface with silicone grease. Trypsin (0.5 ml) is then added to the colony within the cylinder and after cell detachment the suspension is removed with a sterile pasteur pipette. The population can then be expanded (further growth) in any appropriate tissue culture vessel.

Attachment-independent cells can be cloned by plating at low density (~ 10 cells/dish) into semi-solid agar as described later in this chapter. After 7–10 days of incubation the individual colonies can be isolated from the agar using a sterile pasteur pipette. It is usually possible to see well-grown colonies with the naked eye but if desired a low power microscope (×10) can be used. After isolation in this way colonies are added to liquid growth-medium in screw-cap test tubes to allow further growth.

If a cell line will not form colonies in agar or after plating onto a glass or plastic surface it can be cloned by limiting dilution either in microtiter wells or in screw cap test tubes. This is done by diluting the cell suspension to 1 cell/2ml prior to distribution of 2 ml aliquots into, for example, each of 10 culture tubes, or to 4 cells/ml prior to distribution of 250 $\mu$l aliquots into microtiter wells. Tubes or plates are then incubated for 10–14 days and grown colonies can be identified and isolated.

To ensure that the population really has grown from a single cell in all cases it is necessary to reclone by the same method as soon as enough cells from the first isolated colony are available.

### Assay of colony survival in anchorage-dependent cells

The choice of medium for a particular cell line is often somewhat arbitrary and some cells are so 'well trained' that they will grow in a variety of commercially available media with the addition of horse, calf or fetal-calf serum (5–10%). Others, e.g. L5178Y and P388 cells, are more fastidious and require supplementation of a basal type medium, e.g. Eagles MEM with folic acid and asparagine as elegantly demonstrated by Fischer & Sartorelli (1964). Given a cell line growing under conditions which support clonal growth, the following methods should be adopted for quantitative assay of the effect of, for example, X-irradiation on colony-forming ability.

Stock cultures are normally maintained at relatively high density in glass or plastic bottles. It is important for most cell lines that the maximum cell density never exceeds $5 \times 10^5$/ml to ensure exponential growth. This prevents accumulation of polyploid cells, and loss of colony forming ability which occurs in many cell lines if kept at high density. Many cell lines (V79, CHO) can be subcultured routinely from very low inocula e.g. 10 cells/10ml, and where growth from such low inocula is possible, it has the great advantage of reducing the number of subcultures required. It also helps maintain the genetic stability of the cell line. Even under these conditions, cloning, i.e. growth of stock cultures from a single isolated cell, is advisable at approximately three-monthly intervals. For details of methods see above.

Before performing a clonal assay it is important to ensure that cells are in mid-exponential growth, i.e. have reached $(4 \times 10^4) - (2 \times 10^5)$ cells/ml. Under these conditions, plating efficiency, i.e. number of colonies formed from the number of cells plated, is usually maximal, and the age-distribution of cells within the population is likely to be similar from one occasion to the next. This is important as the sensitivity of many cell lines to the cytotoxic effects of drugs and radiation is cell age dependant. For a detailed discussion of age-dependent responses to radiation see Elkind & Whitmore (1967). Preparation of a *single-cell* suspension prior to clonogenic assay is very important and microscopic examination should always be undertaken to ensure that this is achieved after detachment of cells from the stock culture-bottles and prior to plating into test plates. Cells are normally detached by trypsinisation with the minimum possible concentration of trypsin in buffered saline, e.g. Saline A (Table 21.1) which will detach cells after 5–7 minutes incubation at 37°C (0.05–0.25% trypsin is the normal range). There is considerable variation in the concentrations of trypsin but exposure to 0.025 per cent for 5–10 min is normally used for mouse and hamster cells and 0.05–0.1% for 15–20 min for human cells. Prior to trypsinisation the growth medium should be decanted and monolayers of attached cells washed at least twice with serum-free medium or a buffered saline, e.g. PBS or Hanks (Table 21.1), to remove traces of serum which will inhibit the action of the trypsin. Two to three ml of trypsin solution should be sufficient to remove confluent monolayers from a T60 flask. When cells have detached, which should be ascertained microscopically, the action of trypsin should be stopped by addition of medium containing serum (normally 8 ml to give a final volume of 10 ml of cell suspension). After gentle agitation to break up cell clumps (pipetting is usually sufficient) the cell number/ml should then be counted either by haemocytometer or by using an electronic particle counter, e.g. a Coulter counter. If a cell line clumps badly after trypsinisation, suspending in calcium-free medium should eliminate this. Once the cell concentration which should range between $(1 - 5) \times 10^5$ ml, has been determined, appropriate serial dilutions are made to achieve the final required plating-density. For cell lines with a good plating efficiency, greater than 60 per cent, it is usual to plate at a density of 100–200 cells/60 mm diameter dish and 500 cells/90 mm diameter dish in 5 and 10 ml growth medium respectively. This means dilution to a final cell concentration of 20–40 or 50 cells/ml respectively.

**Table 21.1** Composition of saline solutions (g/1)

| Compound | Earle's BSS | Dulbecco's $PO_4$ buffered saline (PBS) | Hank's | Saline A |
|---|---|---|---|---|
| NaCl | 6.8 | 8.0 | 8.0 | 8.0 |
| KCl | 0.4 | 0.2 | 0.4 | 0.4 |
| $NaH_2PO_4H_2$ | 0.15 | – | – | |
| $NaHCO_3$ | 2.2 | – | 1.0 | 0.35 |
| Dextrose | 1.0 | – | 1.0 | 1.0 |
| $CaCl_2$ | 0.2 | 0.075 | 0.2 | – |
| $MgCl_26H_2O$ | – | 0.1 | – | – |
| $MgSO_47H_2O$ | – | 1.15 | 0.06 | – |
| $KH_2PO_4$ | – | 0.2 | 0.06 | – |
| Phenol red | 0.02 | – | 0.02 | 0.02 |

The choice of the number of cells plated per dish of given size must be something of a compromise. The error in estimating the number of colonies per dish decreases as the number of colony-forming units increases. In general, sampling uncertainty can be estimated as the square root of the number of colonies. However, as the initial number of cells per dish is increased there is an increased probability that two cells will be sufficiently close together that the resultant colonies are indistinguishable. We and others (see discussion in Elkind & Whitmore, 1967) have found the figures given above to be a reasonable compromise.

If the effects of radiation or a cytotoxic drug are to be measured, 3–5 plates are normally required for control cultures and 5 for each dose point. Normally 5–6 dose points are needed for each experiment. Once aliquots of the cell suspension have been distributed, the inoculated dishes are incubated in an atmosphere of 5 per cent $CO_2$ in air at 37°C to allow time for cell attachment. The amount of time necessary will depend upon the cell line used. Some cells, e.g. V79, attach within 4 h whereas others, e.g. human diploid fibroblasts, require up to 24 h. It is important to allow cells to attach before further manipulation, particularly if a drug is to be added followed by the subsequent removal of the drug containing medium and washing to remove traces of unreacted drug.

The period allowed between plating of single cells and treatment is important in another respect: if too short a time is allowed, uncontrolled variation in plating efficiency in treated cultures can be a problem, and if it is too long, cells will start to divide and the resultant survival curve will no longer reflect that of an homogeneous single-cell suspension. Attachment can be checked by inspection using a ×10 phase-contrast objective on an inverted microscope. An alternative method of exposure to drugs or radiation is to treat cultures at high cell density, $5 \times 10^5$/90 mm plate, and subsequently to trypsinise and plate at clonal densities. The potential problem with this method is that exposure to drugs may damage cell membranes and subsequent exposure to trypsin can destroy the cells' capability for attachment. Thus, the cytotoxic effects of the treatment may be artificially amplified. After treatment, control and test cultures should be incubated undisturbed in an atmosphere of 5 per cent $CO_2$ in air for a period of time depending on the doubling time of the untreated population, to allow colonies to form. For rapidly proliferating cultures, e.g. V79, CHO cells, this is usually 7 days, whereas for cells with longer doubling-times, e.g.

human tumour cells, the incubation period can be 14–17 days. Considerations of the number of doublings required for scoring a clone as developing from a surviving cell will be discussed in a later section.

**Anchorage-independent cells**

Assessment of the effects of exposure of anchorage-independent cells to cytotoxic agents can be made by plating into semi-solid agar, agarose, carboxy-methyl cellulose or more recently, into multi-well plates.

The same considerations described above for monolayer cultures apply to suspension-grown cells with respect to the maintenance of stock cultures in exponential growth. Exposure to toxic chemicals will require a centrifugation and/or dilution step to remove drug containing medium, therefore cultures are usually exposed to the agent at relatively high density (approx $5 \times 10^5$ cells/ml) and after gentle centrifugation (no greater than 1000 rpm for 10 mins (800 ×G) at room temperature), are diluted in growth medium to the appropriate plating density. One gentle centrifugation followed by a large dilution (1 → 1000) is preferable to repeated centrifugation to remove traces of drug. Centrifugation can damage cell membranes and reduce cloning efficiency. In addition, are cells lost at each centrifugation step and if low cell numbers ($5 \times 10^4$/ml or less) are used initially, it is possible to end up with too few cells when the final plating stage is reached.

*Plating in semi-solid agar*

We use the following method for plating cells in semi-solid agar. A large batch (500 ml) of Difco Bacto Noble agar is prepared as a concentrated solution (5 per cent in double distilled water) distributed in 10 ml aliquots in universal bottles and autoclaved. These aliquots can then be stored at 4°C and used as required. Aliquots of 30 ml of growth medium are then prepared and incubated at 37°C, and the previously-sterilised concentrated agar is heated in a water bath at 90°C until melted. Once melted, 1.5 ml of the hot concentrated agar is added to each 30 ml aliquot of pre-warmed growth medium and after gentle mixing, the appropriate volumes of control and treated cell suspension are added. It is important to keep everything at 37°C to prevent the agar setting, to use wide bore pipettes and to add the hot agar concentrate *before* adding the cells; if done the other way round, the stream of hot agar may cause local cell killing. The agar containing the cell suspension is then distributed into 60 mm diameter plastic petri dishes (5 ml/dish) and the dishes are placed in a tray containing cracked-ice for approximately 5 min to allow setting. This is important because the agar will never set if the plates are put into an incubator at 37°C whilst it is still liquid. The exact concentration of agar which is optimum for clonal growth of a particular cell line has to be determined empirically but is normally in the range 0.15–0.4 per cent. The incubation time for colony formation depends on the cell line, as for monolayer cultures. Colonies are usually scored under a low power dissection microscope (×10).

*Use of carboxymethylase cellulose*

A minicolony assay for CHO cells cloned in suspension in medium containing carboxymethyl cellulose (CMC) was described by Walters et al (1970). It apparently has a number of advantages, but does not appear to have been widely used. Briefly the method is as follows. Equal volumes of autoclaved (120°C, 10 min) carboxymethyl cellulose solution (in distilled water) and double-strength F10 medium were prepared, inoculated with appropriate numbers of cells, dispensed in 17 ml aliquots in disposable culture tubes (18 × 150 mm), and maintained as roller cultures (0.5 rpm) at 37°C. The optimal concentration of CMC was found to be 0.93 per cent w/v. If the CMC concentration was lowered by 0.1 per cent the colonies formed were fragile and irregular, while increasing the CMC concentration resulted in slower growth. After 5–6 days incubation the colonies were counted using an electronic particle counter with an aperture of 127 $\mu$m wide × 270 $\mu$m long, calibrated so that only colonies containing 50 or more cells were counted. Prior to counting the cloning medium was made 0.025M with respect to Na citrate and its pH was adjusted to 4.6 with 0.2N HCl before adding cellulase (3 mg/ml) to reduce the viscosity of the medium. One to two minutes' treatment was found to be sufficient. The cloning efficiency was 90 per cent and the number of colonies formed was proportional to the number of cells inoculated over the range $(5-11) \times 10^3$ cells/ml. Comparable survival of X-irradiated cells using this method and conventional plating of CHO cells on plastic surfaces in liquid medium was obtained. The main advantage of this method is that the numbers of colonies scored are orders of magnitude higher than can be conveniently plated and scored by eye, thus a considerable increase in accuracy can be obtained for high levels of kill.

*Plating on agar surfaces*

A large number of cell lines are also capable of forming colonies when plated on the surface of agar or agarose in petri dishes as described by Kuroki (1973). Plating efficiencies were affected by the source and concentrations of agar or agarose used. Under optimal conditions, plating efficiencies on the agar surface were comparable with those obtained by plating in liquid medium or by agar suspension culture. For L-929 cells and JTC-16 cells, which normally grow attached to the substratum, plating efficiencies were higher on agar surfaces than in liquid medium. A higher plating efficiency on agar plates than in agar suspension was also obtained for primary cultures of Yoshida sarcoma cells.

Briefly, the method used was as follows. a 1.1 per cent solution of agar in water was prepared and after autoclaving and cooling to 45°C was mixed with an equal volume of double-strength growth medium and 10 per cent serum. Aliquots (10 ml for 90 mm diameter dish, 5 ml for 60 mm diameter dish) were then distributed and after setting at 4°C were allowed to stand overnight at room temperature to reduce excess moisture. Sealed plates can be stored several days in a refrigerator. The following day aliquots of cell suspensions (0.2 ml for 90 mm diameter and 0.1 ml for 60 mm diameter dish) were spread evenly over the agar surface. 100 and 500 cells were plated on 60 and 90 mm diameter dishes respectively. Incubation was at 37°C for 2.4 weeks in an incubator gassed with 5 per cent $CO_2$.

*Microtiter plate assay*

Recently clonal assays in microtiter plates have been described for human and mouse lymphomas (Thilly et al, 1980; Cole et al, 1983). Cells harvested from exponentially growing cultures are diluted to either 10 or 3 cells/ml in appropriate growth medium and distributed at 200 $\mu$l/well into 96-well (0.33 ml capacity) micro-titre plates with flat bottoms (Sterilin Ltd) using an Oxford Microdoser-pipette fitted with a 200 $\mu$l piston-barrel assembly and tip. Plates are incubated at 37°C in 5 per cent air for the appropriate time, 7 days for mouse lymphoma cells, 14 days for human lymphoblastoid cell lines. To prevent the outside wells drying out the lid should be sealed with tape. At the end of the appropriate incubation time, positive wells can be scored using a low power disection microscope. Alternatively, a titerteck multiscan automated photometer (Flow Labs) may be successfully used to distinguish positive and negative wells on the basis of pH changes of the medium. One point of caution must be considered when this method is used. Some wells may contain more than one colony forming unit depending on the accuracy of the original dilution and of dispersion into individual wells. This is not a problem with the agar cloning method providing that a good single cell suspension is obtained initially. The method of calculation of plating efficiency (PE) in control and treated samples from microtiter plates will be described in a later section.

*Growth of Chinese hamster cells as spheroids*

A technique has been described by Sutherland et al (1971) in which V79 Chinese hamster cells are cultivated as multicellular 'spheroids' in stirred suspension culture. Exponentially growing cells are harvested from monolayer cultures and $1.5 \times 10^6$ cells put into 250 ml spinner flasks containing 75 ml Eagle's Basal Medium and 5 per cent serum. The spinner was rotated at 180 rpm and 50 ml medium replaced on the 3rd day and on subsequent alternate days. Small aggregates of 5–10 cells were evident during the initial hours of culture which over the next 4 days grew into multicellular spheroids of 150–200 $\mu$m in diameter. The doubling times of the cells were similar to those of cells grown on plastic surfaces. The mitotic index was found to decrease in spheroids greater than 200 $\mu$m in diameter. Individual cells can be released from spheroids by trypsinisation and their survival assessed by conventional plating techniques. Spheroids thus allow the study of the effects of cellular interactions on the survival of individual cells after drug or radiation exposure.

## CONSIDERATIONS ON THE APPROPRIATE DURATION OF EXPOSURE OF CELLS TO DRUGS

The most appropriate duration of exposure of cells in vitro to cytotoxic drugs often depends on the particular question to be asked. However, a number of general principles should be followed. Cytotoxic drugs have varying chemical stabilities in vitro and some exert their cytotoxic effects only on cells in one particular phase of the cell cycle. Thus, if an exponentially growing population is exposed for a short period of time (60 min) to a cytotoxic drug which has its effect only on S phase cells, then irrespective of the chemical half-life of the drug only S phase cells will be killed and a plateau type response will be obtained. Increasing the duration of exposure to one cell cycle or longer will result in exposure of all cells as they progress into S phase, provided that the drug is *chemically* stable under the incubation conditions used. Exposure of an exponentially growing population for a short period (60 min) to a non-cycle-specific drug with a short chemical half-life (less than 30 min) will result in an X-ray-type dose-response curve, and no further cytotoxic effect will be evident if the exposure time is extended. However, if the chemical has a long half-life it may be necessary to compare a short (30 or 60 min exposure) with long (24 h) exposure to determine maximum cytotoxic effects. In general the cytotoxic effects of a non-cycle-specific chemical, with a long half-time for hydrolysis (greater than 120 min), will increase with increasing duration of exposure and analyses of the effects of several exposure times at different external concentrations are necessary for a full understanding of the toxic effects (Fox et al, 1969). Overall, short exposure times are desirable since cells may proliferate during longer exposure times thus complicating the interpretation of survival data. It should also be borne in mind that to produce equitoxic effects short exposures to a drug with a long half-life will require higher concentrations rather than long exposures.

## CONSTITUTION OF A COLONY

It has generally been accepted that the minimum clone size which constitutes a survivor after cytotoxic exposure contains 32 cells, i.e. if a treated cell is able to complete 5 post-treatment cell divisions within a given time then it is likely to have retained an infinite ability to proliferate. In practice the minimum size of clone visible to the naked eye in stained monolayer plates is generally around this size, or even slightly larger, and rigorous examination of smaller colonies for inclusion in total counts is often not undertaken. The effect of dose on the range of colony sizes is also visible to the naked eye in stained plates. In a recently-cloned control culture there is usually little variation in colony size but exposure to cytotoxic agents not only decreases the number of surviving colonies but markedly increases the range of colony sizes. This aspect of cytotoxicity induced by radiation and alkylating agents has been quantitatively analysed by scoring the numbers of cells in individual colonies in plates fixed at various times after treatment (Nias et al, 1965; Nias, 1968). This type of clone size analysis has indicated that the distribution of colony sizes, e.g. after X-irradiation, is not bimodal and therefore that the choice of the 32 cell colony as a cut-off point is rather arbitrary. The distribution of colony sizes will depend on the time chosen for fixation and scoring. This will also influence the number of colonies deemed to be survivors and hence the apparent sensitivity of the exposed cell population. This is illustrated by the data in Figures 21.1 and 21.2. UV-irradiated and EMS-treated V79 cells were fixed either 62 h after treatment or at 7 days. The plates fixed at 62 h were then scored for the colony-size distribution after various UV doses. The data clearly indicate that the absolute sensitivity of the population is dependent on the limit for the cut-off colony size which is chosen. If the 32-cell criterion is applied at 62 h, a time when the untreated population would have undergone 5 cell divisions, then the population appears considerably more sensitive to both EMS treatment and UV-irradiation then when scored at 7 days, and the effect is dose-modifying. Obviously there are a considerable number of slow-growing colonies induced by the treatment which manage to complete 5 or more cell divisions by 7 days.

These data also illustrate the importance of using a number of fixation times together with a standard-size colony cut-off limit when comparisons are being made

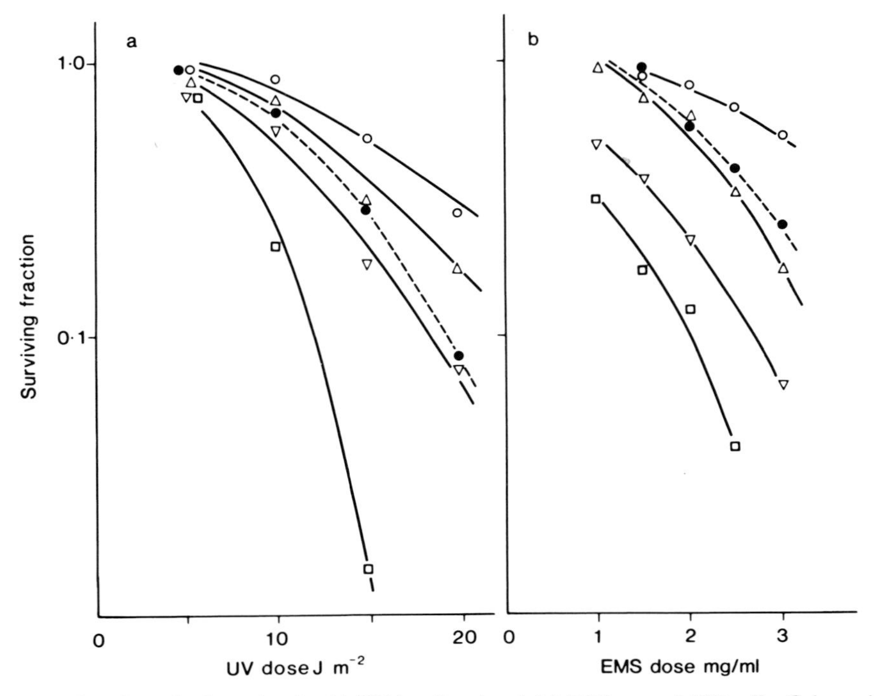


**Fig. 21.1** Survival as a function of colony size for (a) UV irradiated and (b) EMS-treated V79 cells. Colony size distribution was scored on plates fixed 72 h after treatment using a low-power (×10) microscope. Colony numbers in plates fixed at 7 days, were scored by naked eye. Results are expressed as number of colonies scored in treated plates relative to number scored in untreated plates. Symbols: ○——○ > 4 cells, △ > 8 cells, ▽ > 16 cells, □ > 32 cells, all at 72 h. ●——● > 32 cells at 7 days. Data points represent the number of cells in a minimum of 200 colonies on each of two plates at each dose point. M. Fox, unpublished data.

between cell lines suspected to be of differential sensitivity, particularly if they differ markedly in growth rate.

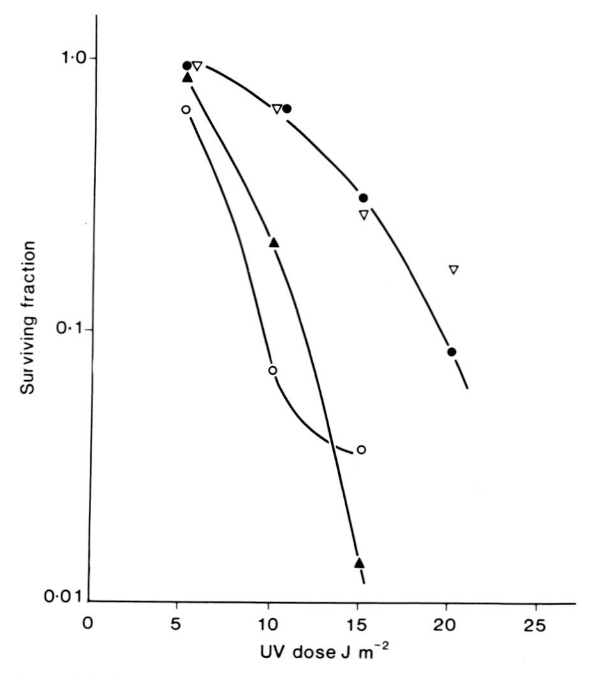


**Fig. 21.2** Relationship between surviving fraction and fixation time for UV irradiated V79 Chinese hamster cells. Colonies containing > 32 cells were scored as survivors on plates fixed at 48 h (○), 62 h (▲), 72 h (▽), or 7 days (●) after treatment. Data points represent a minimum of 200 colonies scored on each of 2 plates for each dose point. M. Fox, unpublished data.

Given that an appropriate fixation time has been chosen, there are several other factors which may influence the actual numbers of colonies scored but which can be controlled with careful attention to methodology.

1. It is important not to inoculate too many cells. This can result in overlapping colonies which makes counting difficult. For cell lines with high plating efficiency (greater than 70 per cent) a maximum of 200 colonies/60 mm diameter and 500/90 mm diameter dish is recommended. For cell lines known to have lower plating efficiencies, the numbers can be increased accordingly. If cells form diffuse colonies, as do human fibroblasts, then the numbers plated may also have to be adjusted.

2. The cells of some cell lines detach easily from the substratum, e.g. from the middle of existing colonies, and form satellite colonies if the cultures are disturbed.

It is possible, however, to select against cells with such undesirable properties, e.g. some V79 cells form colonies which are very convex in vertical section and cells easily detach from their centres. However, clones with a flatter morphology have been selected from such populations by repeated cloning. In some V79 cell lines, daughter cells have been reported to migrate at cytokinesis to a much greater extent than in other cell lines. This phenotype has been deliberately selected in one instance (Newbold et al, 1975), but in most cases it is undesirable.

Satellite colony formation can be a problem in the agar colony-forming assay if the agar concentration is too low and/or the plates are disturbed during incubation, thus care must be taken to ensure that the agar concentration is appropriate for the particular cell line.

## BACK EXTRAPOLATION OF GROWTH CURVES: COMPARISON WITH ASSAYS FOR COLONY-FORMING ABILITY

It is also possible to obtain estimates of toxicity by back extrapolation of growth curves. This method was used widely prior to the development of agar cloning methods (Alexander & Mikulski, 1961) for cells growing in suspension and is still a useful technique for cell lines which have low CE, or which form very loose colonies in semi-solid medium. Briefly, the method is as follows. Cells growing in suspension are treated at high density (1 − 5) × $10^5$/ml then diluted to 1 × $10^4$/ml to allow cell proliferation. Cell counts are made on control and treated populations at 2 day intervals and cultures diluted (1 → 5 or 1 → 10) at appropriate times to maintain the cell density between 1 × $10^4$ and 1 × $10^5$/ml. It is important not to overdilute the treated cultures as this may result in their failure to resume exponential growth. Cell proliferation should be monitored in this way over a period of 10–12 days, or longer if cultures treated with higher doses have not resumed exponential growth. The log of the cumulative cell number (control and treated) is then plotted against time as shown in Figure 21.2 which represents typical data for a human lymphoblast cell line treated with nitrogen mustard (HN2). When treated cultures have recovered and have maintained exponential growth over several days, at a rate comparable to that of untreated cultures, the outgrowth curve can be extrapolated back to zero time and an estimate of the surviving fraction obtained (Fig. 21.3).

It is important to bear in mind when using this technique that it measures only the survival of cells which retain control proliferation rates and hence it is a more stringent end-point than a colony assay. For discussion see Nias & Fox (1968). Slow-growing cells will be either overgrown or diluted out by repeated sub-culture. Also, meaningful back-extrapolation can only be made when outgrowth is parallel in control and treated cultures and when at least four data points are obtained for this part of the growth curve.

Survival as measured by back-extrapolation has been compared with that measured by the colony-forming assay. The data indicated that if HeLa colonies containing 129 cells or more were scored as survivors, good agreement was obtained between the two techniques. For P388 lymphoma cells grown in semi-solid agar, good agreement with back extrapolated growth curves was obtained when the normal colony-scoring procedure was used (Nias & Fox, 1968).

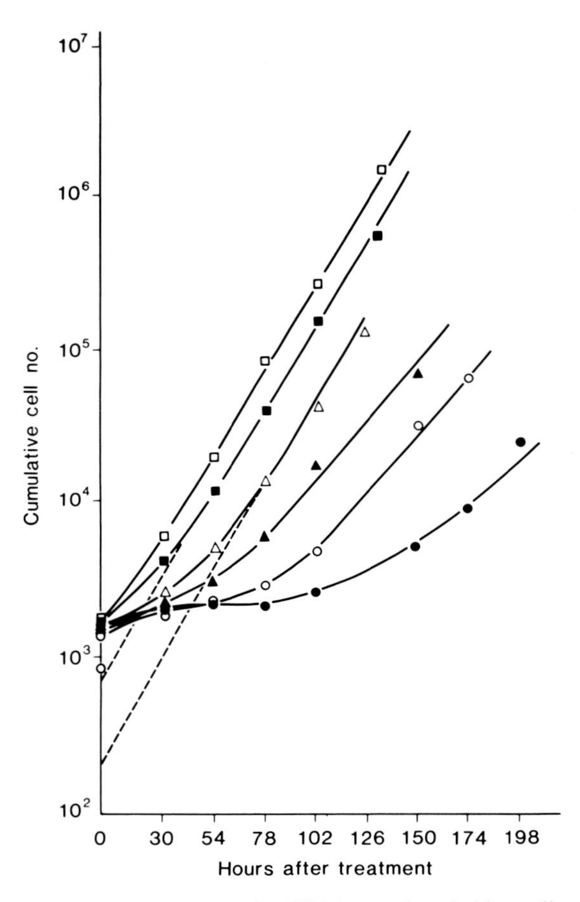


**Fig. 21.3** Growth curves for TK6 human lymphoblast cells untreated or exposed to increasing concentrations of HN2 for 1 h followed by resuspension in drug-free growth medium. Data represent mean cell numbers in three replicate tubes at each time point. Symbols: □ untreated cells ■ 0.02 μg/ml, △ 0.04 μg/ml, ▲ 0.06 μg/ml, ○ 0.08 μg/ml, ● 0.1 μg/ml. Dotted lines represent back extrapolation to zero dose for doses of 0.02 μg/ml (□) and 0.04 μg/ml (△). This yielded values for SF of 0.54 and 0.13 respectively when expressed relative to the zero time for the untreated cells. *N.B.* The non-parallel outgrowth of cells after exposure to higher concentrations of HN2. Back extrapolation of such curves will lead to an over-estimate of SF. S. Dean and M. Fox (unpublished data).

## CALCULATION OF RESULTS AND DATA PRESENTATION

Attached colonies are normally fixed (approx 30 min) with 0.9 per cent saline containing 10 per cent formaldehyde and stained (30–60 min) with 0.04 per cent crystal violet, methylene blue or haematoxylin. Agar and microtiter plates are generally scored unfixed under a low-power microscope. After scoring the mean number of colonies/sample, control and treated, the results are calculated as:

$$\text{Surviving fraction (SF)} = \frac{\text{(no. of colonies on treated plates)}}{\text{(no. of colonies on control plates)}}$$

This assumes that equal numbers of cells have been plated for both groups. If higher numbers of cells have been plated in treated groups then it is necessary to correct for this.

For microtiter plates calculations of the surviving fraction are based on P(O), the proportion of wells in which a colony has *not* grown (Cole et al, 1983):

$$P(O) = \frac{\text{no. of empty wells}}{\text{total number of wells plated}}$$

$$PE = \frac{-\ln P(O)}{\text{no. of cells plated/well}}$$

$$SF = \frac{\text{PE in treated sample}}{\text{PE in control sample}}$$

Classically, data obtained in this way are presented graphically in the form of a semi-log plot of surviving fraction (log) against dose (linear). From this type of plot the conventional parameters of 'n' the extrapolation number, '$D_o$' the slope of the exponential region of the curve and '$D_q$' the quasi-threshold dose, can be derived. Curves describing cell survival data are frequently fitted by eye, and procedures for computer fitting of survival data are described in Chapter 1.

An alternative way of presenting survival data which may be particularly useful when comparisons between cell lines of different sensitivity are made has been recently discussed by Williams & D'Arpa (1981), but was first used by Kao & Puck (1969) and later by Fox (1974). Data were reduced from units of dose to a standardised dose by first estimating $D_o$ or MLH (mean lethal hit) (Williams & D'Arpa, 1981). The abscissa is then transformed (see Ch. 1) into these units by dividing dose by $D_o$. The number of mean lethal hits ($D/D_o$) is then plotted on a linear scale (abscissa) against log survival (ordinate). Once curves for different cell lines, or the same cell line manipulated under different experimental conditions, have been so transformed, an effects ratio (E) can be calculated. This is the ratio between survival of normal cells to that of non-normal or manipulated cells. The data can then be further transformed by plotting log E (effects ratio) against the number of $D_o$s. Using this type of analysis the responses of normal human fibroblasts to various DNA damaging agents have been compared, with the responses of cell lines derived from ataxia telangiectasia, xeroderma pigmentosum and Fanconi's anaemia patients measured after the same spectrum of agents. Survival data for a series of mouse lymphoma cell lines of different X-ray sensitivity are plotted conventionally and after using the $D/D_o$ transformation in Figure 21.4.

## MODIFICATIONS OF MATERIALS AND METHODS

The survival parameters measured in colony-forming assays in vitro should not be considered to be absolute values. They may only represent the response of a particular cell line at the particular point in time when the assay was performed, particularly in the case of exposure to cytotoxic drugs. It is notoriously difficult to reproduce exact survival levels from one experiment to the next after exposure to cytotoxic drugs, even when conditions are controlled as carefully as possible. The variation with time in survival seen in repeat experiments over an approximately 2 year period in which P388 cells were exposed to MMS for 1 h is shown in Figure 21.5. At least 3 different serum batches were used but the variation in response showed no particular relationship with serum batch, in fact the greatest variation was seen between 1:12 and 14.12 when the same serum batch was used. The reasons for the variation are not known but could be ascribed to clonal evolution or genetic drift within this continuously-cultured cell line even when cultures are grown weekly from low dilution and recloned at approximately three monthly intervals. A similar variation in response with time which *was* serum-related was observed when comparisons were made between the responses of the mutagen-sensitive V79/79 cell line and wild-type cells as illustrated in Figure 21.6. Survival of the sensitive cell line after MNU exposure was considerably affected by alteration of the serum batch, but that of wild type cells less so. This could be due to variations in NAD or thymidine content of the serum to which the sensitive cell line responds more markedly. Commercially-available formulations of media vary in their NAD content from 0.2–6.0 mg/l and therefore the choice of medium could be important.

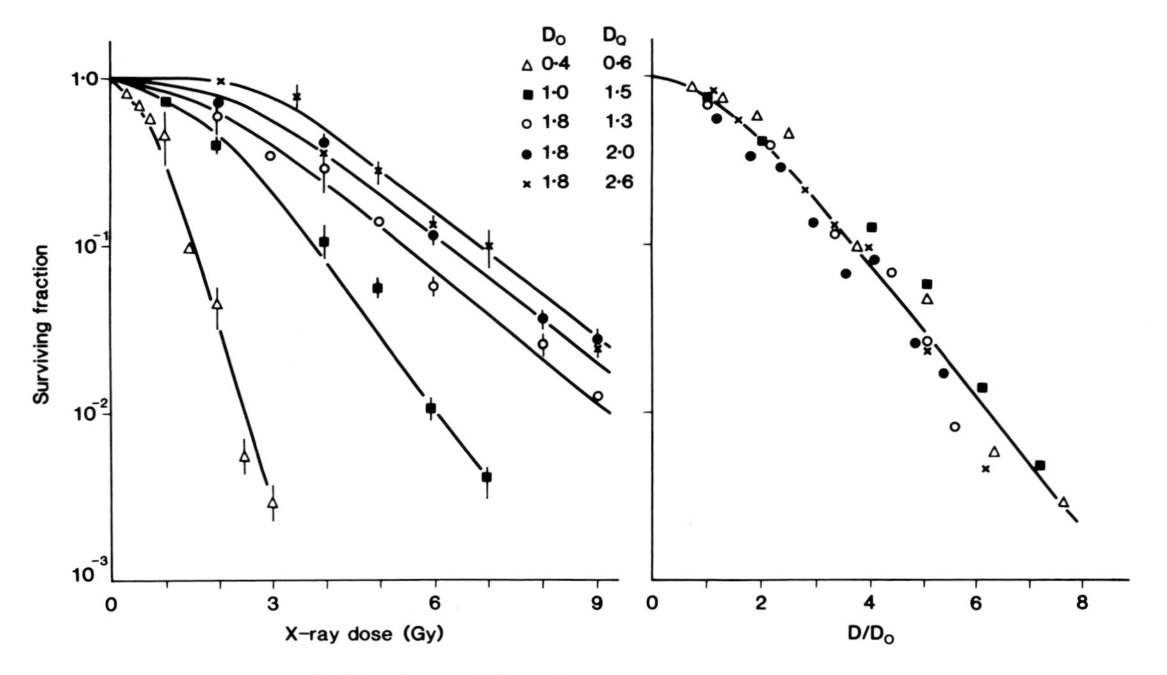


**Fig. 21.4** **a** Dose response curves for 5 lymphoma cell lines after exposure to X-rays. Results are means of at least three experiments for each cell line. Error bars are ± SE of mean. Symbols: △ LS, ■ P388, ○ A11, ● A111, X A1V. **b** Survival data as above, plotted in terms of $D/D_o$ where D is dose and $D_o$ is the dose required to reduce the surviving fraction to 0.37 on the exponential part of the dose-response curve. Data from Fox (1974) reprinted by permission of Elsevier, North Holland.

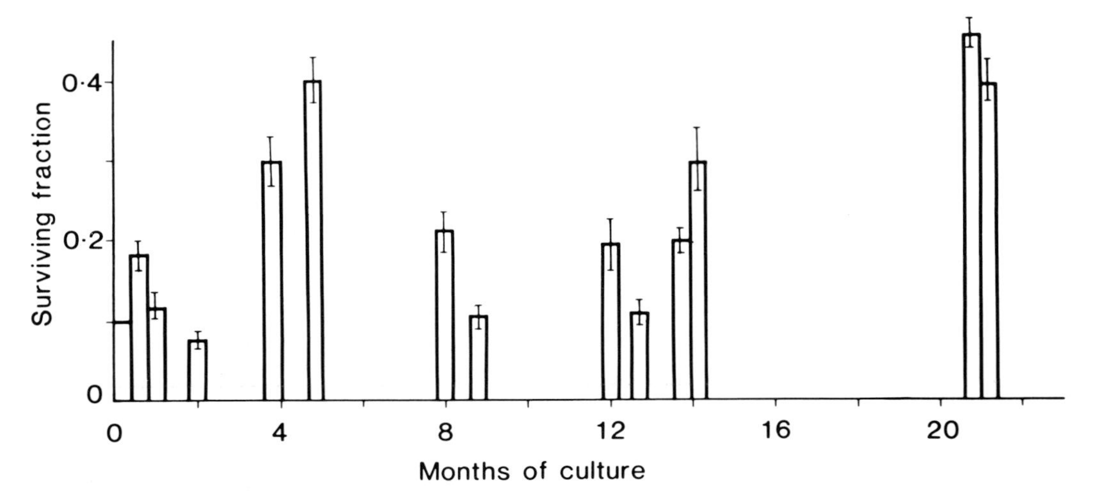


**Fig. 21.5** Variation in survival of P388 mouse lymphoma cells exposed to a single dose of 0.6 mM MMS at various times over a period of 24 months. Survival was measured by plating in 0.25% agar. The data plotted represent the mean colony count on at least five plates expressed as a percentage of the number of colonies on untreated plates and illustrate the variability discussed in the text. Error bars are ± one SD of mean.

There is little possible variation in methods except to alter the duration of incubation prior to fixation which has already been discussed in the section on the constitution of a colony, or to vary the number of cells or number of dishes plated/dose point. Bedford & Hall (1966) demonstrated that if more than $5 \times 10^4$ cells are plated/60 mm dish the surviving fraction decreased as the cell number increased due to overcrowding and competition for nutrients. They recommended therefore that if measurements are to be made at low survival levels, $1 \times 10^{-3}$ or less, then the number of plates should be increased, not the number of cells/plate, and the number of plates used should be adjusted so that 150–200 colonies are scored for each of the lower dose points. Trypsinisation has also been shown transiently to affect the radiation response, largely the $D_o$ value (Berry et al, 1966), so it is important to keep this interval constant to avoid variation but to allow long enough for the cells to both attach and flatten and to recover from trypsin damage to cell membranes. Obviously if long intervals are allowed between trypsinisation and plating, cells may divide and the resultant survival curve will not reflect that of single cells but pairs or clumps of cells.

The increased number of cells (cellular multiplicity) at the time of irradiation influences mainly the size of the extrapolation number which tends to increase as cellular multiplicity increases (Elkind et al, 1961). The relationship between increased cellular multiplicity, $D_o$, and n, of resultant survival curves is however complex, and involves considerations as to whether individual cells are able to survive independently when exposed to cytotoxic drugs in groups. For a detailed discussion the reader is referred to Elkind & Whitmore (1967). However, the problem is best avoided if at all possible.

## INFORMATION TO BE OBTAINED FROM THE TECHNIQUE

The technique has been used extensively since its inception for the study of the responses of a variety of rodent and human cell lines to radiation and cytotoxic drugs. Much effort has gone into the mathematical analysis of survival curve shapes. Information as to the mechanism of cell killing can be inferred from this type of analysis, particularly of radiation survival curves. Care however is needed in the extrapolation of such models to the interpretation of survival curve shapes after drug exposure, as other factors such as drug half-life and cell permeability to the drug will affect the dose reaching the target site.

A variety of closely-related cell lines which show differential responses to the cytotoxic effects of radiation and drugs are now available, and these have been used extensively in the study of repair mechanisms in mammalian cells. The colony-forming assay has also been used extensively in the study of anti-cancer drugs and in recent years in the assay of the responses of mammalian cells to a wide variety of mutagens. Isolation of mutant clones resistant to a wide variety of metabolic inhibitors has greatly increased our knowledge of the biochemistry and genetics of mammalian cells.

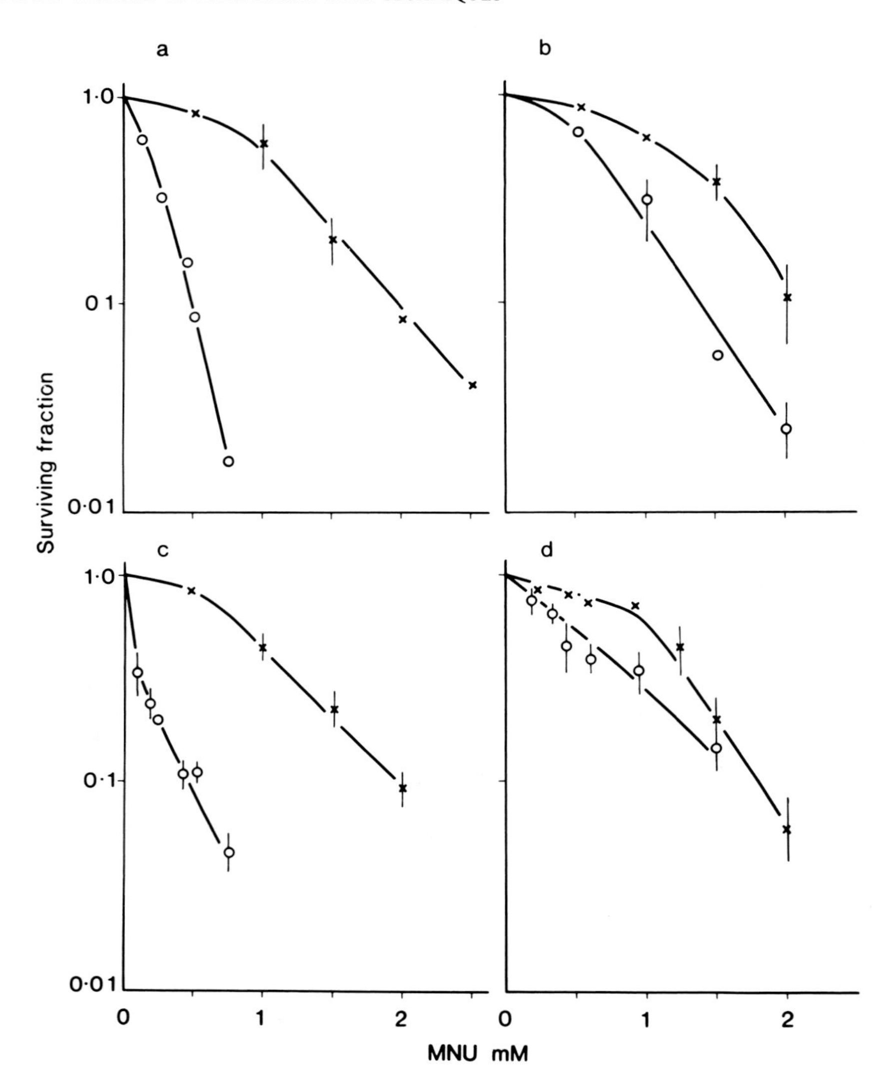


**Fig.21.6** Variation in cellular sensitivity of V79 and V79/79 cells to MNU exposure in relation to serum batch. Symbols: X V79, ○ V79/79. Error bars are ± 1 SE of mean of a number of replicate experiments each one consisting of 5 plates at each dose level. **a**15% donor calf serum 294907, **b** 5% fetal bovine serum 2906109, **c** 5% fetal bovine serum 2927089, **d** 5% fetal bovine serum 2913110.

## REFERENCES


Alexander P, Mikulski Z B 1961 Differences in the response of leukaemia cells in tissue culture to nitrogen mustard and dimethyl myleran. Biochemical Pharmacology 5:275

Bedford J S, Hall E J 1966 On the shape of the dose-response curve for HeLa cells culture in vitro and exposed to gamma-radiation. Nature 209:1363

Berry R J, Evans H J, Robinson D M 1966 Perturbations in X-ray dose response in vitro with time after plating. A pitfall in the comparison of results using asynchronous cell systems. Experimental Cell Research 42:512

Burrows M T 1910 The cultivation of tissues of chick embryo outside the body. Journal of the American Medical Association 55:2057

Carrel A 1912 On the permanent life of tissues outside the organism. Journal of Experimental Medicine 15:516

Carrel A 1913 Artificial activation of the growth in vitro of connective tissue. Journal of Experimental Medicine 17:14

Cole J, Arlett C F, Lowe J, Muriel M 1983 A comparison of the agar cloning and microtitration techniques for assaying cell survival and mutation frequency in L51781Y mouse lymphoma cells. Mutation Research 111:371

Eagle H 1955 Nutritional needs of mammalian cells in tissue culture. Science 122:501

Elkind M M, Whitmore G F 1967 The radiobiology of cultured mammalian cells. Gordon & Breach, New York, p 615

Elkind M M, Sutton H, Moses W B 1961 Post irradiation survival kinetics of mammalian cells grown in culture. Journal of Cellular and Comparative Physiology 58 (suppl 1):113

Fischer G A, Sartorelli A G 1964 Development, maintenance and assay of drug resistance. Methods in Medical Research 10:247

Fisher H W, Puck T T 1956 On the functions of X-irradiated 'feeder' cells in supporting growth of single mammalian cells. Proceedings of the National Academy of Sciences (USA) 42:900

Fox M 1974 Repair synthesis and induction of thymidine resistant variants in mouse lymphoma cells of different radiosensitivity. Mutation Research 23:129

Fox M, Gilbert C W, Lajtha L G, Nias A H W 1969 The interpretation of 'split dose' experiments with mammalian cells after treatment with alkylating agents. Chemical-Biological Interactions 1:241

Kao F T, Puck T T 1969 Genetics of somatic mammalian cells IX Quantitation of mutagenesis by chemical and physical agents. Journal of Cellular Physiology 14:245

Kuroki T 1973 Colony formation of mammalian cells on agar plates and its application to Lederburg's replica plating. Experimental Cell Research 80:55

Morgan J F, Morton H J, Parker R C 1950 Nutrition of animal cells in tissue culture 1. Initial studies on a synthetic medium. Proceedings of the Society for Experimental Biology and Medicine 73:1

Newbold R F, Brookes P, Arlet C F, Bridges B A, Dean B 1975 The effect of variable serum factors and clonal morphology on the ability to detect hypoxanthine guanine phosphoribosyl transferase (HGPRT) deficient variants in cultured Chinese hamster cells. Mutation Research 30:143

Nias A H W 1968 Clone size analysis: A parameter in the study of cell population kinetics. Cell and Tissue Kinetics 1:153

Nias A H W, Fox M 1968 Minimum clone size for estimating normal reproductive capacity of cultured mammalian cells. British Journal of Radiology 41:468

Nias A H W, Gilbert C W, Lajtha L G, Lange C S 1965 Clone size analysis in the study of cell growth following single and during continuous irradiation. International Journal of Radiation Biology 9:275

Puck T T, Marcus P J 1955 A rapid method for viable cell titration and clone production with HeLa cells in tissue culture: The use of X-irradiated cells to supply conditioning factors. Proceedings of the National Academy of Sciences (USA) 41:432

Puck T T, Marcus P I 1956 Action of X-rays on mammalian cells. Journal of Experimental Biology and Medicine 103:653

Puck T T, Marcus P I, Cieciura S J 1956 Clonal growth of mammalian cells in vitro. Growth characteristics of colonies from single HeLa cells with and without a 'feeder' layer. Journal of Experimental Medicine 103:273

Sanford K K, Earle W R, Likely G D 1948 The growth in vitro of single isolated tissue cells. Journal of the National Cancer Institute 9:229

Sutherland R M, McCredie J A, Inch W R 1971 Growth of multicell spheroids in tissue culture as a model of nodular carcinomas. Journal of the National Cancer Institute 46:113

Thilly W G, DeLuca J G, Furth E E, Hoppe H, Kaden J J, Krolewski J J, Liber H L, Skopek T R, Slapikoff R J, Tizard R J, Penman B W 1980 Gene-locus mutation assays in diploid human lymphoblast lines. In: de Serres F J, Hollander A (eds) Chemical mutagenesis. Plenum, New York, vol 6

Walters R A, Hutson J Y, Burchill R R 1970 A new method of cloning mammalian tissue culture cells in suspension. Journal of Cellular Physiology 76:85

Williams J R, D'Arpa P 1981 Epigenetic and genetic factors in cellular respone to radiations and DNA-damaging chemicals. Journal of Investigative Dermatology 77:125

# Primary cultures from tumours

## HISTORICAL BACKGROUND

### Monolayer culture

Attempts to grow tumour cell colonies in monolayer culture by methods appropriate for established cell lines are rarely successful. Apart from their more demanding growth requirements there are a number of reasons for failure. Some types of tumour cell e.g. small cell tumour of lung, may fail to attach to the surface of culture dishes. Growth may be very slow and discrete colonies may not be formed. Cell suspensions from disaggregated tumours contain a mixture of tumour cells and non-tumour cells that form the stroma. The latter may overgrow the tumour cells or sometimes interfere with tumour cell colony formation through cell-mediated cytotoxicity. Stromal cells often form a significant proportion, or even the majority, of cells in the suspension and it is necessary to have some means of distinguishing between tumour and stromal cell colonies.

### Soft agar culture

Virtually all colony assays systems for tumour cells in primary culture depend on the use of soft agar or agarose. Macpherson & Montagnier demonstrated in 1964 that transformed cells had on ability to form colonies in agar not found in the normal cells from which the transformed cells were derived. Hence, colony formation in agar is now one of the accepted criteria of tumorigenicity. Growth in soft agar therefore provides a valuable tool for discriminating between tumour cells and stromal cells such as fibroblasts which are anchorage dependent and unable to form colonies in agar. The double layer agar technique using 35 mm Petri dishes was developed by Bradley & Metcalf in 1966 for the growth of colonies from normal bone marrow. Marrow cells were suspended in an upper layer of 0.3 per cent agar above a thicker layer of 0.5 per cent agar. The method was later modified by Park et al (1971), for the growth of mouse myeloma cells, by incorporating a feeder layer of kidney tubules in the under layer and replenishing the nutrients in the agar by allowing liquid medium placed on top of the agar layers to percolate down through the agar and escape via holes in the bottom of the dish. Since then double layer methods without replenishment of the medium have been used by many workers, including Thomson & Rauth (1974) and Courtenay (1976) for the growth of colonies from animal tumour cells, and Hamburger & Salmon (1977) who obtained colonies from human tumour biopsies. The Petri-dish method has the advantage that colonies can be counted in situ and this may be the method of choice for some rapidly growing cells capable of forming 50 cell colonies within about 12 days. However, after this time colony growth tends to slow down or stop due to deterioration and depletion of the medium. Most tumour cells in primary culture grow less rapidly, particularly those obtained directly from human biopsy material, and in order to maintain colony growth up to 50 cells per colony it is necessary to replenish the nutrients in the agar. This is difficult to achieve in Petri dishes and a simple method using test tubes was developed by Courtenay (1976). With this technique (Courtenay & Mills, 1978) tumour cells are suspended in a hemisphere of agar in the bottom of the test tube and liquid medium is pipetted on top and changed as necessary without breaking the agar. The method, which will be described in detail in this chapter, has been used in a variety of studies on xenografted human tumour cells and on human tumour biopsies. It has been found to give plating efficiencies (PE) higher than were attainable by previous methods (Tveit et al, 1981b).

### Culture medium

In colony assays, the necessity for initiating cultures from suspended cells, separated from each other by large distances, imposes highly artificial growth conditions on cells taken directly from a solid tissue and previously in 3-dimensional contact with their neighbours. To enable such cells to survive and grow it is necessary that culture conditions should, as far as possible, simulate the biochemical and gaseous environment of the tissue from which they were derived.

The more comprehensive media such as Ham's F12m or McCoy's 5A medium can supply the principal nutrients and cell metabolites present in tissue fluid, while the addition of fetal calf serum provides growth factors and other substances lacking in culture medium. Feeder layers including red blood cells (RBC) may provide other possibly labile substances lacking in stored serum. Other additives, including hormones, may be necessary for some types of tumour. Bicarbonate-buffered media are generally used in preference to HEPES buffer which may be toxic to some tumour cells. The correct pH is maintained by incubating with a gas phase containing 5 per cent $CO_2$.

**Oxygen concentration**

Using standard culture conditions, it has been a general observation, both in monolayer and in agar, that primary cultures of tumour cells fail when plated out at low cell densities, but may survive when the number of cells is increased. A density of between $2 \times 10^5$ and $5 \times 10^5$ cells/ml has therefore been commonly used. However, for colony assays such high cell numbers may lead to difficulties due to early exhaustion of the medium, limiting the growth of potentially clonogenic cells. The need for high cell densities has been thought to be related to the conditioning of the medium by the release of various normal metabolites by the cells thus modifying the biochemical constitution of the medium. But another consequence of metabolic activity is the uptake of oxygen by the cells. In static cultures at high cell density this may lead to a local reduction of $O_2$ within the medium. $CO_2$, which diffuses more rapidly in culture medium than $O_2$, is also produced.

Studies by Courtenay (1976) on primary cultures of Lewis lung mouse carcinoma cells showed that they could be grown at lower cell densities if they were incubated with a gas phase containing 5 per cent $O_2$ instead of air. Increased $CO_2$ concentration or conditioned medium alone did not improve survival. Subsequent studies on many other tumours have consistently shown improved colony formation when cultures are gassed with mixtures containing 3–5 per cent $O_2$. The importance of low $O_2$ concentrations for the growth of some normal cells was demonstrated by Richter et al (1972), studying mouse embryo cells in primary culture. The use of low oxygen concentrations provides a physiological condition closer to that normal for cell growth since the oxygen concentration in venous blood is equivalent to 5 per cent $O_2$, much lower than the 21 per cent $O_2$ in air. In solid tumours the majority of tumour cells are in a microenvironment with $O_2$ concentrations ranging down to hypoxic levels.

**Rat red blood cells (RBC)**

The growth-promoting action of lysed RBC was first demonstrated by Bradley et al (1971) who showed that the growth of colonies from normal mouse marrow in agar was enhanced by the addition of washed RBC from rats or mice but not sheep. Using RBC from the August rat strain, we have found a consistent enhancement of colony formation by tumour cells (Courtenay, 1976; Tveit et al, 1981a). The growth-promoting substances in RBC are evidently partly labile since suspensions of whole RBC which lyse after addition to the agar are more effective than lysates. The nature of the growth factors involved has been studied by Bertoncello & Bradley (1977) and Kriegler et al (1981) and it is apparent that two or more different growth factors may be released.

From our observations the RBC factors aid in the initiation of colony formation and to be effective it is necessary that lysis occurs within the first 5–7 days of setting up the cultures. The action of RBC does not appear to be related to the ability of haemoglobin to take up oxygen because intact RBC are ineffective. Conditioned medium from tumour cell cultures has proved to be ineffective as a possible alternative to RBC and in our hands has most often reduced the plating efficiency (PE).

## DESCRIPTION OF THE TECHNIQUES

**Cell suspensions**

The preparation of good viable single-cell suspensions is essential. The most appropriate method will vary depending on the nature of the particular tumour concerned. Mechanical disaggregation may be suitable for a small proportion of tumours. Here the specimen is forced through a mesh or passed through syringe needles of decreasing size. However, for most tumours cautiously applied enzyme methods (Courtenay, 1983) give higher yields of viable cells. With some experimental animal tumours, trypsin may give good preparations but for human tumours other enzymes, particularly collagenase, alone or in combination with pronase and DNase or hyaluronidase, are less destructive to the cell membrane. Cell suspensions should be prepared as soon as possible after excision of the tumour. In the case of human biopsies, these should be transported in medium at 4°C and set up in agar within 24 h.

**Agar colony assays**

The two agar colony methods in vitro to be described here differ from previous methods using tumour cells in the use of low oxygen concentrations in the gas phase and the admixture of rat RBC with the tumour cells.

*Method I* (Petri dishes)

This is a double-layer agar technique (Fig. 22.1A) and was developed for assaying cell survival in rapidly growing animal tumours capable of forming colonies of over

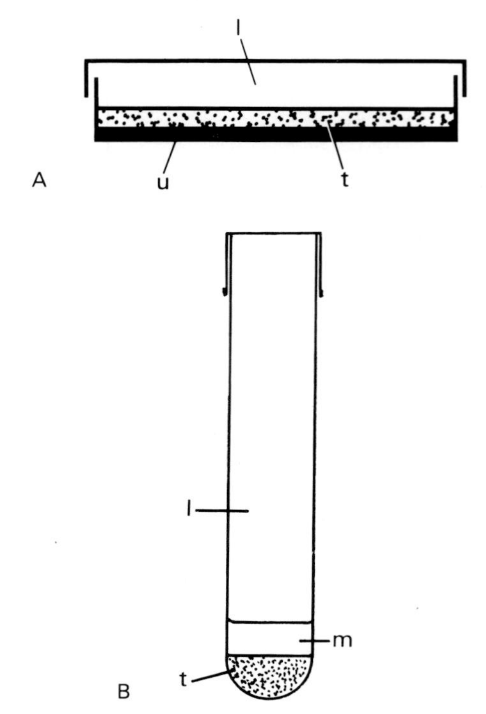


**Fig. 22.1** Agar colony techniques. **A** Petri dish method, **B** test tube method; l, gas phase of 3–5 per cent $O_2$ + 5 per cent $CO_2$ + 92–90 per cent $N_2$; t, tumour cells + RBC in 0.3 per cent agar; u, underlayer of 0.5 per cent agar; m, liquid medium.

50 cells within two weeks of incubation (Courtenay, 1976). It has been extensively used in studies with the Lewis lung mouse carcinoma and the B16 mouse melanoma.

To form the underlayer, 0.6 ml molten 0.5 per cent agar in Ham's F12m medium containing 15 per cent fetal calf serum is pipetted into a 35 mm Petri dish and allowed to set. The upper agar layer is mixed separately. One volume of washed August rat RBC, diluted $\frac{1}{8}$, is mixed with one volume of $5 \times 10^4$ tumour cells (including lethally irradiated (HR) cells when necessary) and warmed to 37°C. Three volumes of molten 0.5 per cent agar medium at 44°C are added and mixed. 1 ml aliquots are immediately pipetted on top of the underlayers in replicate dishes. After the agar has set, the dishes are incubated at 37°C in clear plastic boxes (humidified, with dilute $CuSO_4$ solution) and gassed with 5 per cent $O_2$ + 5 per cent $CO_2$ + 90 per cent $N_2$. The detailed procedure for making up the agar and preparing the RBC is given under Method II.

Colonies are counted in situ under a dissecting microscope.

*Method II* (test tubes)
No underlayer is required in this method as shown in Figure 22.1B. Tumour cells and RBC are mixed and suspended in molten 0.3 per cent agar as for method I. Aliquots of 1 ml are pipetted into each tube and gelled rapidly in crushed ice. This method is easier and quicker to set up than method I and the danger of mould growth is greatly reduced since each tube is individually capped.

The tubes are incubated in a gas mixture of 3 per cent $O_2$ + 5 per cent $CO_2$ + 92 per cent $N_2$. At one and two weeks, 1 ml of liquid medium is added and then at three weeks (and four weeks if necessary) about 1 ml is removed and replaced with a further 1 ml of medium. When the majority of colonies in untreated control tubes exceed 50 cells they are counted under the microscope by decanting the agar onto a slide or Petri dish and placing a cover glass on top, thus bringing colonies into the same focal plane. Details of the practical procedures are given below.

## Preparation of RBC

*a. Materials*
August rats; 10 ml syringe with No. 12 needle; heparin (preservative free); PBS; Ham's F12m medium with 15 per cent serum and antibiotics; universal containers.

*b. Procedure*
1. With sterile precautions take up sufficient heparin into the syringe to fill the needle.
2. Anaesthetise the rat with ether and soak the fur of the thorax with methylated spirit.
3. Withdraw blood by cardiac puncture using the syringe with heparin and eject into a universal container.
4. Mark the level of the meniscus and centrifuge at 3000 rpm for 15–20 min.
5. Remove the buffy coat together with the supernatant and discard them.
6. Rinse the RBC twice with PBS and resuspend to the original blood volume in medium with serum.
7. The RBC may be stored for up to three weeks at 4°C.
8. Blood used within seven days of collection must be heat-inactivated for one hour at 44°C.

## Setting up agar cultures

*a. Materials*
17 × 100 mm sterile test tubes (Falcon 2051); metal racks in tray containing iced water; water baths at 37°C and 44°C; culture medium; Ham's F12m + 15 per cent fetal calf serum; 5 per cent agar (Difco, Noble Agar) in PBS; washed August rat RBC suspended in culture medium to original blood volume; tumour cell suspension.

*b. Procedure*
1. Dilute the tumour cell suspension with culture

medium to five times the required final concentration.

2. Dilute the washed RBC $\frac{1}{8}$ with medium.
3. Heavily irradiated tumour cells (HR) (omitted when there are more than $5 \times 10^3$ tumour cells per tube) are exposed to 40 Gy of X or $\gamma$ radiation and diluted to $1 \times 10^5$ cells/ml.
4. To set up five replicate tubes, measure out in a test tube 1.2 ml tumour cell suspension + 0.6 ml RBC + 0.6 ml HR cells (A).
5. Prepare agar
   (i) To make 50 ml of 0.5 per cent agar medium, weigh 250 mg agar in a bijou bottle and add 5 ml PBS.
   (ii) Warm 45 ml culture medium to 44°C in a water bath.
   (iii) Sterilise five per cent agar solution by boiling in a water bath for 10 mins (with the caps loose to permit escape of steam).
   (iv) Remove from the bath, tighten the cap and swirl to mix, and cool to 60°C.
   (v) Stand in a water bath at 44°C* for one min and then add to medium at 44°C and mix thoroughly.
6. Warm the mixed cells (A) to 37°C.
7. Add 3.6 ml of 0.5 per cent agar medium and mix by inverting the capped tube.
8. Pipette out 1 ml aliquots into five vertical Falcon tubes in a rack in an ice bath.
9. When the agar has set (five min) the loosely capped tubes in racks are placed in a gassed incubator at 37°C with 3 per cent $O_2$ + 5 per cent $CO_2$ + 92 per cent $N_2$.
10. 24 h later the caps are tightened to prevent evaporation.

*c. Alternative gassing systems*

Culture tubes may be incubated in gas-tight containers such as polystyrene lunch boxes (Stewart Plastics) or modular incubator chambers (Flow Laboratories). For this method, both individual culture tubes and containers are gassed.

1. Uncap the vertical tubes in racks and gas the individual tubes at a rate of 2 l/min for 4 s using a vertical pasteur pipette.
2. Immediately press on caps in the 'gas tight' position.
3. Transfer tubes, which must be kept within 45° of vertical, to lunch boxes and seal box lids with polythene sellotape.
4. Gas the boxes at 2 l/min for 10 mins via two holes on opposite sides of the box and then seal the holes with polythene sellotape.

**Maintenance of cultures**

1. After 5–7 days, gently pipette 1 ml medium on top of the agar.
2. Repeat at 14 days.
3. At 21 days, withdraw 1 ml of 'old' medium and replace with 1 ml 'new' medium.

**Colony counting (usually at 21 to 28 days)**

1. Discard the medium and decant the agar onto a slide or shallow dish.
2. Cover with a 25 mm × 50 mm cover glass and count colonies containing more than 50 cells under an inverted microscope.

**Cell survival**

In each experiment the control PE is calculated as a percentage from the number of control colonies produced per 100 cells plated out. Surviving fraction (SF) for each dose is calculated from

$$SF = \frac{\text{number of colonies produced by treated cells}}{\text{number of treated cells plated out}} \times \frac{100}{\text{control PE}}$$

**Agar diffusion chambers**

An alternative method of growing colonies in soft agar employs agar diffusion chambers (ADC) implanted in the peritoneal cavity of mice. The mouse acts as an incubator and nutrients necessary for cell growth are supplied by diffusion from the mouse peritoneal fluid. The method was first developed by Gordon in 1974 for the growth of bone marrow cells and was subsequently used for the growth of tumour cell colonies by Smith et al (1976).

ADC are made as shown in Figure 22.2 by cementing a millipore filter (pore size 0.22 $\mu$m) to an acrylic ring by the solvent action of acetone. A circular glass coverslip is attached to the other side of the ring and the chambers are sterilized by exposure to 25 kGy $^{60}Co$ $\gamma$-rays or an equivalent radiation. A hole in the ring allows injection of the cell suspension into the chamber.

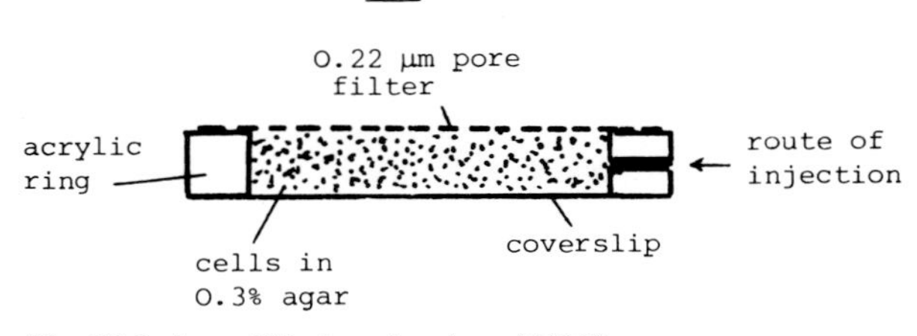


**Fig. 22.2** Agar diffusion chambers (ADC).

* 0.5 per cent is a nominal concentration. A small amount of agar remaining in the bottle can be ignored. Providing the agar is firm enough to remain intact with careful handling, colony growth is generally better at final agar concentrations below 0.3 per cent

Before injection into the ADC, molten agar is added to the cell suspension to a final concentration of 0.3 per cent. The chamber is filled to capacity (350 μl) and the agar is allowed to set at room temperature. The injection hole is sealed with paraffin wax. Single chambers are then inserted into the peritoneal cavity of a mouse. To prevent the immune rejection of the foreign tumour cells it is necessary either to use genetically athymic mice as hosts, or to pretreat normal mice with 9.0 Gy whole body irradiation from a $^{60}$Co source at a standard time of 3 h before implantation. This dose is generally lethal after 10–12 days, requiring transfer of the chambers to fresh hosts after about nine days unless radiation death is prevented by i.v. injection of normal mouse bone marrow immediately after irradiation, or by i.p. injection of 200 mg/kg cytosine arabinoside two days before irradiation (Smith et al, 1976).

With this system colony growth is sustained for up to three weeks. For colony counting the chamber is removed from the peritoneal cavity, wiped clean with a tissue and the millipore filter carefully peeled off the agar. Colonies are counted under the microscope.

The method has the advantage that no tissue culture facility is required and colonies may grow slightly faster.

**Spheroids**

Unlike colonies in agar, spheroids are not clonal in origin but are formed by the aggregation of tumour cells from a suspension. They are maintained in spherical form by the continuous whirling of the culture medium. They can be grown to diameters of up to 1000 μm and may consist of hypoxic and oxic zones and of cycling and non-cycling cells (Sutherland & Durand, 1976). They have been used as models of solid tumours in vitro in radiotherapeutic and chemotherapeutic studies.

## CONSTITUTION OF A COLONY AND PROBLEMS ASSOCIATED WITH ITS MEASUREMENT

The starting premise for a valid colony assay is that the initial cell suspension should consist of single cells held in position in the agar so that the subsequently-formed colonies are derived from the repeated division of one initial cell. With few exceptions tumour cell colonies are compact as shown in Figure 22.3 with each cell in direct contact with its neighbour. The tightness of contact in colonies of different tumour cell types varies but it is usually possible to distinguish individual cells under a good microscope, with a x20 objective. Occasionally diffuse colonies are formed by normal blood cells, particularly macrophages, which can be identified under the microscope from the fact that the cells tend to move apart through the agar with rarely more than two cells in direct contact.

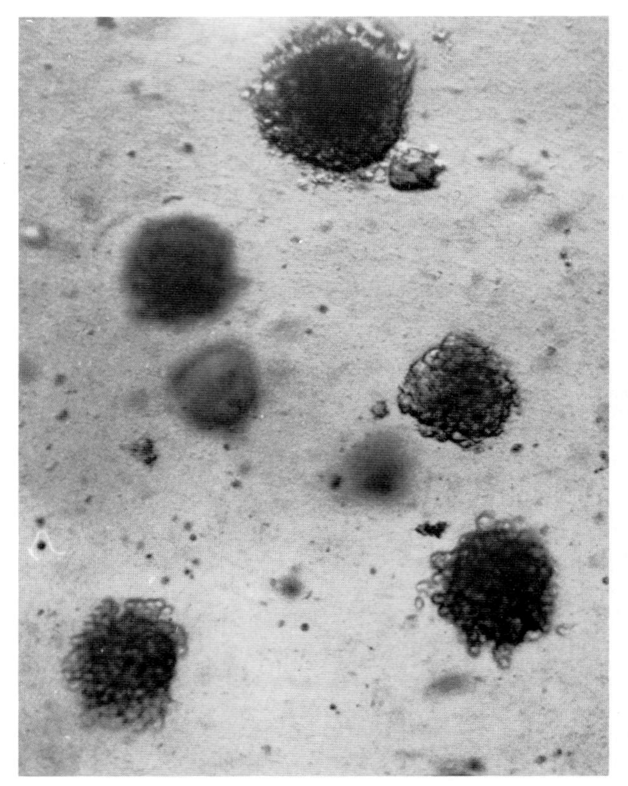

**Fig. 22.3** Human melanoma colonies in agar (test-tube method).

For a given number of cells per colony the diameter of the colony will depend on the average size of the cells and before starting a count on an unknown tumour it is necessary to determine the diameter corresponding to 50 cells per colony. Tumour cell colonies originating from a single cell are approximately spherical and only the cells on the upper or lower surface can be readily counted under the microscope (about 18 cells for a 50 cell colony). The size criterion selected must be rigidly adhered to throughout the counting procedure.

## PITFALLS IN EXISTING PROCEDURES

**Cell suspensions**

One of the major problems in growing colonies from primary tumour material is that of obtaining good single-cell suspensions that are representative of the original tumour and that are not irreparably damaged in the process of disaggregation. Mechanical methods which rely on cell spill-off from cut-up pieces of tumour tend to give an unrepresentative number of dead cells (Pavelic et al, 1980) together with a high proportion of blood cells. Forcing tumour pieces through hypodermic needles of decreasing bore exposes the cells to considerable trauma and can be successfully used on only a small proportion of soft tumours. Teasing also tends to disrupt cells. For most tumours, enzyme methods are to be preferred and generally give much higher cell yields but overtreatment can damage cell membranes and may alter drug sensitivity (Barranco et al, 1980; Rasey & Nelson, 1980).

Viability should always be assessed by examination using phase microscopy assisted by a dye exclusion method.

Whatever method of disaggregation is employed some clumps must be expected to remain. Filtration through a 30 $\mu$m mesh excludes large clumps but small clusters and occasional 'strings' of cells may pass through and clusters of up to eight cells may still be found on inspection of the resulting suspension. Such clusters may 'seed' colonies.

### Cell clumps

The growth of cells in agar has been observed by Agrez et al (1982) taking serial photomicrographs of agar cultures from 30 different human tumours grown by the method of Hamburger & Salmon (1977) omitting conditioned medium. They found that cell aggregates of 50 cells or more were indistinguishable from colonies and when counted one to three weeks after plating were found only in locations in which a cluster of cells had been seen on day one. Their results demonstrate that when the plating efficiency is low it is difficult to exclude the possibility that the clonogenic assay may in fact be a 'clumpogenic' one. Where large numbers of cells are available it may be possible to remove most of the clumps by some kind of gravity sedimentation or cell elutriation technique. The finding that single cells did not form 50 cell colonies within three weeks, although some had gone through two to four divisions (Agrez et al, 1982), stresses the importance of maintaining active colony growth for four weeks or more.

### Abortive colonies

The 50 cell criterion for colony counting was set by Puck & Marcus (1956) when they first introduced colony assay techniques, and it has been used in experimental studies ever since. Nias et al (1965) demonstrated that 50 cells was a minimum number since some lethally — irradiated HeLa cells were shown to be capable of completing six divisions before death and so producing abortive colonies of up to 64 cells. Abortive colonies may also be produced after treatment with some drugs. In work on human tumour cells there has been a tendency, because of this slow growth rate, to count 30-cell colonies, but there is a danger that cell survival may be overestimated in some cases.

### Cell number

If viable cells are plated out at too high a cell density, nutrients in the medium may become rapidly depleted, inhibiting cell growth. Also, particularly in the test tube method, cells towards the bottom of the agar may become hypoxic due to the metabolism of oxygen by the cells above them (Fig. 22.4). Generally the maximum number of viable tumour cells that can be safely plated out is about $1 \times 10^4$. If more than 200 colonies are formed, colony size and plating efficiency may be reduced. For tumour cells which have a high plating efficiency additional cell dilutions should be plated out and the total cell number made up to $1 \times 10^4$ with heavily irradiated tumour cells.

**Fig. 22.4** Effect of cell density on colony formation in deep agar gassed with 5 per cent $O_2$. Cells from the HX34 xenografted melanoma plus RBC were set up in 0.3 per cent agar in 6 mm diameter tubes. Colony formation was optimal with $4 \times 10^3$ cells/ml at a depth of 4.5 to 9.0 mm. With $4 \times 10^4$ and $4 \times 10^5$ cells/ml colonies were small and overcrowded and there were no colonies below 2.3 and 1.5 mm due to hypoxia.

### Gelling of agar

If the agar does not set or becomes broken by rough handling, few colonies will be obtained. When making up the agar medium it is important that the five per cent agar does not begin to set before it is added to the medium or the final mix may not set properly.

## ACCURACY OF THE TECHNIQUES

Agar assay techniques are capable of giving accuracies comparable to those from methods in vivo. Method I was first used by Shipley et al (1973) in radiation studies on the Lewis lung mouse carcinoma. Tumours were irradiated in situ in the air breathing mouse using $^{60}$Co γ rays. Tumours were immediately excised and assayed either in vitro or by the in vivo lung colony assay (Hill & Stanley, 1975). The $D_o$ of hypoxic cells in the tumour was measured from the slope of the survival curve measured below 10 per cent survival. The $D_o$, determined by linear regression analysis, was 3.09 Gy and 3.22 Gy for the in vitro and lung colony assays respectively, a difference that was well within the calculated range of experimental error of 2.90 and 3.47 Gy. Selby & Thomas (1980) measuring the drug sensitivity of the melanoma HX34 obtained similar results by the ADC and lung colony methods.

For any particular tumour the accuracy of results is mainly dependent on the plating efficiency that can be obtained. This may be in the range of 20–50 per cent for certain animal tumours and a few selected human tumours grown as xenografts in immune deficient mice. In these tumours a representative proportion of the tumour cells can be assayed and survival accurately measured down to two or three decades of kill. Such tumours provide valuable experimental models. Figure 22.5 shows radiation survival curves (Courtenay et al, 1976) obtained from a human pancreatic tumour xenograft irradiated in situ in the mouse and assayed either within 10 minutes or 18 h after irradiation. A PE of approximately 30 per cent was obtained from this tumour. Cells that remained in the intact tumour for 18 h showed in situ repair of potentially lethal damage giving a higher survival than those from tumours that were disaggregated directly after irradiation. In these studies parallel measurements using ADC gave similar results to the agar test-tube method (Fig. 22.4). From the hypoxic region of the survival curve using the two sets of data a value for $D_o$ of 3.05 ± 0.16 Gy was obtained. Drug sensitivity has also been studied in the same xenografted tumour treated in vitro (Bateman et al, 1979) or in vivo (Courtenay et al, 1982).

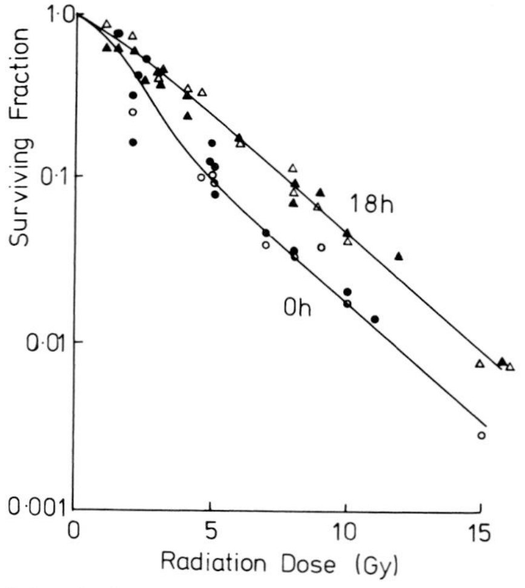


**Fig. 22.5** Survival curves from a human pancreatic tumour HX32 irradiated in situ in the air-breathing mouse and excised and assayed either 10 min (labelled 0 h) or 18 h later. Results from the test-tube assay are shown by closed symbols and from ADC by open symbols.

**Biopsy specimens**

Tumour cells from patient biopsies have generally given low plating efficiencies in soft agar. Table 22.1 shows results obtained using both ADC and the test-tube method when the tubes were gassed with 5 per cent $O_2$. Out of 40 different tumours examined (Courtenay et al, 1978), 25 per cent gave a PE > 1.2 per cent using the test tube method and 21 per cent with ADC (from 29 tumours) gave a similar PE. No colonies were obtained from the six breast tumours by either method. Melanomas are among the most readily disaggregated human tumours and good single cell suspensions can be obtained using a collagenase-pronase enzyme cocktail (Howell & Koch, 1980). In subsequent studies on melanoma biopsies cultured with three per cent instead of five per cent $O_2$, seven out of 12 tumours gave a PE ≥1.2 per cent. The response in vitro to single doses of γ radiation from a $^{60}$Co source was measured using these tumours. Oxic cell suspensions in culture medium were exposed to doses up to approximately 6 Gy γ rays and then assayed in agar. The results in Figure 22.6 from five different tumours with PE between 1.2 per cent and 1.5 per cent indicate the accuracy that can be expected from the melanomas. The values obtained fit a single survival curve. The dose range that could be studied was limited by the PE obtainable. With a PE of 1.2 per cent, from $1 \times 10^4$ cells given 6 Gy, an average of only about four colonies was obtained in each tube and it is obvious that statistical error becomes unacceptable at higher

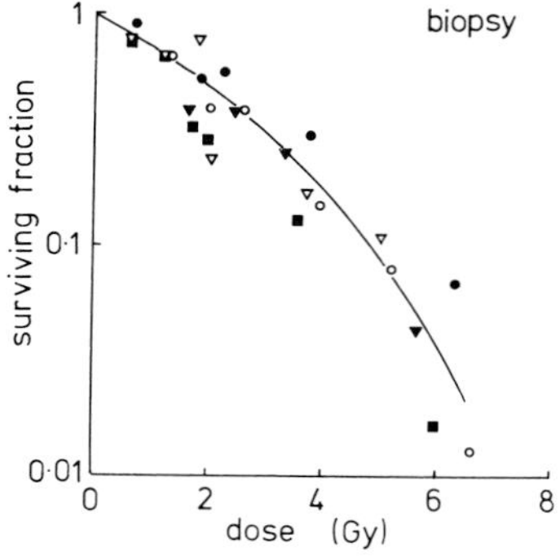


**Fig. 22.6** Radiation survival measured for cell suspensions of five different human melanoma biopsies (various symbols) exposed in vitro to γ-rays from a $^{60}$Co source under fully oxygenated conditions.

**Table 22.1** Plating efficiencies using colonies grown in vitro or in agar diffusion chambers (ADC) from cell suspensions prepared from tumour biopsies

| Tumour type | Form | Treatment of suspension | Plating efficiency, % | |
|---|---|---|---|---|
| | | | in vitro | ADC |
| Melanoma | sc deposit | m | 0 | 0.23 |
| | sc deposit | m | 15 | >2.5 |
| | sc deposit | m | 0.5 | 0 |
| | sc deposit | m | 3.0 | 0.92 |
| | sc deposit | m | 5.6 | 11.5 |
| | sc deposit | m | 0 | 0.066 |
| | sc deposit | m | 0.2 | – |
| | Node deposit | m | 0 | 2.0 |
| | Node deposit | m | 0 | – |
| | Node deposit | m | 0.5 | – |
| | Node deposit | m | 0.25 | – |
| | Primary | m | 0 | 0.6 |
| | Ascites | u | 0 | 0 |
| | Pleural effusion | u | – | 0.072 |
| Ovarian carcinoma | Ascites | u | 2.7 | 0.13 |
| | Ascites | u | – | 0.01 |
| | Ascites | u | 0.25 | – |
| | Ascites | u | 0.02 | – |
| | Ascites | u | 0.2 | – |
| | Ascites | u | 1.0 | – |
| | Ascites | u | 1.0 | 1.7 |
| | Primary | t | 0.4 | – |
| | Secondary deposits | t | 1.3 | – |
| | Peritoneal deposits | m | 4.5 | 2.2 |
| Breast cancer | Primary | t | 0 | 0 |
| | Primary | t | 0 | – |
| | Primary | t | 0 | – |
| | Pleural effusion | u | 0 | 0 |
| | Pleural effusion | u | 0 | 0 |
| | Pleural effusion | u | 0 | – |
| Colorectal cancer | Primary | t + c | 0 | – |
| | Primary | t + c | 0 | 0 |
| | Primary | t | 0.03 | 0 |
| | Primary | t | 0 | – |
| | Primary | t | – | 0.25 |
| | Secondary deposits | t + c | 0.3 | 2.0 |
| Teratoma testis | Primary | t + c | – | 0 |
| | Primary | t | 0 | – |
| Seminoma | Primary | t | 0 | 0.10 |
| | Primary | m | – | 0 |
| Pancreatic carcinoma | Primary | t + c | 0.17 | 1.0 |
| Uterine Ca (body) | Secondary dep. | m | 12 | – |
| (cervix) | Ascites | | – | 0 |
| Oat cell Ca bronchus | Secondary dep. | m | – | 0.1 |
| Hypernephroma | Secondary dep. | t | 0.1 | 0 |
| Orchioblastoma | Primary | t | 0.04 | – |
| Osteosarcoma | Local recurrence | m | – | 0 |
| Leiomyosarcoma | Primary | m | 0 | – |

u = untreated
m = mechanical separation
t = trypsin
t + c = trypsin + collagenase
sc = subcutaneous
Ca = carcinoma
– = not attempted

doses. For a number of other tumours, e.g. ovarian carcinomas, which are difficult to disaggregate, errors are further increased at low PE by the difficulty of excluding cell clumps.

The idea of testing radiation or drug sensitivity in vitro from human biopsy material is an attractive one. However, until higher PEs can be obtained, it is doubtful whether (with the possible exception of the melanomas) results based on a single set of measurements can provide an adequate basis for predicting the response of individual patients' tumours (Bertoncello et al, 1982; Selby et al, 1983).

## EFFECT OF VARYING THE METHODS AND MATERIALS

### Oxygen concentration

As has been shown, the control of $O_2$ concentration at physiological levels is an important requirement for the growth of tumour cells in primary culture and we have conducted experiments using different $O_2$ concentrations.

*Test tubes in gassed boxes*

Figure 22.7 shows the effect of using different $O_2$ concentrations in the gas phase on colony formation by HX34 human melanoma cells from xenografts. Between 100 and 500 tumour cells were set up in test tubes with $10^4$ HR cells and gassed at weekly intervals with gas mixtures containing 0, 2, 3, 5 or 20 per cent $O_2$. The gassed and tightly capped tubes were incubated in sealed plastic boxes gassed with the same mixtures. PE was highest from boxes gassed with 0–3 per cent $O_2$. The fact that colonies were obtained from 0 per cent $O_2$ illustrates the problem of maintaining low $O_2$ levels in plastic boxes and tubes, which may adsorb and release oxygen and from which there is always a small gas leak. With the use of an incubator which can mix three gases, $O_2$ concentration can be more accurately controlled than in boxes, as well as saving the time taken in gassing individual boxes.

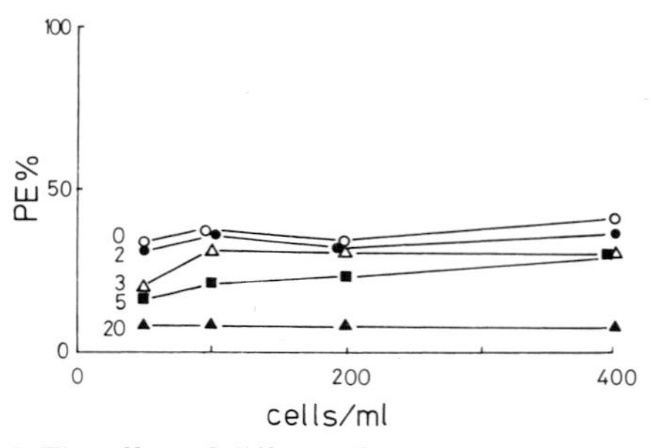


**Fig. 22.7** The effect of different $O_2$ concentrations (0, 2, 3, 5, 20%) on PE of cells from the HX34 melanoma xenograft. Test tubes were incubated in gassed boxes in an air incubator.

*Agar depth in tubes and multiwell plates*

The effect of agar depth on PE was shown in an experiment comparing tubes and 17 or 24 mm multiwell plates (Linbro): 100 HX34 cells with or without HR cells were plated out in volumes of 1.0 or 0.5 ml of 0.3 per cent agar in the wells over an 0.5 mm underlayer of 0.5 per cent agar. At the same time 100 HX34 cells were set up as usual in test tubes. All were maintained in an incubator gassed with 3 per cent $O_2$. The results in Table 22.2 show that PE depended on the depth of the agar. With added HR, PE ranged from 35 per cent in a 10 mm depth of agar in tubes down to 2.1 per cent in 1 mm deep agar in 24 mm wells. In tubes without HR, PE was always lower than at corresponding depths with HR. The effect was most marked in the 24 mm wells and with only 1 mm depth of agar without HR, PE fell to 0.16 per cent.

It can be shown that the $O_2$ concentration at the bottom of the dishes and tubes depends on the depth of the agar (McLimans et al, 1968). The rate of flow of oxygen in agar medium can be calculated from Fick's Law, which states that the rate of flow of $O_2$ across a plane is directly proportional to the gradient of $O_2$ concentration and the area of the plane, thus

$$\text{flow rate (Q)} = -D\frac{dg}{dx} \qquad \text{equation (1)}$$

where D = diffusion coefficient of $O_2$ in agar medium
g = $O_2$ concentration
x = distance

and

$$Q = -D\left(\frac{g_l - g_s}{l}\right) \qquad \text{equation (2)}$$

where l = depth of agar
$g_l$ = $O_2$ concentration at bottom of agar medium
$g_s$ = $O_2$ concentration at surface of agar medium

In the steady state, $O_2$ flow rate equals $O_2$ consumption by cells

so

$$Q = nC \qquad \text{equation (3)}$$

where n = number of cells/cm$^2$
C = $O_2$ consumption per cell

Substituting from (2):

$$\frac{nCl}{D} = g_s - g_l$$

$$g_l = g_s - \frac{nCl}{D} \qquad \text{equation (4)}$$

Thus for a given $O_2$ concentration in the gas phase the $O_2$ concentration at the bottom of the agar depends on the depth of the agar as well as on the cell number.

**Table 22.2** HX34 Melanoma: effect of agar depth and HR cells on PE

| Containers | Vol. of agar (ml) | Depth (mm) | No. of tumour cells | PE with $1 \times 10^4$ HR cells (%) | PE without HR (%) |
|---|---|---|---|---|---|
| Falcon test tubes | 1.0 | 10 | 100 | 35.0 | 27.5 |
| 17 mm wells | 1.0 | 5 | 100 | 22.2 | 22.0 |
| | 0.5 | 2.5 | 100 | 20.5 | 17.3 |
| 24 mm wells | 1.0 | 2.2 | 100 | 15.8 | 11.3 |
| | 0.5 | 1.1 | 100 | 2.1 | 0.16 |

PE = plating efficiency
HR = heavily irradiated

The low PE obtained in the shallow dishes in which the $O_2$ (Table 22.2) concentration at the bottom more nearly approaches that of the gas phase, indicates that the optimal $O_2$ level for the growth of HX34 cells is considerably lower than three per cent. The improved PE obtained with HR cells would appear to be dependent on their ability to metabolise $O_2$.

The test-tube method gave the best results and provides a more easily-used technique. It can be used with a wider range of $O_2$ concentrations which could be an advantage since optimum $O_2$ concentrations may differ for different tumours. The results from the 17 mm wells show PEs almost as high as from the test-tube method and the wells have the advantage that colonies can be counted directly without decanting. However, they are rather more difficult to handle and more liable to contamination with moulds. With shallower dishes further improvements in PE might be obtained by further reducing the $O_2$ concentration. However for routine use lower $O_2$ gas mixtures are more difficult to maintain and further work is needed to determine the optimum $O_2$ level. The effect of cell number on colony formation at different depths in deep agar is shown in Figure 22.4. HX34 cells were set up in agar in open narrow tubes gassed with five per cent $O_2$. With only $4 \times 10^3$ cells/ml, colony formation was good between 4.5 mm and 9.0 mm below the surface. Above 4.5 mm, where $O_2$ levels approached five per cent, colonies were fewer and smaller. With $4 \times 10^4$ cells/ml and $4 \times 10^5$ cells/ml no viable colonies were seen below 2.3 mm and 1.5 mm respectively due to hypoxia produced by the cells above.

### RBC

In the test tube assay, RBC are added to give a final concentration 1/80 of that in normal blood. At a 1/40 concentration their growth enhancing effect is increased but breakdown products have sometimes been found toxic after about 10 days. In our hands RBC from August rats have given the best results. Human RBC or RBC from Marshall or Wistar rats do not normally lyse in agar until two or more weeks after plating out and fail to promote growth. There is evidence that RBC have little or no effect on colony formation in ADC (Selby et al, 1980).

### Agar

Above a concentration of 0.3 per cent, PE tends to decline. Provided the agar can be handled without fracture, we and others, including Thomson & Rauth (1974), have found improved PE at concentrations down to 0.2 per cent agar. We examined one batch of agarose as a possible alternative to agar but RBC failed to lyse and PE was low as compared to that in agar. A comparison of agar and agarose by Neugut & Weinstein (1979) using 13 different tumour cell lines showed that agarose was superior to agar for the growth of three of the tumours but for the other 10 there was no difference.

Methyl cellulose over an agar underlayer has been used as an alternative to agar and has the advantage that colonies can be picked out more readily (Buick & Fry, 1980).

### Medium

Ham's F12m medium, originally developed for the growth of cells at low concentrations, was chosen for use in both the liquid and agar phases. We have found it superior to basal MEM $\alpha$ medium or RPMI 1640 but little different from McCoy's 5A in a limited number of tests. No other additions have been found useful apart from serum and RBC.

The fact that PE in agar cultures was as high as in ADC, in which nutrients required for cell growth were supplied by diffusion from the peritoneal fluid of the living mouse, suggests that at least for cells from secondary tumours, further additives are not very likely to produce a marked improvement in colony formation. Less success has been obtained with primary tumours. This could be related to a slower growth rate or to special requirements for hormones or perhaps for unstable or rapidly metabolised substances specific to the tissue of origin.

## PRESENTATION OF DATA

Results obtained from the colony assays after treatment with drugs or radiation are best presented in the form of survival curves using a semilog plot with the surviving fraction plotted on the vertical logarithmic axis as in Figure 22.5. In general, survival curves are exponential, often with an initial shoulder. Curves terminating in a plateau may be obtained with phase specific drugs, when there is a resistant cell population or when colony-sized clumps are present in the cell suspension. The shape of the curve may also be changed in drug exposure in vitro when there is some interaction between the drug and the medium (Takahashi et al, 1980) or when cells are allowed to settle out in the bottom of a test tube during treatment, so that those underneath may not receive the calculated dose.

## NATURE OF THE CELL BEING ASSAYED

That cells forming compact colonies in agar are tumour cells has been demonstrated by a number of techniques including chromosome preparations from aneuploid tumours and the use of monoclonal antibodies. The most direct indication of colony identity has been obtained from melanotic melanomas which produce dark colonies of cells containing granules that stain for melanin. Amelanotic melanomas have also been identified from the presence of melanosomes using electron microscopy (Selby et al, 1980).

Cells other than tumour cells may form diffuse colonies in agar. Stephens et al (1978), studying the Lewis lung mouse carcinoma, identified diffuse colonies of macrophages interspersed with compact colonies of tumour cells. Also from human marrow suspensions we have found diffuse colonies of red cell precursors and granulocytes as well as macrophages.

Not all tumours are capable of forming colonies in agar. In some cases this may simply be due to a long cell cycle time. On the other hand there is increasing evidence that some tumour cells, capable of growing in monolayer culture, may retain a degree of anchorage dependence as found by Hastings & Franks (1981) for human bladder carcinomas, and Dodson et al, (1981) studying a squamous cell carcinoma. These tumours formed xenografts in immune-deficient mice and grew well in monolayer culture but failed to give colonies in agar.

### Tumour stem cells

It has been suggested that the low PEs generally obtained from human biopsy material reflect a small stem-cell compartment (Buick & McKillop, 1981). This may be true for some tumours, but the wide variations in PE that we have obtained from the same xenografted tumour line depending on the oxygen tension in the medium and also on the presence of substances released from rat RBC, shows that PE is strongly dependent on culture conditions. Thus PE is a poor indicator of the size of the stem-cell compartment. In primary cultures, measured values of PE are reduced by the inclusion of non-tumour cells in the cell counts, by lethal damage to cells sustained during the preparation of cell suspensions and in some cases, the inhibitory effect of agar on cell growth. With further investigation and development of these aspects of cell culture and with the careful control of oxygen concentration, very much higher PEs eventually may be attainable.

## REFERENCES


Agrez M V, Kovacs J S, Lieber M M 1982 Cell aggregates in the soft agar 'human tumour stem-cell assay'. British Journal of Cancer 46:880

Barranco S C, Bolton W E, Novak J K 1980 Time-dependent changes in drug sensitivity expressed by mammalian cells after exposure to trypsin. Journal of the National Cancer Institute 64:913

Bateman A E, Peckham M J, Steel G G 1979 Assays of drug sensitivity for cells from human tumours: in vitro and in vivo tests on a xenografted tumour. British Journal of Cancer 40:81

Bertoncello I, Bradley T R 1977 The physicochemical properties of erythrocyte derived activity which enhances murine bone marrow colony growth in agar culture. Australian Journal of Experimental Biological and Medical Sciences 55:281

Bertoncello I et al 1982 Limitations of the clonal agar assay for the assessment of primary human ovarian tumour biopsies. British Journal of Cancer 45:803

Bradley T R, Metcalf D 1966 The growth of mouse bone marrow cells in vitro. Australian Journal of Experimental Biological and Medical Sciences 44:287

Bradley T R, Telfer P A, Fry P 1971 The effect of erythrocytes on mouse bone marrow colony development in vitro. Blood 38:353

Buick R N, Fry S E 1980 A comparison of human tumour-cell clonogenicity in methylcellulose and agar culture. British Journal of Cancer 42:933

Buick R N, McKillop W J 1981 Measurement of self renewal in culture of clonogenic cells from human ovarian carcinoma. British Journal of Cancer 44:349

Courtenay V D 1976 A soft agar colony assay for Lewis lung tumour and B16 melanoma taken directly from the mouse. British Journal of Cancer 34:39

Courtenay V D 1983 The Courtenay clonogenic assay. In: Dendy P P, Hill B T (eds) Human tumour drug sensitivity testing in vitro: techniques and clinical applications. Academic Press, London

Courtenay V D, Mills J 1978 An in vitro colony assay for human tumours grown in immune-suppressed mice and treated in vivo with cytotoxic agents British Journal of Cancer 37:261
Courtenay V D, Mills J, Steel G G 1982 The spectrum of chemosensitivity of two human pancreatic carcinoma xenografts. British Journal of Cancer 46:436
Courtenay V D, Smith I E, Peckham M J, Steel G G 1976 In vitro and in vivo radiosensitivity of human tumour cells obtained from a pancreatic carcinoma xenograft. Nature 263:771
Courtenay V D, Selby P J, Smith I E, Mills J, Peckham M J 1978 Growth of human tumour cell colonies from biopsies using two soft-agar techniques. British Journal of Cancer 38:77
Dodson M G, Slota J, Lange C, Major E 1981 Distinction of the phenotypes of in vitro anchorage-independent soft-agar growth and in vivo tumorigenicity in the nude mouse. Cancer Research 41:1441
Gordon M Y 1974 Quantitation of haemopoietic cells from normal and leukaemic RFM mice using an in vivo colony assay. British Journal of Cancer 30:421
Hamburger A W, Salmon S E 1977 Primary bioassay of human tumour stem cells. Science 197:461
Hastings R J, Franks L M 1981 Chromosome pattern, growth in agar and tumorigenicity in nude mice of four human bladder carcinoma cell lines. International Journal of Cancer 27:15
Hill R P, Stanley J 1975 The lung-colony assay: extension to the Lewis Lung tumour and the B16 melanoma — radiosensitivity of B16 melanoma cells. International Journal of Radiation Biology 27:377
Howell R L, Koch C J 1980 The disaggregation, separation and identification of cells from irradiated and unirradiated EMT6 mouse tumours. International Journal of Radiation Oncology Biology and Physics 6:311
Kriegler A B, Bradley T R, Hodgson G S, McNiece I K 1981 Identification of the 'factor' in erythrocyte lysates which enhances colony growth in agar cultures. Experimental Haematology 9:11
McLimans W F, Blumenson L E, Tunnah K V 1968 Kinetics of gas diffusion in mammalian cell culture systems II theory. Biotechnology and Bioengineering X:741
MacPherson I, Montagnier L 1964 Agar suspension culture for the selective assay of cells transformed by polyoma virus. Virology 23:291
Neugut A I, Weinstein I B 1979 The use of agarose in the determination of anchorage-independent growth. In Vitro 15:351
Nias A H W, Gilbert C W, Lajtha L G, Lange C S 1965 Clone size analysis in the study of cell growth following single or during continuous irradiation. International Journal of Radiation Biology 9:275
Park C H, Bergsagel D E, McCulloch E A 1971 Mouse myeloma tumour stem cells: a primary cell culture assay. Journal of the National Cancer Institute 46:411
Pavelic Z P, Slocum H K, Rustum Y M, Creaven P J, Karakousis C, Takita H 1980 Colony growth in soft agar of melanoma sarcoma and lung carcinoma cells disaggregated by mechanical and enzymatic methods. Cancer Research 40:2160
Puck T T, Marcus P I 1956 Action of X-rays on mammalian cells. Journal of Experimental Medicine 103:653
Rasey J S, Nelson N J 1980 Response of an in vivo-in vitro tumour to X-rays and cytotoxic drugs: effect of tumour disaggregation method on cell survival. British Journal of Cancer 41:217
Richter A, Sanford K K, Evans V J 1972 Influence of oxygen and culture media on plating efficiency of some mammalian tissue cells. Journal of the National Cancer Institute 49:1705
Selby P J, Thomas J M 1980 Clonogenic cell survival curves for human melanoma xenografts using agar diffusion chamber and lung colony assays. British Journal of Cancer 41:(Suppl IV) 150
Selby P J, Courtenay V D, McElwain T J, Peckham M J, Steel G G 1980 Colony growth and clonogenic cell survival in human melanoma xenografts treated with chemotherapy. British Journal of Cancer 42:438
Selby P J, Tannock I, Buick R N 1983 A critical appraisal of the 'human tumour stem cell assay'. New England Journal of Medical 308:129
Shipley W U, Stanley J A, Courtenay V D, Field S B 1975 Repair of radiation damage in Lewis Lung carcinoma cells following in situ treatment with fast neutrons and $\gamma$ rays. Cancer Research 35:932
Smith I E, Courtenay V D, Gordon M Y 1976 A colony forming assay for human tumour xenografts using agar in diffusion chambers. British Journal of Cancer 34:476
Stephens T C, Currie G A, Peacock J H 1978 Repopulation of $\gamma$-irradiated Lewis Lung carcinoma by malignant cells and host macrophage progenitors. British Journal of Cancer 38:573
Sutherland R M, Durand R E 1976 Radiation response of multicell spheroids — an in vitro tumour model. Current Topics in Radiation Research Quarterly 11:87
Takahashi I, Ohnuma T, Kavy S, Bhardwaj S, Holland J F 1980 Interaction of human serum albumin with anticancer agents in vitro. British Journal of Cancer 41:602
Thomson J E, Rauth A M 1974 An in vitro assay to measure the viability of KHT tumour cells not previously exposed to culture conditions. Radiation Research 58:262
Tveit K M, Fodstad O, Pihl A 1981a Cultivation of human melanomas in soft agar. Factors influencing plating efficiency and chemosensitivity. International Journal of Cancer 28:329
Tveit K M, Endresen L, Rugstad H E, Fodstad O, Pihl A 1981b Comparison of two soft agar methods for assaying chemosensitivity of human tumours in vitro: malignant melanomas. British Journal of Cancer 44:539

# The assay of tumour colonies in the lung

## HISTORICAL BACKGROUND

Monitoring the growth of nodules in the lung following intravenous injection of tumour cells was first used as an approach to studying various aspects of the metastatic process (Zeidman et al, 1950) and has continued to be used in this context (Fisher & Fisher, 1975; Poste & Fidler, 1980). In 1966, Williams & Till reported that the approach could also be used as a basis for a useful quantitative assay for determining the malignancy of cell lines transformed with polyoma virus, and in 1969, Hill & Bush described the use of the assay to determine the viability of KHT Sarcoma cells following treatment of the tumour cell population with radiation. Subsequent to this report it was demonstrated that the technique can be used to assay cell viability in a variety of animal tumours: mouse fibrosarcomas FSAI and II (Grdina et al, 1975; Rice et al, 1980), mouse mammary tumour KHJJ (Kallman, 1973), mouse B16 Melanoma and Lewis lung tumour (Hill & Stanley, 1975a), and rat tumours Y–P 388 and Walker 256 (Van den Brenk et al, 1973). The assay has been used for studies of the effect of both radiation and chemotherapeutic drugs, and to study various aspects of tumour cell antigenicity (Boone et al, 1973; Janik & Szaniawska, 1978; Peters et al, 1977). Recently its use to determine the radiosensitivity and chemosensitivity of two human malignant melanoma xenografts has been reported (Thomas, 1979).

## DESCRIPTION OF THE TECHNIQUE AND METHODOLOGY

The basic procedure in the assay is to prepare a suspension of tumour cells and to inject a known number of morphologically viable (dye excluding) cells, mixed with heavily radiated HR tumour cells and/or 15 μm plastic microspheres, into one of the tail veins of a number of recipient animals. A convenient volume for the injection is 0.2–0.5 ml. In the animals the tumour cells are arrested in the lung capillaries and a proportion of them grow to form discrete nodules (colonies) over a period of a few weeks (see Fig. 23.1). At an appropriate time the animals are killed and their lungs removed and fixed. The nodules can then be counted either by eye, or more easily, using a low power dissection microscope. The number of lung colonies per cell injected, the lung colony forming efficiency (LCE) can then be calculated. The fraction of cells surviving a given treatment can be determined by taking a ratio of the LCE for the treated population to that for an untreated control population. An alternative approach (Shaeffer et al, 1973, 1974) is to inject tumour cells intravenously so that they will be arrested in the lungs and then to treat the recipient mice with drugs or thoracic irradiation at a later time. The cells, which survive the treatment, are then allowed to grow to form lung nodules which can be counted and the surviving fraction calculated as above. Such a calculation may be influenced by a multiplicity factor if the single cells arrested in the lung have time to proliferate

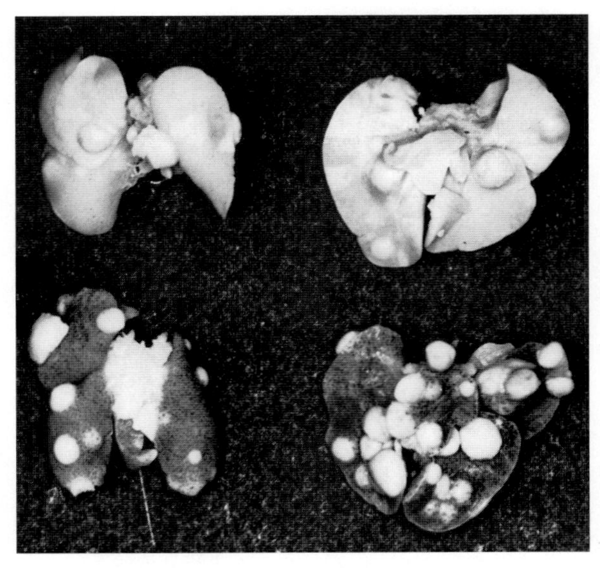

**Fig. 23.1** Photograph of mouse lungs containing nodules growing from injected KHT Sarcoma cells. The lungs were fixed in Bouin's solution. The lower two lungs are from mice injected with tumour cells admixed with plastic microspheres.

before the initiation of treatment. Also the calculation may be affected if cells seed to the lung from other parts of the body after the thoracic irradiation. Such as problem was reported by van den Brenk (1973), who found that, after i–v injection of Walker 256 carcinoma cells into rats, a small fraction (~0.01%) of the cells could be trapped in the tail capillaries and, over a period of a few hours, some could seed to the lung.

It is necessary for the lung nodules to be a certain size before they can be detected macroscopically; consequently the time between injecting the cells and removing the lungs will depend on the growth-rate of the tumour being studied. For the KHT Sarcoma it was demonstrated (Hill & Bush, 1969) that the mean number of colonies, observed in groups of mice injected from the same suspension of untreated cells, was constant between 14 and 20 days after injection, implying that all the colonies were large enough to be detected in this interval. In their studies with the $C_3$HBA tumour Shaeffer et al (1973) did not observe a plateau in colony number until 4–6 weeks after cell injection. Thus, for each tumour, an appropriate time range for colony assessment must be established. The optimal time for observation of the colonies may be influenced by treatment. Cells from irradiated tumours have been found to grow more slowly in the lung with the result that a constant level of colonies is not reached until a later time than after injection of untreated cells (Hill & Bush, 1969; Hill & Stanley, 1975a; Hill, 1980b). Thus in studies with irradiated populations of cells, a policy of removing the lungs from mice injected with cells from control and treated populations at different times was adopted, so that the colonies are a similar size when counted (see Fig. 23.2). This phenomenon did not occur in studies with cytotoxic drugs (Hill & Stanley, 1975a, b — see Fig. 23.2) but it should be investigated for each treatment under study.

Lungs may be fixed and prepared for colony counting in a number of ways. If the tissue is fixed in Bouin's solution and transferred to 70 per cent ethanol after 24 hours, then the lung parenchyma is stained yellow and the tumour nodules are seen as a very pale yellow against this background (see Fig. 23.1), unless they are derived from melanoma cells when they are often clearly distinguished by their black pigment. For counting, the individual lobes of the lung are separated from one another and the trachea and mediastinal tissue discarded. If black plastic microspheres (obtainable from 3M Products Ltd, Minnesota, USA) are included in the injected suspension, the lung traps them, leading to a better contrast between the dark lung parenchyma and the light tumour nodules which do not contain microspheres (see Fig. 23.1). An alternative approach (Wexler, 1966) is to inject India ink into the trachea immediately after killing the animals, tie off the trachea and remove the lung en

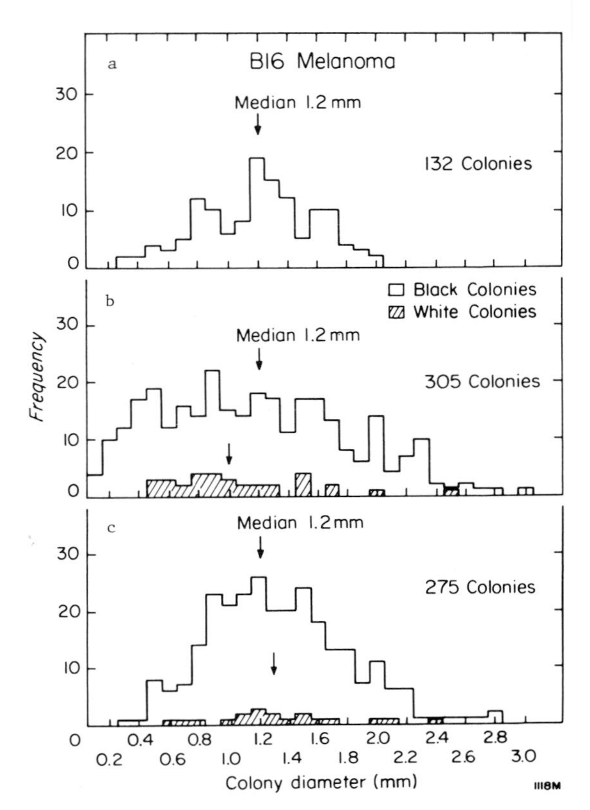


**Fig. 23.2** A size (diameter) distribution of colonies growing from B16 Melanoma cells, which were obtained from subcutaneously-growing tumours treated in situ with (a) 300 mg/kg cyclophosphamide (CY), (b) 20 Gy $\gamma$-irradiation, (c) no treatment. For (a) and (c) the mice were killed 22 days after cell injection but for (b) they were killed 24 days after cell injection. Melanotic and amelanotic colonies are shown separately.

bloc. Then immerse the tissue in a bleaching agent such as Fekete's fixative. This results in white nodules standing out against a black background. A third approach, described by Boone et al (1973), involved fixing the lungs in formalin and then flattening each lobe to about 1 mm thickness before embedding in a paraffin block. Two sections were then cut at equidistance through the thickness of the lungs and the colonies counted on histological slides using a low-power microscope. While, in this latter method, colonies growing entirely inside the lung will be detected, it is possible using the first two methods that such colonies might be missed during the counting. In practice, this is not a serious problem with Bouin's-fixed mouse lungs, if the colonies are allowed to grow to a median size of 1–2 mm. The lungs are quite thin (< 5 mm) and can be cut with a scalpel to investigate any suspicious lumps and colonies inside the lung. Occasionally infected lungs may show regions on their surface which appear whitish because they are filled with air or clear fluid and are slightly raised from the surface.

These regions can be mistaken for tumour colonies by inexperienced observers but can be distinguished from colonies by slicing with a scalpel revealing their lack of solidity.

The addition of HR cells and/or microspheres to the suspension of viable cells is an important part of the assay because the LCE is often quite low when small numbers of cells are injected alone and may be dependent on the number of cells injected (i.e. non-linearity occurs). However it was shown by Hill & Bush (1969) that the LCE could be substantially increased and stabilized by admixing a large number ($>10^6$) of HR cells and/or microspheres with the injected cells. For the KHT sarcoma it was found that either HR cells or microspheres were equally effective in increasing the LCE to its maximum level but, for the Lewis lung tumour and B16 melanoma, HR cells were not as effective as microspheres (Hill & Stanley, 1975a). In contrast to this, in his studies with human malignant melanoma cells injected into immune deprived mice, Thomas (1979) found that addition of $10^6$ HR cells or microspheres reduced the LCE. However, studies with human bladder cell lines, injected into similar mice, have found that both HR cells and microspheres increase the LCE (Kovnat & Tannock, private communication). The most likely reason for these findings of an increase in the LCE when using HR cells or microspheres is that the retention of the viable cells in the lung is increased under these conditions. Studies with labelled cells have demonstrated such an effect when a large number of cells is injected (Peters et al, 1978). The reason why the effect did not occur with the human melanoma cells remains unclear. However, if different numbers of cells are being injected for the treated and control groups, these results make it desirable to equalise the cell numbers injected by the addition of HR cells to the appropriate groups. The use of black microspheres has additional practical value because they increase the contrast between the lung tissue and the colonies and they allow detection of any badly-injected animals, since such animals will have no, or only a few, microspheres in their lungs at sacrifice.

For the calculation of the surviving fraction of treated cells it is necessary to know the relationship between the number of viable cells injected and the number of colonies observed. Ideally one would wish the LCE to be constant for different numbers of cells injected. Hence most workers have investigated the linearity of their assay. An example of such linearity for KHT Sarcoma cells is shown in Figure 23.3. Up to a mean of about 70

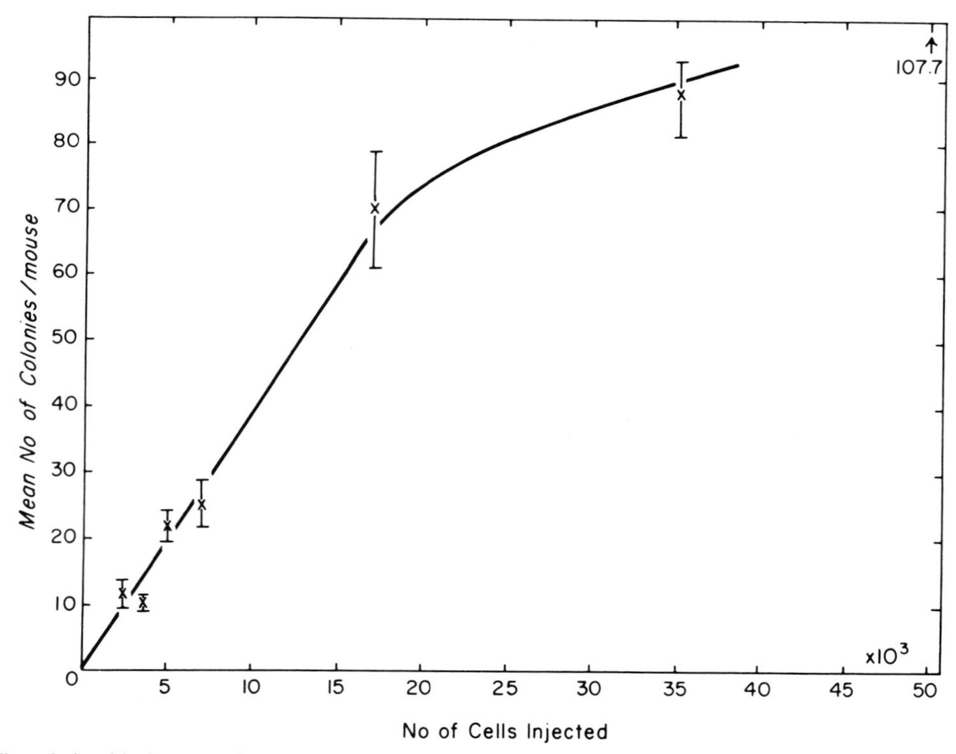


**Fig. 23.3** The relationship between the mean (+SE) number of colonies observed per mouse and the number of cells injected. The cells were irradiated with 3 Gy before injection and were admixed with $2 \times 10^6$ HR cells. Results from Hill & Bush (1969), reproduced from the International Journal of Radiation Biology with permission.

colonies/lung there is a linear relationship but this is lost at higher colony numbers. The most likely reason is that it becomes increasingly difficult to count all the colonies in mouse lungs when the numbers exceed about 80–100 and many colonies may coalesce when they originate very close together. The cell numbers to be injected should be adjusted for the treated and untreated groups to maintain the number of colonies obtained within the linear region.

## FACTORS AFFECTING THE ASSAY

A factor which can increase the colony efficiency is pretreatment of the lungs of the recipient animals with a dose of irradiation (5–20 Gy) (Brown, 1973; van den Brenk et al, 1973; Withers & Milas, 1973; Hill & Stanley, 1975a; Janik, 1976) and some groups of workers use this procedure routinely (Grdina et al, 1975). A similar effect can be obtained by pretreatment with a large dose of cyclophosphamide (van Putten et al, 1975; Carmel & Brown, 1977; Steel & Adams, 1977) while pretreatment with anticoagulants or other agents, such as prostacyclin, will reduce the number of colonies formed Brown, 1973; Brown & Parker, 1979; Honn et al, 1981; Maat & Hilgard, 1981). The exact reason for these effects remains uncertain but it is thought that they are a result of damage to the lung and to inhibition of thrombus formation.

An important requirement for the lung colony assay is that a good single cell suspension be available for injection which, if the suspension is to be obtained from a solid tumour, may require both mechanical and enzymatic treatments. Appropriate treatments vary from tumour to tumour and hence need to be determined for each type of tumour studied. A significant amount of debris in the cell suspension reduces the maximum number of cells which can be injected without killing the recipient animals and clumps of cells probably have a higher LCE than single cells. Thompson (1974) found for $C_3H$ mouse mammary tumours that the LCE depended on clump size. It is usual when cell suspensions are prepared from solid tumours to assess the morphological viability of the cells recovered, by a technique such as dye-exclusion, and to use the number of such cells as a base for calculations. When enzyme-treatments are used as part of the separation procedure the dye-excluding cells are usually more than 90 per cent of the total cells recovered.

It has also been shown that a number of factors including cell size can influence the LCE (Grdina et al, 1977; Suzuki & Withers, 1979), which potentially could have a bearing on the interpretation of results obtained, particularly from irradiated cell populations. However, as discussed later the good correlation of results with those obtained using other assays suggests that it is not a serious problem.

### Sample size

Another aspect of the assay is the number of mice which

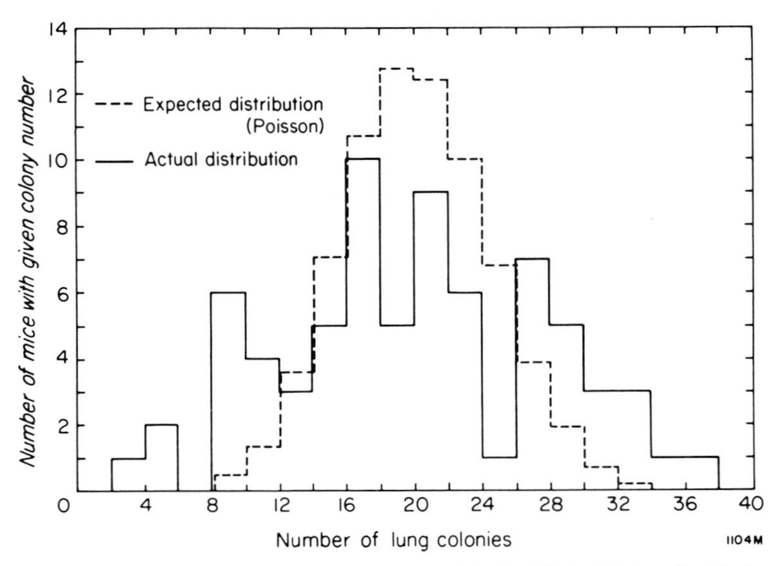


**Fig. 23.4** The distribution of colony numbers among 72 $C_3H$ mice (weight 24–25 g) all injected with the same volume (0.2 ml) of the same suspension of untreated KHT sarcoma cells (admixed with $2 \times 10^6$ HR cells + $10^6$ microspheres) is shown in the solid line. A Poisson distribution for the same number of animals and based on the same mean is shown with the broken line. The mean of the distribution is 19.7 with a standard deviation of 7.8; the median is 19.0 with quartile values of 14 and 26.

should be injected from each cell suspension. It would be expected that the variation in colony number from animal to animal, injected with the same cell suspension, would be described by a Poisson distribution. Thus, if an accuracy of a standard error (SE) of 10 per cent of the mean (M) is required, then, for a mean of 10 colonies per mouse, 10 mice should be injected for each group (standard deviation (SD) = $\sqrt{M}$, SE = SD/$\sqrt{N}$). In practice the spread of values among individual mice is greater than expected (see Fig. 23.4). When a group of 96 mice were all injected with the same volume from the same cell suspension of KHT Sarcoma cells the standard deviation was 7.7 with a mean of 13.1 colonies/mouse (Hill & Bush, 1969). In another group of 72 mice the mean was 19.7 and the standard deviation 7.8 (see Fig. 23.4). In both cases the standard deviation is approximately twice that expected from the Poisson distribution, presumably because of inter-animal differences. Since the variance is much greater than expected an alternative to using the mean value is to calculate the median value and to use non-parametric statistics for error calculations and comparative analysis. In practice, the number of mice which can be injected per group will probably be governed by expense and injecting 5–7 mice per group gives satisfactory results unless the mean colony number is less than about five (Hill, 1980a). Since in in vivo studies of tumour response to drugs or radiation there is usually substantial inter-experiment variation in cell survival (about 2–10 fold) the degree of accuracy for any individual survival value is rarely limiting in defining an average cell survival curve.

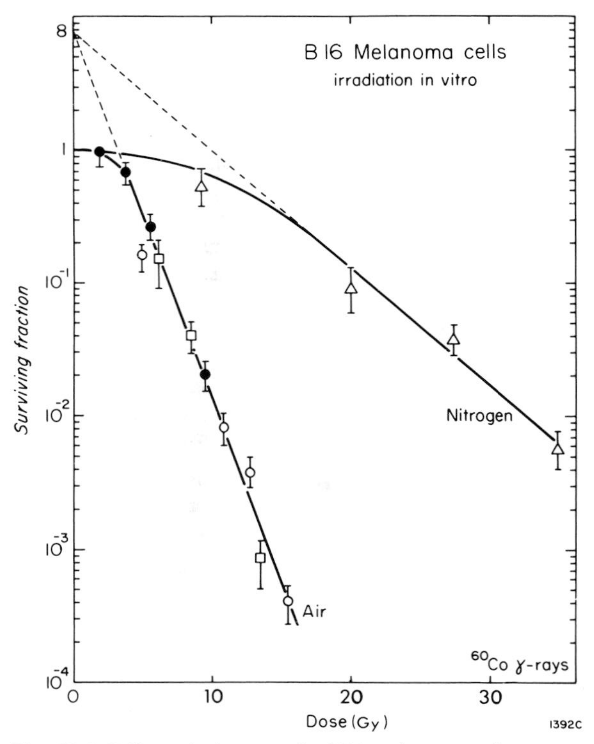


**Fig. 23.5** Cell survival curves for B16 melanoma cells irradiated in vitro with $^{60}$Co γ-rays. The different symbols represent results obtained in four independent experiments with irradiation given under oxic (air) or anoxic (nitrogen) conditions. Errors shown are 1 SE. $D_o$ (air) = 1.58 Gy. $D_o$ (nitrogen)= 4.96 Gy. Results from Hill & Stanley (1975a), reproduced from the International Journal of Radiation Biology with permission.

## PRESENTATION OF RESULTS

The results obtained from the assays are calculated as surviving fractions of cells following treatment as noted earlier. These results can be plotted as a function of treatment dose to give a survival curve. It is usual to plot surviving fraction on a log scale for such a presentation because, particularly for radiation treatment, surviving fraction is usually found to be some exponential function of dose (see Fig. 23.5). If multiple data points are obtained at the same dose level then it is appropriate to obtain an arithmetic mean of their logarithmic values with corresponding standard error for the purpose of plotting in this type of presentation. If the assay procedure is being used to examine the response of a tumour cell population to treatment over a prolonged time period (i.e. > few hours) then a potential problem in presenting the results, which must be addressed, is that of cell yield. Loss of killed cells or proliferation of surviving cells in the population between the initiation of treatment and the assay procedure will affect the calculated level of survival. In such a situation the results can be presented in terms of surviving cells per tumour by taking into account cell yield from the cell suspension procedure (Hill & Bush, 1973). Since cell yield per tumour in the controls will increase during a prolonged treatment period, this factor will also influence the survival calculation, and may need to be corrected.

The range of survival levels which can be measured using a lung colony assay is controlled by two factors, first the LCE which can be achieved with untreated cells from the particular tumour being studied, and second, the total number of cells which can be injected without causing respiratory distress. In $C_3H$ mice, a maximum of (2–5) × $10^6$ KHT Sarcoma cells can be injected and the LCE is about three per cent when HR cells and/or microspheres are admixed with the cells injected (Hill & Stanley, 1975a). These numbers mean that for an average of 10 colonies per mouse, growing from the treated cell population, a survival level of about $10^{-4}$ can be detected. Lower levels of survival will produce smaller numbers of colonies and consequently increased errors, unless very large numbers of mice are injected. It should be noted however that there is significant variation in the LCE of untreated cells from one experiment to another (Hill & Bush, 1969; Hill & Stanley, 1975a), and consequently it is important to use controls for each experiment.

The results obtained using a lung colony assay have been compared with those obtained using other in vivo or in vitro assays for a number of different animal tumours treated with drugs or radiation (Steel et al, 1977; Shipley et al, 1975; Rice et al, 1980; Hill, 1980a). In general the results have demonstrated very good agreement between the lung colony assay and the endpoint dilution assay or the agar colony assay.

## ACKNOWLEDGEMENTS

The work of the author is supported by the National Cancer Institute of Canada and the Ontario Cancer Treatment and Research Foundation.

## REFERENCES

Boone W E, Lundberg E, Orme T 1973 Brief communication: quantitative lung colony assay for tumour immunity in mice. Journal of the National Cancer Institute 5:1731

Brown J M 1973 The effect of lung irradiation on the incidence of pulmonary metastases in mice. British Journal of Radiology 46:613

Brown J M, Parker E T 1979 Host treatments affecting artificial pulmonary metastases: Interpretation of loss of radioactively labelled cells from lungs. British Journal of Cancer 40:677

Carmel R J, Brown J M 1977 The effect of cyclophosphamide and other drugs on the incidence of pulmonary metastases in mice. Cancer Research 37:145

Fisher E R, Fisher B 1975 Circulating cancer cells and metastases. International Journal of Radiation Oncology, Biology and Physics 1:87

Grdina D J, Basic I, Mason K A, Withers H R 1975 Radiation response of clonogenic cell populations separated from a fibrosarcoma. Radiation Research 63:483

Grdina D J, Hittelman W N, White R A, Meistrich M L 1977 Relevance of density, size and DNA content of tumour cells to the lung colony assay. British Journal of Cancer 31:659

Hill R P 1980a An appraisal of in vivo assays of excised tumours. British Journal of Cancer 41 (Suppl. IV):230

Hill R P 1980b Radiation-induced changes in the in vivo growth rate of KHT Sarcoma cells: Implications for the comparison of growth delay and cell survival. Radiation Research 83:99

Hill R P, Bush R S 1969 A lung-colony assay to determine the radiosensitivity of the cells of a solid tumour. International Journal of Radiation Biology 5:435

Hill R P, Bush R S 1973 The effect of continuous or fractionated irradiation on a murine sarcoma. British Journal of Radiology 46:167

Hill R P, Stanley J A 1975a The lung-colony assay: extension to the Lewis lung tumour and the B16 melanoma — radiosensitivity of B16 melanoma cells. International Journal of Radiation Biology 4:377

Hill R P, Stanley J A 1975b The response of hypoxic B16 melanoma cells to in vivo treatment with chemotherapeutic agents. Cancer Research 35:1147

Honn K V, Cicone B, Skoff A 1981 Prostacyclin: a potent antimetastatic agent. Science 212:1270

Janik P 1976 Lung colony assay in normal, irradiated and tumour bearing mice. Neoplasma 5:495

Janik P, Szaniawska B 1978 Search for an influence of natural immunity on the lung colony assay of a syngeneic transplated murine tumour. British Journal of Cancer 37:1083

Kallman R F 1973 Oxygenation and reoxygenation of a mouse mammary tumor. In: Duplan, Chapiro (eds) Advances in radiation research, biology and medicine, vol. 3. Gordon & Breach, Paris, p 1195

Maat B, Hilgard P 1981 Anticoagulants and experimental metastases, evaluation of antimetastatic effects in different model systems. Journal of Cancer Research and Clinical Oncology 101:275

Peters L J, Mason K A, McBride W H 1977 Pitfalls in the use of the lung colony assay to assess T-cell function in irradiated mice. British Journal of Cancer 36:386

Peters L J, Mason K A, McBride W H, Patt Y Z 1978 Enhancement of lung colony forming efficiency by local thoracic irradiation: interpretation of labelled cell studies. Radiology 126:499

Poste G, Fidler I J 1980 The pathogenesis of cancer metastases. Nature 283:139

Rice L R, Urano M, Suit H D 1980 The radiosensitivity of a murine fibro-sarcoma as measured by three cell survival assays. British Journal of Cancer 41 (Suppl. IV):240

Shaeffer J, El-Mahdi A M, Constable W C 1973 Radiation control of microscopic pulmonary metastases in $C_3H$ mice. Cancer 32:346

Shaeffer J, El-Mahdi A M, Constable W C 1974 Treatment of metastatic osteosarcoma by cyclophosphamide and radiotherapy. Radiology 111:467

Shipley W R, Stanley J A, Courtenay V D, Field S B 1975 Repair of radiation damage in Lewis lung carcinoma cells following in situ treatment with fast neutrons and $\gamma$-rays. Cancer Research 35:932

Steel G G, Adams K, Stephens T C 1977 Clonogenic assays in the B16 melanoma: response to cyclophosphamide. British Journal of Cancer 36:618

Steel G G, Adams K 1977 Enhancement by cytotoxic agents of artificial pulmonary metastases. British Journal of Cancer 36:653

Suzuki N, Withers H R 1979 Lung colony formation: A selective cloning process for lung-colony-forming ability. British Journal of Cancer 39:196

Thomas J M 1979 A lung colony clonogenic cell assay for human malignant melanoma in immune-suppressed mice and its use to determine chemosensitivity, radiosensitivity and the relationship between tumour size and response to therapy. British Journal of Surgery 66:696

Thompson S C 1974 The colony forming efficiency of single cells and cell aggregates from a spontaneous mouse mammary tumour using the lung colony assay. British Journal of Cancer 30:332

Van den Brenk H A S 1973 Measurements of tumour-cell radiosensitivity in vivo using lung macrocolony assay: dose-survival artefact due to 'tail trapping'. International Journal of Radiation Biology 6:631

Van den Brenk H A S, Burch W M, Orton C, Sharpington C 1973 Stimulation of clonogenic growth of tumour cells and metastases in the lungs by local X-radiation. British Journal of Cancer 27:291

Van Putten L M, Kram L K J, Van Dierendonck H H C, Smink T, Fuzy M 1975 Enhancement by drugs of metastatic lung nodule formation after intravenous tumour cell injection. International Journal of Cancer 15:588

Wexler H 1966 Accurate identification of experimental pulmonary metastases. Journal of the National Cancer Institute 36:641

Williams J F, Till J E 1966 Formation of lung colonies by polyoma transformed rat embryo cells. Journal of the National Cancer Institute 37:177

Withers H R, Milas L 1973 Influence of preirradiation of lung on development of artificial pulmonary metastases of fibrosarcoma in mice. Cancer Research 33:1931

Zeidman I, McCutcheon M, Cowan D R 1950 Factors affecting the number of tumor metastases: experiments with a transplantable mouse tumor. Cancer Research 10:357

# Measurement of tumour clonogens in situ

## HISTORICAL BACKGROUND

The survival of tumour cells following treatment in vivo can be measured by excision and transfer of the treated cells into a new environment. This may be done in vitro, where the surviving colony forming units can give rise to visible colonies (Ch. 22), or also by injecting defined numbers of cells into recipients where the presence of at least one surviving clonogen is proven by the production of a tumour (Ch. 25). It is realized, however, that the excision assays have their limitations, partly because the measurable survival level is restricted, but also more fundamentally because they require a disruption of the histological structure.

A valuable addition to existing methods therefore would be an assay where the surviving clonogens could be quantified directly in situ by their formation of microcolonies, as has been established for a number of normal tissues (e.g. Ch. 5). In view of the simplicity and technical feasibility of this approach one is faced with a striking scarcity of published communications. Negative findings were mentioned by Mendelsohn (1967) on a mouse mammary carcinoma. In a more recent paper, Marsden et al (1980) described for the first time identifiable tumour clones in an irradiated chondrosarcoma. These could, to a limited extent, be used to derive cell survival data. In our opinion, the absence of any further reports indicates negative results rather than a lack of attempts. We conclude from our own experience with a number of tumour systems that several reasons usually contribute to the failure of such an approach. Clearance of sterilized cells is often delayed for long periods of time in solid tumours; this process overlaps and thus masks the proliferation of survivors. Even more seriously, the growth of surviving cells tends to be invasive and ill delineated in most tumours, thus making regenerating foci difficult to distinguish. While there is little hope of overcoming this inherent problem with a particular technique, the chance is that among the increasing number of transplantable tumour systems more than one may become suitable for study. Recently we found that adenocarcinoma AT 17, a slow growing mammary tumour induced by radiation in our C3H mouse colony, has a histology before and after treatment that lends itself to a quantitative assessment of microcolonies.

## DESCRIPTION OF THE TECHNIQUE

A detailed knowledge of the histological changes following higher radiation doses is essential to understand the evolution of colonies and to select an optimal time for their evaluation. It may also serve as a guideline for future screening studies. As shown in Figure 24.1, the untreated adenocarcinoma has a mixed histological structure. Parenchymal tissue is organized into roundish or irregular nodules, with proliferation at the periphery; larger nodules have a tendency to keratinize. A similar structure with 'cell nests' or 'pearls' is not uncommon in human epithelial tumours, e.g. squamous cell carcinoma (Willis, 1967). The interstitial areas contain stroma and vessels as well as tumour cells which in many parts seem to 'exude' from the parenchyma and form satellites. Following a higher radiation dose (e.g. 32 Gy given under local hypoxia) the two components express their damage at different rates. Two days after radiation the nodules show advanced decline, partly by lysis of the proliferating layers but mainly because the more central cells become keratinized. (Fig. 24.2A). As this continues, the tumour loses all its epithelial features by day six (Fig. 24.2B). The remaining interstitial cells are swollen and subsequently undergo more severe degeneration and a gradual removal over the next few weeks. The first foci of regeneration are detectable at day eight or nine when small nests of densely-packed cells begin to emerge from the debris (Fig. 24.2C). These nests increase in size and number and by day 18 have grown into well demarcated spheroids with an average diameter of 200 to 300 μm. They are comprised of parenchymal, chromophilic, viable-looking cells and they stand out clearly against the otherwise cell-depleted tumour mass (Fig. 24.3A).

Tumour regeneration does not immediately proceed

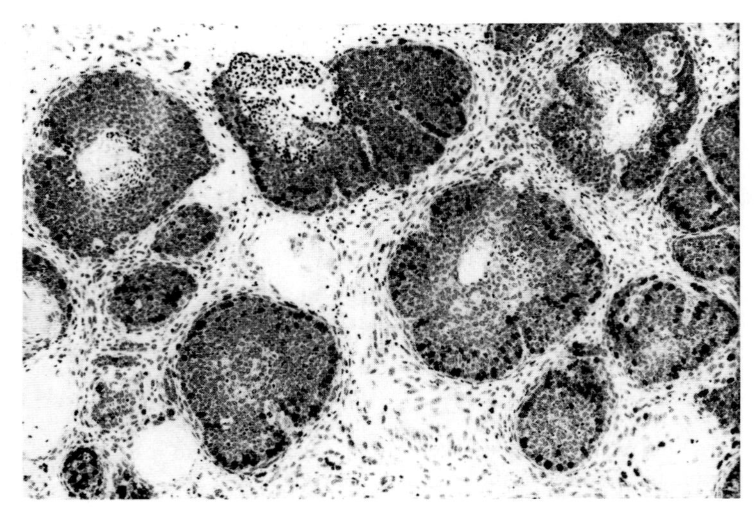

**Fig. 24.1** Histology of adenocarcinoma AT 17, showing the mixed structure of epithelial and interstitial areas. This picture is also a pulse-label autoradiograph revealing the sites of cell proliferation. Methacrylate embedding, H & E × 113.

from this stage. For reasons still unclear, the colonies develop keratin centres in an almost synchronous wave and within another two weeks differentiate into keratin pearls that at best maintain a shell of flattened cells (Fig. 24.3B). After a delay that is apparently dose-dependent, the rims begin to proliferate again. However, in contrast to the primary regeneration pattern where only parenchymal nodules were formed, the cells now also repopulate the interstitium, giving rise to secondary nodules. This process usually starts in several areas of the tumour and eventually reinstates the original histology (Fig. 24.3C).

## DEFINITION OF A COLONY

A schematic representation of this evolution and involution of colonies as seen at two dose levels is given in Figure 24.4. This diagram also illustrates a few points that are relevant for the quantitation of the response:

1. Developing colonies pass through a relatively short initial stage of compact growth but subsequently undergo differentiation, a feature seen also with small colonies developing from normal progenitor cells committed to a differentiation pathway (Ch. 3). Whether this decline reflects a true state of induced differentiation or a microenvironmental (vascular) problem is yet unclear. The crucial question is whether it is justified to score colonies in their compact peak stage as a measure of surviving clonogens. Evidence is presented below which shows that even at high doses most declining colonies will eventually regrow. From this evidence and as the compact colonies meet the common criterion for viability, i.e. contain more than 50 cells, it is concluded that they arise from cells that have retained their reproductive integrity.

2. The choice of the excision interval can play an essential role in quantitating the response. Too long an interval makes it difficult to distinguish between totally effete colonies and remnants of those parenchymal structures that existed at the time of irradiation. This is particularly true for high doses, and for paraffin sections. Too short an interval, on the other hand, will result in a shift towards small colonies with an increasing probability of missing a colony altogether in a given section or of obtaining more cross sections below a critical scoring threshold. The choice is facilitated by the fact that the initial growth period and in particular the peak stage of compact colonies, though transient, is surprisingly little influenced by dose. The optimum sampling stage is at a compact peak stage, and should include a proportion of colonies with incipient central degeneration.

3. The threshold for scoring a colony is a diameter of 40 $\mu$m which corresponds to about 20 nuclei in a cross section. Considering that such small cross sections constitute a minority in the spectrum, and that some of them are tangential sections, the error made by including abortive small colonies should be minimal. For morphometric reasons it would be desirable to know the spectrum of colony sizes, both for various excision intervals and dose levels. Attempts in that direction are being made. If a decrease in colony size with dose exists, it is not very conspicuous.

## RESULTS

A limited number of experiments has been done so far,

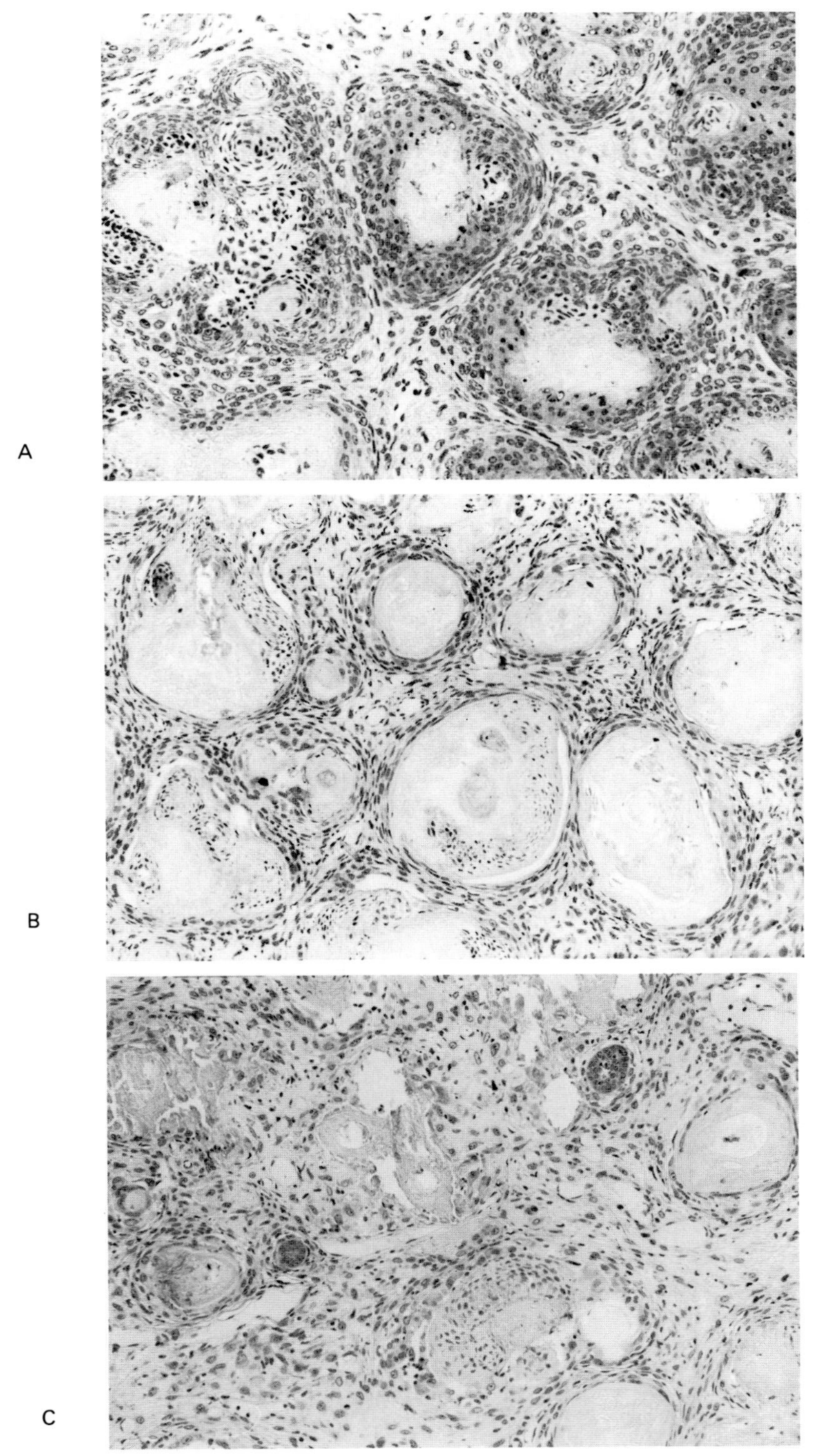


**Fig. 24.2** Tumour histology following a single dose of 32 Gy given under local hypoxia induced by clamping, × 113. **A** two days after irradiation. Rapid cell depletion from nodules by lysis and bulk keratinization. **B** six days after irradiation. Epithelial features are lost from the tumour as the nodules are totally depleted of cells. **C** eight days after irradiation. Beginning of clonal regrowth is indicated by isolated buds of chromophilic epithelial cells.

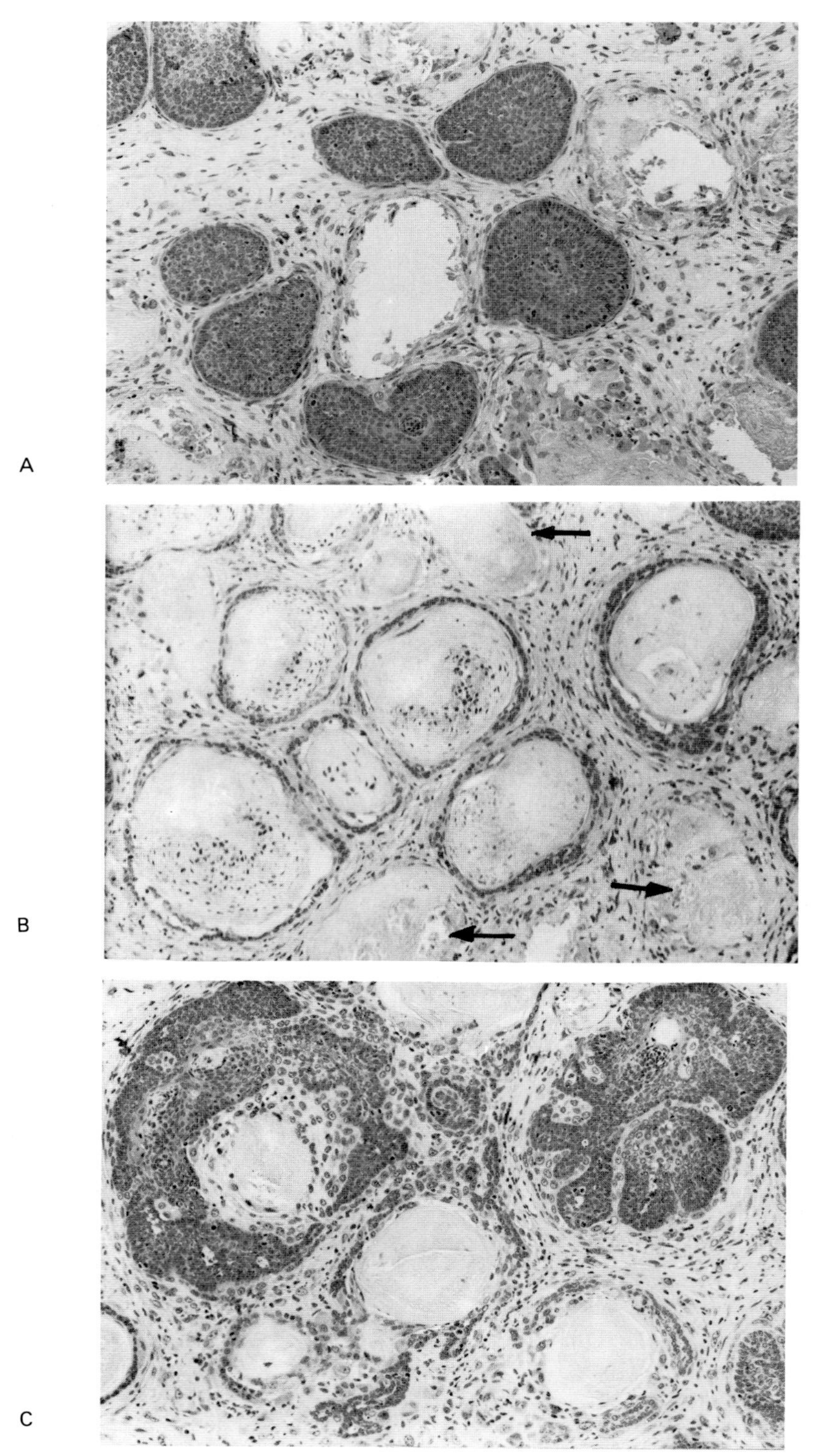


**Fig. 24.3** Tumour histology after 32 Gy given under local hypoxia, × 113. **A** 15 days after irradiation. Fully developed colonies of parenchymal cells grouped around the keratinized remnant of a former parenchymal nodule (lost from the section). **B** 29 days after irradiation. Progressive degeneration ends up with effete clones with only a rim of cells left. Arrows indicate the keratin ghost of an original parenchymal nodule. **C** 48 days after irradiation. Secondary outgrowth from remaining cellular rims eventually reinstitutes the original histology.

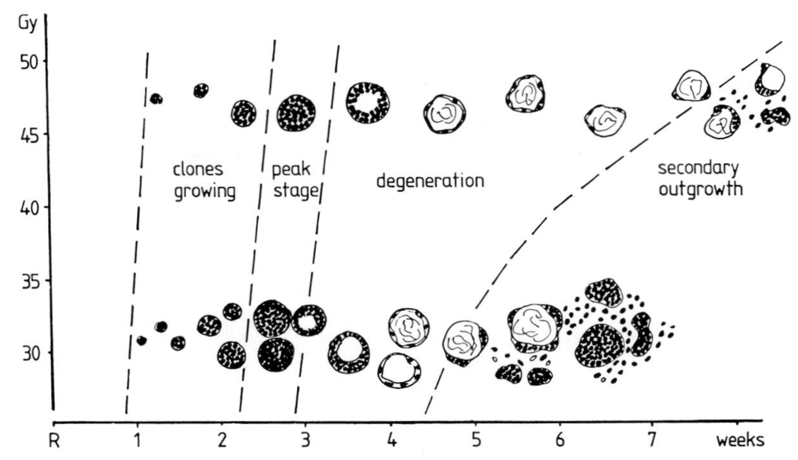


**Fig. 24.4** Schematic representation of the fate of colonies in adenocarcinoma AT 17 irradiated at two dose levels (single doses given under local hypoxia). Developing colonies pass through a short peak stage of growth, then degenerate and remain 'dormant' during a dose-dependent delay period until they resume proliferation. Quantitative assessment of the number of colonies has to be done during the peak stage of compact colonies preceding degeneration, around day 18 to 20. R = time of irradiation.

with the main objective of establishing a 'survival curve' for irradiation. Figure 24.5 shows the results of an experiment where tumours were given single doses with their blood supply clamped. This was done to render the cells uniformly hypoxic and resistant so that the response was governed by the total clonogenic population and not by a naturally hypoxic fraction of variable size. The range of doses given was from 33 to 54 Gy, i.e. up to the curative dose level. Tumour cross sections are shown in Figures 24.6 A, B and C and 24.7 to demonstrate the decline in colony frequency with dose but also to illustrate how a single surviving clone after 54 Gy may be constituted from perfectly viable-looking cells.

For reasons of interpretation and comparison, colony number should be calculated per whole tumour, provided tumour size at treatment is sufficiently standardized. The dose-response curve can then immediately be related to the size of the target-cell population, and also to relevant responses in situ, particularly the probability of local control of tumours. To obtain an absolute count of colonies in a 'crowded' tumour, however, is nearly impossible. For the routine sampling procedure we follow, tumours are sectioned (5 $\mu$m sections) at sequential levels of 500, 250, 100 or 50 $\mu$m, according to the colony frequency expected. The sum of colonies counted in all sequential levels is normalized to a standard separation of sections of 500 $\mu$m and denoted $\Sigma$ 500 $\mu$m. Values obtained from closer spacing than 500 $\mu$m are simply divided by the ratio of the separations (e.g. a $\Sigma$ 100 $\mu$m is divided by five to get the $\Sigma$ 500 $\mu$m. Considering that the average colony diameter is 200 to 300 $\mu$m, values obtained at, or corrected for a 500 $\mu$m separation must necessarily underestimate the absolute colony number by a sampling factor of two to three. The $\Sigma$ 500 $\mu$m therefore is a relative figure which, however, is operationally convenient as it gives comparable dose response data and is independent of individual tumour

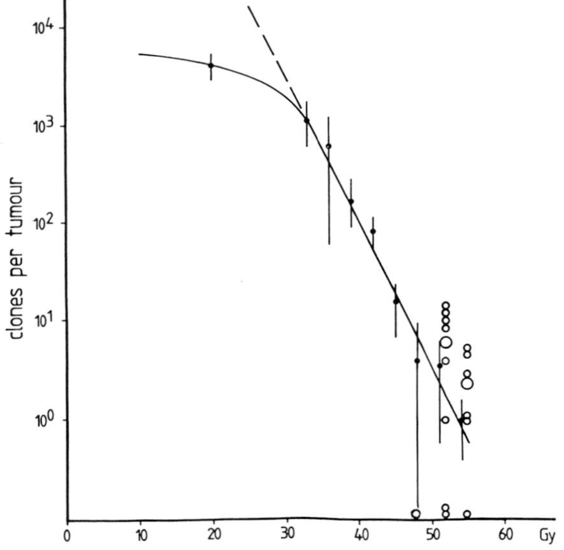


**Fig. 24.5** Dose response of clonogens to single radiation doses of 300 kV X-rays delivered under local hypoxia. Solid points give the mean of six to eight tumours ($\pm$ SD); these data are based on cumulative counts of sequential sections at 500 $\mu$m intervals and underestimate the absolute number of colonies by a factor of two to three (see text). Open symbols are *absolute* numbers of colonies in individual tumours, established from sequential sections at 50 $\mu$m intervals. Symbols drawn at the $10^{-1}$ level represent tumours with zero colonies. Large open circles = arithmetical mean.

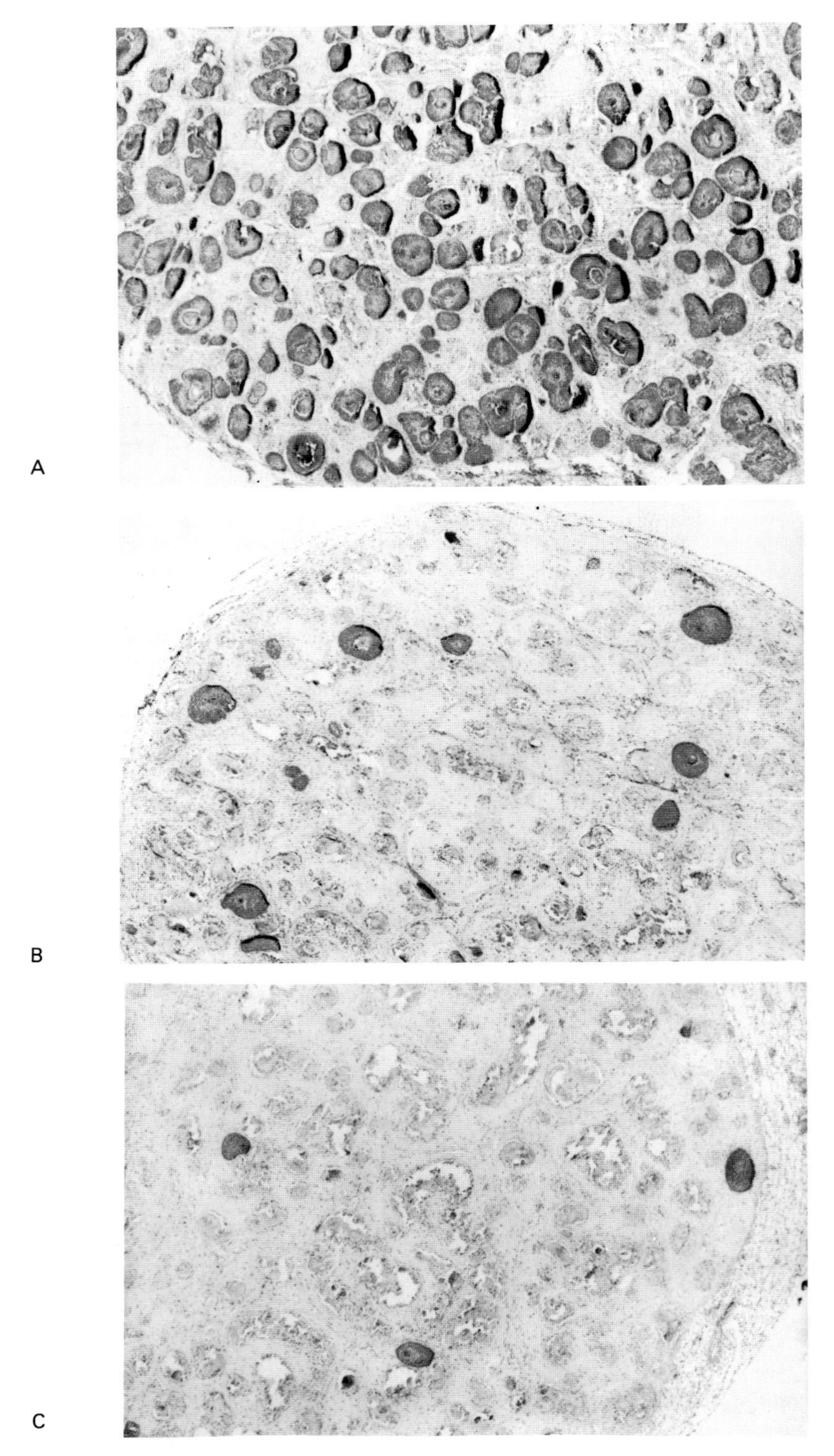


**Fig. 24.6** Dose-dependent decrease in colony frequency in tumours excised 18 days after single doses given under hypoxia. Paraffin embedding, × 29. Doses given were **A**: 33 Gy, **B**: 42 Gy, **C**: 51 Gy.

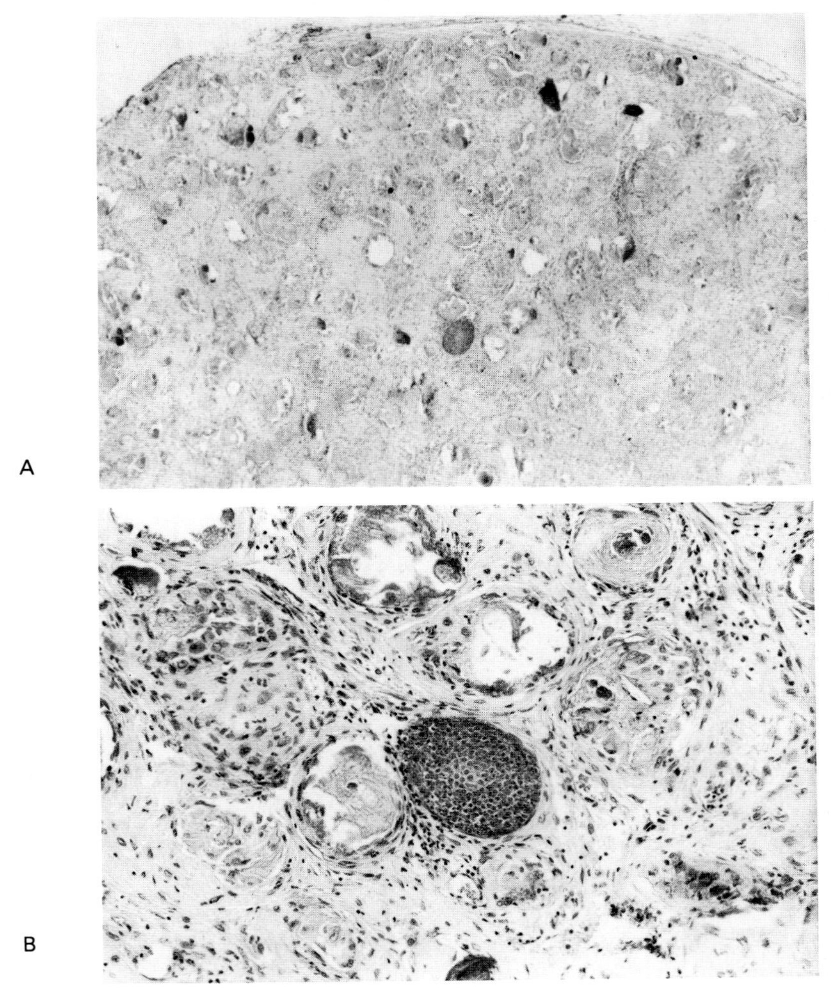


**Fig. 24.7 A** Tumour cross section after a single dose of 54 Gy (under hypoxia) showing a solitary colony, × 29. **B** Higher magnification of this colony to show morphological viability, × 113.

volume at the time of excision. In addition to this procedure, absolute colony counts were performed in tumours sectioned at 50 μm levels, i.e. tumours with very low colony frequency at high dose levels.

Figure 24.5 shows the response measured in the single-dose experiment expressed in both ways. The full symbols and the drawn curve represent the relative values (Σ 500 μm). This dose-response curve bears a general similarity to other survival curves assessed in situ. At the low dose of 20 Gy (data obtained in a separate experiment) the system demonstrates a clear saturation effect; histologically this is obvious from the irregular and coalescent colony pattern and also from areas of advanced regeneration that no longer meet the scoring criteria. Separate colonies are observed from a dose of 33 Gy upwards when the curve extends downwards exponentially over three decades. There is also good theoretical evidence that at about this dose of 33 Gy the majority of the colonies is formed from single surviving clonogens. Assuming a critical population of $5 \times 10^7$ clonogens in a 100 mg tumour, the average density of surviving cells at 33 Gy can be estimated to be one cell in a tumour volume of 0.5 $mm^3$. However, this density becomes quickly diluted at higher doses. In other words, as most colonies arise from single parental cells, the curve can be regarded as a true cell survival curve. Over the dose range tested, the radiosensitivity of the clonogenic population can be characterized by a $D_0$ of 3.0 Gy. If the exponential curve is extrapolated backwards it gives an intercept at slightly more than $10^8$ targets per tumour. Taking into account the sampling factor and a conventional cellular extrapolation number (between 4 and 20),the true number of target cells cannot be far off the $5 \times 10^7$ parenchymal cells estimated as present in the untreated tumour. The region of very low survival has particularly interesting functional implications. When the mean number of clones per tumour reaches ln 2 (= 0.693), Poisson statistics would predict that half the tumours are left without a clone at all and should therefore be locally controlled. The absolute clone counts available for the top doses make it just possible to check this, although they do not fully reach the curative range. If a line with the slope of the survival curve is drawn thorugh their mean values (large open circles in Fig. 24.5) it meets the ln 2 survival level at about 60 Gy. This is in very good agreement with the TCD 5O value obtained in a previous local control experiment which was 59 Gy (unpublished). A further interesting finding was the indication of an over-dispersion in the absolute clone counts (e.g. one tumour with no clones at 48 Gy), compared to the predictions using Poisson statistics. More detailed experiments in the curative dose range will show whether this additional heterogeneity can explain the shallow slopes of many dose-cure relationships (Wheldon, 1980).

## REFERENCES


Marsden J J, Kember N F, Shaw J E H 1980 Irregular radiation response of a chondrosarcoma. British Journal of Cancer 41 (suppl. IV):88

Mendelsohn M L 1967 Cell cloning experiments as models for radiotherapy: a critical appraisal. In: del Regato J A (ed) Conference on radiobiology and radiotherapy. National Cancer Institute Monograph 24:157

Wheldon T E 1980 Can dose-survival parameters be deduced from in situ assays? British Journal of Cancer 41 (Suppl IV):79

Willis R A 1967 Pathology of tumours. Butterworth, London, p 268

Withers H R, Elkind M M 1970 Microcolony survival assay for cells of mouse intestinal mucosa exposed to radiation. International Journal of Radiation Biology 17:261

# The $TD_{50}$ assay for tumour cells

## HISTORICAL BACKGROUND

Early work describing the development of a quantitative assay for the transplantation of two different mouse tumours, Sarcoma 37 and a spontaneous sarcoma of $C_3H$ mice, was reported by Hewitt (1953). In this work, he examined the transplantability of the tumours using inocula of different numbers of tumour cells and he introduced the term '$TD_{50}$', defined as the number of tumour cells required to give rise to tumours in 50 per cent of the inoculated sites. Following this early work the experimental technique was refined and in 1959, Hewitt & Wilson reported on its use to determine a radiation survival curve for leukaemic cells irradiated in vivo. Powers & Tolmach (1963, 1964) used the approach to examine the radiosensitivity of the 6C3HED lymphosarcoma growing subcutaneously and published the first biphasic survival curve, directly demonstrating the presence of hypoxic cells in a solid tumour. These results stimulated other workers to examine the radiosensitivity of the cells of a number of animal tumours using the $TD_{50}$ assay (for review see Kallman, 1968). Meanwhile Revesz (1956, 1958) demonstrated that the addition of a large number of lethally-irradiated tumour cells to the inoculated suspension can reduce the number of viable tumour cells required for transplantation. This so-called 'Revesz effect' will be discussed in more detail later.

The $TD_{50}$ assay was the first assay available for quantitative assessment of the capability of tumour cells to initiate tumour growth. It has been widely used both to study the radiosensitivity and chemosensitivity of tumour cell populations (e.g. Kallman et al, 1967; Reinhold, 1966; Hewitt et al, 1967; Bush & Bruce, 1965; Bruce & Lau, 1975; Urano et al, 1973; Steel et al, 1977; Steel & Adams, 1975) and to quantitate the in vivo immunogenicity of tumour cell populations (e.g. Silobrcic & Suit, 1967; Porteous et al, 1979; Hewitt et al, 1976; Jurin & Suit, 1972; Hewitt, 1978; Kovnat et al, 1982). A modification of the assay has also been used recently to estimate the frequency of tumourigenic variants within transformed cell populations (Thomassen & De Mars, 1982). While the basic methodology is similar in all these studies, in the discussion which follows, the emphasis will be placed on studies using the $TD_{50}$ assay to determine cell survival following radiation or drug treatment of a tumour cell population.

## METHODOLOGY AND DESCRIPTION OF TECHNIQUE

The basic procedure of the $TD_{50}$ assay is to prepare a suspension of tumour cells and to inject different numbers of these cells into groups of recipient animals. It is usual to admix large numbers of heavily-radiated tumour (HR) cells with the viable cells, just prior to injection, to equalise the total number of cells injected at each site. The injections can be given at many different sites; for leukaemias and lymphomas they are often given intravenously (i.v.) or intraperitoneally (i.p.) when the tumour may disseminate throughout the animal; for solid localised tumours they can be done subcutaneously (s.c.) or intradermally (i.d.) often into the inguinal and axillary regions or intramuscularly (i.m.) into the rear leg muscles. Particularly for s.c., i.d. or i.m. injections, it is usual to use multiple sites on the same animal, to reduce the number of test animals required in the an experiment. Each site on the same animal is usually injected with the same number of viable cells, but the different numbers of cells, injected into the different groups of animals, are chosen with the aim that tumours will grow in 0–100 per cent of the injection sites (s.c., i.d. or i.m.), or the injected animals (i.v. or i.p.), over the range of cell numbers used. The injection sites (or animals) are examined frequently for tumour growth for a period which is long enough to ensure that all the tumours are detected. The fraction of tumours arising in the sites (or animals) injected with a given inoculum is then plotted as a function of the number of cells injected (see Fig. 25.1). The number of cells required to produce a tumour in 50 per cent of injection sites (or animals), i.e. the $TD_{50}$, is determined either directly from the plot or more usually by an analytic procedure to be discussed later. If the extent of cell killing following a cytotoxic

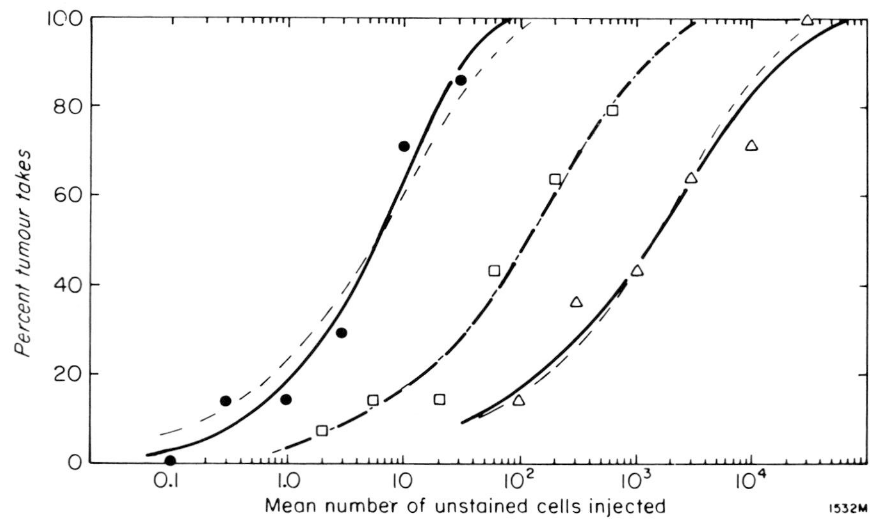


**Fig. 25.1** Percent tumour takes plotted as a function of the cell number injected. Each point was obtained for a group of seven mice each inoculated i.m. into both rear legs (i.e. 14 injection sites/group). The three sets of data are for KHT Sarcoma cell populations, (1) untreated (circles), (2) irradiated with 10 Gy in vivo (squares) or (3) irradiated with 20 Gy in vivo (triangles), prior to the preparation of the cell suspension. The solid lines through the data were obtained by fitting each set of data separately to the modified Poisson model discussed in the text. The broken lines were obtained by determining a mean slope, since the individual slopes were not significantly different, and fitting each set of data with a line of this slope. The $TD_{50}$ values calculated using these analyses are given in Table 25.1.

treatment is to be determined, this is given by the ratio $TD_{50}$(control)/$TD_{50}$ (treated). It is advisable in such studies to run control $TD_{50}$ determinations in every experiment since $TD_{50}$ values have been found to fluctuate from experiment to experiment (Kallman, 1968).

The detection of a growing tumour depends on the site (or sites) of inoculation of the tumour cells. When the injections are i.p. or i.v. and/or the tumour disseminates to the whole body, the decision concerning tumour growth is usually made when the animal becomes sick and is killed or dies, when an autopsy can be performed. For inoculations done s.c., i.d. or i.m., when local tumour growth is expected, the injection sites must be checked carefully for signs of tumour growth. Most investigators require lumps at the injection sites to be at least a defined size and to demonstrate an increase in size over a period of observation, before they are positively identified as tumours. The assay thus applies a very strict criterion for colony expansion since a tumour of a few mm diameter will contain about $10^7$ cells which will require more than 20 divisions to arise from a single cell.

The time period over which the injection sites should be checked for tumour growth before being assessed as negative depends, of course, on the growth rate of the tumour cells being studied, and an appropriate time must be empirically established. Tumours growing from the inocula with small cell numbers (i.e. at the $TD_{50}$ or lower) take the longest time to reach a detectable size. However, one factor which must be considered when establishing a suitable time period is that tumours growing from irradiated cell populations often do so more slowly than those from untreated populations.

When multiple injection sites on the same animal are used (e.g. inguinal and axillary regions for s.c. injections) then two potential problems arise. The first is whether all the sites are equally receptive to tumour growth. In general it is found that fewer cells are required to obtain tumour takes if the cells from solid tumours are injected s.c. rather than i.p. or i.v. but it has also been shown that the $TD_{50}$ can be lower for i.m. injections than for s.c. injections (Kallman et al, 1967; Steel & Adams, 1975; Steel et al, 1977). This should not create a problem because injections of different types are rarely mixed in $TD_{50}$ assays. However, a more serious difficulty is possible when injections are given in different regions of the body. Hewitt et al (1967) found, in their studies using $TD_{50}$ with a squamous carcinoma which involved giving four s.c. injections (two axillary and two inguinal) per mouse, that there was a slight predominance of tumours growing in the axillary, as compared to the inguinal sites. A similar result was reported by Hewitt (1953) in his early studies on quantitative transplantation of tumour cells. Auerbach and coworkers (1978, 1982) have investigated regional differences in the incidence and growth of tumour cells and have obtained very similar results with a number of different animal tumour systems. Although these workers did not do formal $TD_{50}$ studies, their results indicate that there can be a difference by a factor of two in the number of tumour takes for the same number of cells injected into axillary and inguinal regions. The reason for these differences is not known but such heterogeneity in inoculation sites will influence the analysis of the $TD_{50}$ value, and it is probably advisable to use only two injection sites per animal in either the anterior or posterior region. A very convenient approach is to inject i.m. into both rear legs.

The second problem with multiple injection sites is to ensure an adequate test for tumour growth at all of the inoculation sites. If a tumour arises early at only one of the injection sites, it may grow so large that the animal must be sacrificed for humane reasons before the end of the time period designated for checking tumour growth. In such a situation, the question arises whether the other injection site(s) should be scored as negative for tumour growth or completely eliminated from the analysis. Kallman and coworkers (1967, 1968) examined this question in detail for the KHT Sarcoma using two different approaches. Firstly, they analysed the times of detection of tumours arising from a large number of different sized inocula and estimated the probability of a tumour arising at any given site which was lost from observation by the death of an animal. Secondly, they excised all tumours as they arose at injection sites, so that all the sites could be checked for the entire observation period. They concluded that less than three per cent of injection sites, on average, would require to be corrected due to early animal death and hence they felt it unnecessary to adjust their results to take account of this factor. They cautioned that their results should not be applied indiscriminantly to other tumours. However, if only two injection sites per animal are used, as suggested above, and both sites on the same animal are injected with the same number of viable cells, it seems unlikely that there will be a problem with other tumour systems.

The preparation of a good cell suspension from the tumour being studied is an important part of the $TD_{50}$ assay. For solid tumours this can require a substantial effort. In their initial studies Hewitt (1953) and Kallman et al (1967) used a mechanical procedure followed by differential sedimentation to obtain single cell suspensions. Combined mechanical and enzymatic procedures are now common however (e.g. Reinhold, 1965; Hill & Bush, 1969) and these are generally felt to be more satisfactory both because the cell yield is higher and because there is less debris. Appropriate treatments may vary from tumour to tumour, and hence must be determined for each type of tumour studied. It is usual, when cell suspensions are preparaed from solid tumours, to assess the morphological viability of the cells recovered by a technique such as dye-exclusion, and to use the number of dye-excluding cells as a base for calculations. This requires that cell counting be done by direct observation, e.g. using a haemocytometer chamber. When enzyme treatments are used as part of the separation procedure, the dye-excluding cells are usually more than 90 per cent of the total cells recovered, although it should be noted that the total cells recovered rarely represent more than 10 per cent of the mass of the starting tumour material.

## FACTORS AFFECTING THE ASSAY

One obvious factor which can influence the value of the $TD_{50}$ for a tumour cell population is the immunological compatibility of the tumour cells and the recipient animals. This factor has been discussed at length by Hewitt and his coworkers (Hewitt et al, 1976; Hewitt, 1978), who argue that it can influence the results of studies designed to examine radiation or drug sensitivity of tumour cell populations and should be avoided by using non-immunogenic tumour/host systems.

Another very important factor which can influence the $TD_{50}$ value is the 'Revesz effect', the addition of a large number of heavily-radiated tumour (HR) cells to the injected volume. This effect has been found to influence the $TD_{50}$ for a large number of tumours. The size of the effect depends on the number of HR cells injected but varies substantially from tumour to tumour (Revesz, 1958; Hewitt et al, 1976; Kallman et al, 1967; Hill, 1980). The maximum effect appears to occur when greater than $10^6$ HR cells are mixed with the viable cells. In his early work Revesz (1958) demonstrated that HR cells both reduced the $TD_{50}$ and caused and earlier appearance of the tumours. He found that the HR cells had to be injected at the same site as the viable cells to be effective. Neither heat-killed tumour cells nor a local inflammatory reaction could cause the effect. Revesz further found that the effect only occurred for viable tumour cells injected into genetically compatible hosts and that if genetically incompatible hosts were used, the addition of HR cells often increased the $TD_{50}$. This result implies that the 'Revesz' effect is not due to the abrogation of some localized immunological reaction, a conclusion which is supported by extensive studies reported by Hewitt et al (1976).

Studies by Hewitt et al (1973) and Peters & Hewitt (1974) also confirmed Revesz's other observations and the latter authors proposed that the 'Revesz effect' was due to the induction of fibrin formation at the site of injection, which prevented the viable cells from escaping from that site. Because of the Revesz effect, it is important, in $TD_{50}$ assays designed to examine the sensitivity of tumour cell populations to cytotoxic treatments, to add a large number ($\sim 10^6$) of HR cells to all inocula to equalize the total number of cells injected. This approach has two advantages. It avoids the possibility that the cells killed by the treatment might cause an artifact by producing a Revesz effect which does not occur in the control group and, by reducing the $TD_{50}$, it extends the range of survival values over which the assay can be used.

## ANALYSIS OF RESULTS

The results from $TD_{50}$ assays of two groups of irradiated and one group of control tumours are shown in Figure 25.1, plotted as the percentage of inoculation sites which gave rise to tumours (tumour takes) against the logarithm of the number of viable cells injected at those sites. The shape of such a tumour take versus log cell number curve has been discussed in detail by Porter et al (1973) and various analytical procedures for determining the $TD_{50}$ have been described and compared (Finney, 1964; Porter & Berry, 1963; see Ch. 1).

Some of the methods of analysis suggested for $TD_{50}$ assays start with an assumption concerning the expected shape of the tumour takes versus log cell number curve and then apply a maximum likelihood method to obtain the best fit of the expected curve to the data. The method involves applying a transformation to the data, giving a linear relationship which can be fitted by least squares regression analysis using appropriate weighting coefficients. The parameters of the fit are then used in a series of iterations of the fitting procedure, which are continued until the parameters converge. At this point, a $\chi^2$ value can be determined to indicate the goodness of fit of the data to the expected curve (Finney, 1964). The transformation favoured by Finney (1964) is the probit, on the grounds that the normal distribution accords fairly well with observations of biological traits. An alternative is the logit transformation (based on the logistic distribution) which is very similar to the probit over the range of response rates from 0.01 to 0.99.

Porter (Porter & Berry, 1963; Porter et al, 1973; Porter, 1980) argues strongly that the appropriate transformation is one based on a Poisson distribution if it is assumed that the population of tumour cells to be tested contains a proportion of clonogenic cells, any one of which is capable of initiating tumour growth. The Poisson distribution predicts that the number of sites without tumours should be given by $e^{-a}$, where a is the average number of tumour stem cells in the inoculum. A maximum likelihood approach for calculating $TD_{50}$ values based on this simple formula is described in detail by Porter & Berry (1963).

It is often found, however, that the results from a $TD_{50}$ assay are not compatible with the simple model based on the Poisson distribution (called single cell transplantation kinetics) discussed above (Kallman et al, 1967; Porter et al, 1973), in that the slope of the curve is anomalously shallow (as in the case for the results in Fig. 25.1). Porter et al (1973) suggest that the reason for such findings must be unrecognized variability in the experimental procedure. As noted earlier, variation does exist between inguinal and axillary injection sites but it is unclear whether it is sufficiently large to explain completely the differences in slope that have been observed (Porter et al, 1973). Certainly the shallow slopes of the curves in Figure 25.1 are not due to this effect, since injections were administered only into both rear leg muscles. A modification to the mathematical transformation of the Poisson distribution, which will accommodate changes in the slope of the curve and allow analysis from which the $TD_{50}$ value and the goodness of the fit of the curve to the data can be determined, has been suggested by Porter et al (1973). This is a purely empirical modification, which has no clear biological basis in relation to the Poisson model.

An alternative method of analysis, which does not rely overtly on an assumed distribution, is that ascribed to Spearman and Karber. It is described by Finney (1964), and provides an estimate of the log $TD_{50}$ value and the variance of the estimate. However, it does not permit a test of the validity of the method for the data available. According to Finney's discussion, this method compares very favourably with the maximum likelihood approach based on probits, but nominally it requires that the doses (log cell number) be equally spaced and cover the whole range of tumour takes, since it is implicitly assumed in the method that the next dose above and below the doses tested would give 100 per cent and 0 per cent tumour takes, respectively. However, it seems from Finney's discussion of the methods that the Spearman-Karber approach performs quite well even when the distribution of tumour takes is quite badly skewed. The method does not have the inherent flexibility and efficiency of the maximum likelihood approaches but it does have the virtue of simplicity. An outline of the method is given in the Appendix; for a more detailed discussion, particularly of the limitations of the formulae given, see Finney (1964). It is beyond the scope of this chapter to give a similar treatment for the maximum likelihood methods discussed above but the references cited give all the information necessary to develop a suitable program for a small desk calculator. Such a program, based on Porter's suggested mathematical modification of the Poisson distribution, has been used to analyse the results shown in Figure 25.1 and the $TD_{50}$ values, with error ranges, are given in Table 25.1, for comparison with calculations made using the Spearman-Karber method. The data were not compatible with a fit using the simple Poisson model (i.e. the slopes were anomalously shallow), but the three sets of data could

**Table 25.1**

| | TD$_{50}$ values: number of cells (s.e. range) | |
|---|---|---|
| Method A | Method B | Method C |
| 1 4.8 (6.3–3.6) | 6.1 (8.1–4.6) | 5.8 (8.4–4.3) |
| 2 87 (120–63) | 1.2 (1.7–0.85) × $10^2$ | 1.2 (1.7–0.86) × $10^2$ |
| 3 1.3 (1.8–0.9) × $10^3$ | 1.5 (2.1–1.0) × $10^3$ | 1.9 (2.6–1.4) × $10^3$ |

The results shown in Fig. 25.1 were analysed by two different approaches: the Spearman & Karber technique as outlined in the Appendix (Method A) and the modified Poisson distribution model suggested by Porter et al (1973). The fits to the modified Poisson model were made using the method described by Porter (1980) for dose-cure relationships using ln (cell number) to represent dose and tumour takes as equivalent to tumour recurrences. The slopes of the lines representing the individual fits to the three sets of data (Method B) were not significantly different so it was possible to do a parallel fit to the three sets of data giving the results in Method C

be fitted acceptably by curves of the same slope and a parallel line fit was also done. The two different sets of curves obtained are shown in Figure 25.1 and the $TD_{50}$ values given in Table 25.1. The $TD_{50}$ values obtained by all these methods are very similar.

## EXPERIMENTAL DESIGN

The design of a practical experiment to determine $TD_{50}$ values is discussed by Finney (1964) and Porter & Berry (1963) and depends on a number of factors: the most efficient use of the available test animals; the extent of advanced knowledge concerning the expected $TD_{50}$ value; and the planned method of analysis. If no advance information is available, a wide range of cell numbers must be investigated; four or five different cell numbers spaced by factors of eight or 10, each injected into a group of about four animals, seems appropriate. Such a design should cover the complete range of tumour takes and can be easily analysed using the Spearman-Karber method. When the information from such a preliminary assay is available, greater precision in the estimate can be achieved by reducing the cell number intervals and concentrating on the region of interest. If the maximum likelihood methods of analysis are to be used at this stage, then concentrating on the expected tumour take range of 15–85 per cent gives maximum efficiency, and the use of three or four groups of about eight animals each at cell number intervals of a factor of about four is recommended (Finney, 1964; Porter & Berry, 1963). If the Spearman-Karber method of analysis is to be used at the second stage of the experiment, then a wider range of cell numbers is desirable to cover the complete range of tumour takes (with some insurance for error). An appropriate approach might be to have six groups of mice at cell dose intervals of a factor of three or four centred on the expected $TD_{50}$ value. In both cases, however, the number of groups and the cell number intervals needed, will depend on the slope of the tumour takes versus log cell number curve and on the degree of uncertainty in the estimate of the expected $TD_{50}$ value. As can be seen from Figure 25.1, a design of six groups of seven mice each with cell number intervals of a factor of three was barely adequate to cover the range of tumour takes required, although in this case the expected $TD_{50}$ values were well defined.

## PRESENTATION OF RESULTS

The $TD_{50}$ values obtained from assays of treated and control groups (gp) of tumour cells are commonly used to calculate the fraction of cells surviving the treatment by taking the ratio $TD_{50}$ for control gp/$TD_{50}$ for treated gp. An estimate of the error on this surviving fraction can be obtained by combining the variances of the two values (available from the analysis discussed above) in the conventional manner (see Ch. 1). An implicit assumption in using this approach is that the curves of cell number vs tumour takes are parallel for the treated and untreated tumour cell populations. The $TD_{50}$ value is used for calculation because it is the best defined value (i.e. has the smallest variance) for symmetrical distributions ( the $TD_{63}$ value is the one for the Poisson distribution). If the two curves are not parallel, then the surviving fraction calculated will depend on the level of tumour takes being compared and a unique value will not be defined. One value of using a method of analysis which allows an assessment of the goodness of fit of the raw data to a given model is that it provides for a measure of whether the assumption of parallelism is correct (see Table 25.1).

The surviving fractions can be plotted as a function of treatment dose to obtain a survival curve (see Fig. 25.2). It is usual to plot surviving fractions on a log scale for such a presentation because, particularly for

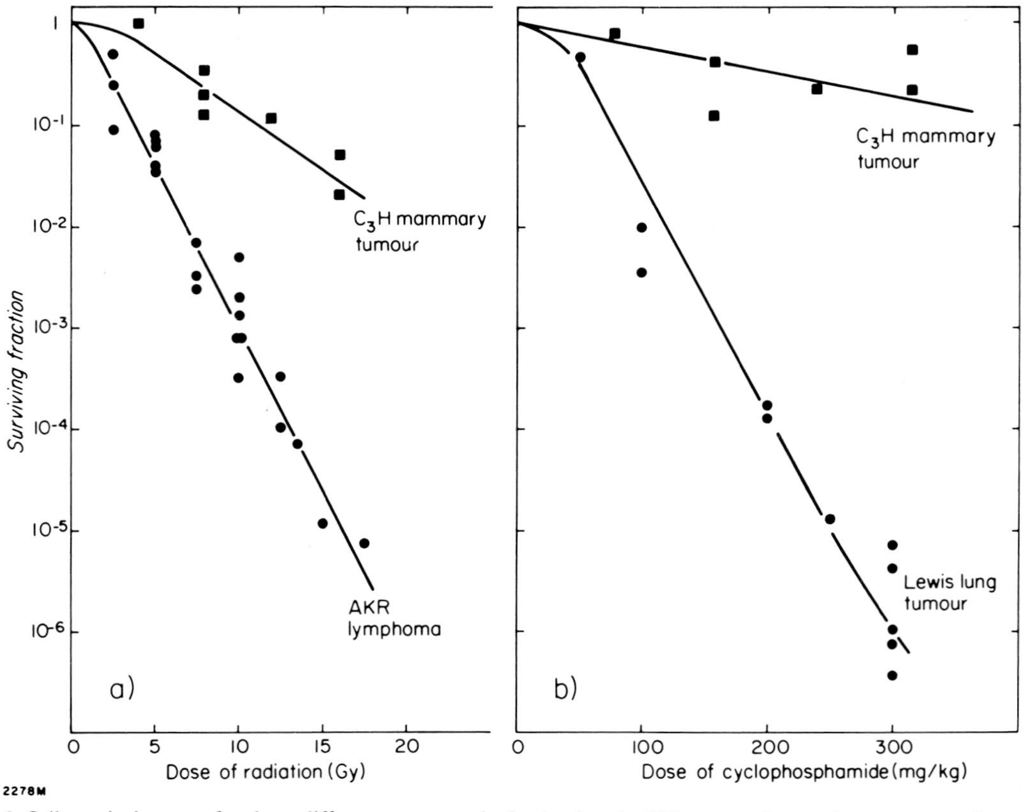


**Fig. 25.2** Cell survival curves for three different tumours obtained using the $TD_{50}$ assay. In panel a curves are shown for AKR lymphoma (redrawn from Bush & Bruce, 1965) and 1st generation transplanted $C_3H$ mouse mammary tumours (redrawn from Bruce & Lau, 1975) treated with $^{60}Co$ γ-rays or 100 kVp x-rays respectively. The difference in the slopes of these lines (lymphoma $D_o = 1.33 \pm 8$ Gy, mammary tumour $D_o \sim 3.9$ Gy) is presumably due largely to the presence of hypoxic cells in the solid mammary tumours. In panel b curves are shown for Lewis lung tumour (redrawn from Steel & Adams, 1975) and 1st generation transplanted $C_3H$ mouse mammary tumour (redrawn from Bruce & Lau, 1975) treated with the drug cyclophosphamide.

radiation treatment, surviving fraction is usually found to be some exponential function of dose. If multiple data points are obtained at the same dose level, then it is appropriate to obtain an arithmetic mean of their log values with corresponding standard error for the purpose of plotting in this type of presentation. Since in studies of tumour response in vivo there is usually substantial inter-experiment variation (see Fig. 25.2), the standard error associated with combining a number of independent survival determinations usually turns out to be greater than that obtained for the individual survival determinations. When the treatment of a tumour cell population extends over a period of time, then the calculated surviving fraction may be influenced by the loss of killed cells or proliferation of surviving cells in the population between the initiation of treatment and the assay procedure. If this aspect is a potential problem in interpreting the data, then the results can be expressed in terms of surviving cells per tumour, by taking into account the cell yield from the cell suspension procedure (Hill & Bush, 1973). Since cell yield per tumour in the control tumours will increase during a prolonged treatment period, this factor may need to be corrected in such a calculation.

The range of survival levels which can be measured using a $TD_{50}$ assay is theoretically controlled by the $TD_{50}$ of the untreated cells and the maximum number of treated cells which can be injected into each inoculation site. In practice, however, because the $TD_{50}$ is not exactly known, a relationship between cell number and tumour takes must be established in order to be reasonably certain of being able to determine the $TD_{50}$ value. Thus the number of untreated cells required to establish tumours in 90–100 per cent of inoculation sites and not the $TD_{50}$ is the relevant value. For the KHT Sarcoma this value is 10–$10^2$ cells (Fig. 25.1). The maximum number of cells which can be satisfactorily injected at a local site is about ($5 \times 10^6$) to $10^7$ (assuming an injection volume of 0.1 ml and a maximum available cell concentration in the suspension of ($5 \times 10^7$) to $10^8$ cells/ml); thus for the KHT Sarcoma a minimum surviving fraction of $10^{-5}$–$10^{-6}$ should be attainable with reasonable precision. As shown in Figure 25.2, such low levels of survival have been measured for transplantable AKR lymphoma and the Lewis lung tumour, both being tumours which have very small $TD_{50}$ values similar to that of the KHT Sarcoma. In contrast, the $TD_{50}$ value for early-generation transplanted $C_3H$ mammary tumours is much higher ($10^4$–$10^5$ cells, Silobrcic & Suit, 1967) which limits the survival levels which can be measured for this tumour.

The results obtained using a $TD_{50}$ assay, to determine the response of a number of different tumour cell populations to radiation or drug treatment, have been compared with those obtained using other in vivo or in vitro assay procedures. In general, good agreement was obtained (Steel & Adams, 1975; Steel et al, 1977; Hill, 1980) but some discrepancies were observed (Bush & Bruce, 1965; Rice et al, 1980). A possible explanation for these discrepancies is that, in these latter two studies, optimum numbers of HR cells were not added to the injected cell suspensions. In both cases, the surviving fraction calculated following large radiation doses was higher, using the $TD_{50}$ assay, than that obtained with the other assay procedures.

An assessment of the practicalities of performing the $TD_{50}$ assay on the KHT Sarcoma in comparison to other assays in vivo such as the lung colony assay, and an assay in vitro, has recently been presented (Hill, 1980). Not surprisingly, in such a comparison, assays in vitro show up as easier and cheaper to perform than a $TD_{50}$ assay and this has meant that for drug and radiation studies the $TD_{50}$ assay is now used less commonly than before. However, the $TD_{50}$ assay does not require the extensive preparative studies necessary to establish an assay in vitro and thus it potentially has a wider applicability. It also usually has a greater range over which survival values can be determined, allowing the examination of the sensitivity of small subpopulations.

## ACKNOWLEDGEMENTS

The work of the author is supported by the National Cancer Institute of Canada and the Ontario Cancer Treatment and Research Foundation.

## APPENDIX

The Spearman-Karber method for calculating $TD_{50}$ values is outlined below. It is condensed from Finney (1964) and can be used for experiments which conform to the designs discussed in the main text. The reader is urged to consult Finney's discussion of the method for the limitations of these formulae, particularly that for the variance.

Suppose that the logarithms of the cell numbers tested are represented by $x_1, x_2 \ldots x_k$ and that $r_i$ of the $n_i$ injections of log cell number $x_i$ give rise to tumours. Then $P_i = r_i n_i$ is an estimate of the tumour take rate at $x_i$. Provided successive cell numbers are fairly close together, the logarithm of the $TD_{50}$ value can be estimated by:

$$m = S\left\{(P_{(i+1)} - P_i)\frac{(x_i + x_{(i+1)})}{2}\right\}$$

where S represents the summation. It is assumed here that log cell numbers less than $x_i$ would have resulted in no tumour takes and log cell numbers above $x_k$ would have resulted in 100 per cent tumour takes. If the doses are equally spaced, so that

$$x_{(i+1)} - x_i = d$$

then

$$m = x_k + \frac{d}{2} - d.Sp_i$$

with a variance

$$V(m) = d^2 . S\left\{\frac{P_i q_i}{(n_i - 1)}\right\}$$

## REFERENCES

Auerbach R, Auerbach W 1982 Regional differences in the growth of normal and neoplastic cells. Science 215:8

Auerbach R, Morrissey L W, Sidky Y A 1978 Regional differences in the incidence and growth of mouse tumours following intradermal or subcutaneous inoculation. Cancer Research 38:1739

Bruce W R, Lau L C 1975 Prediction at the cellular level — a study with the $C_3H$ mammary tumour. In: Pharmacological basis of cancer chemotherapy. Williams & Wilkins, Baltimore, p. 756

Bush R S, Bruce W R 1965 The radiation sensitivity of a transplanted murine lymphoma as determined by two different assay methods. Radiation Research 25:503

Finney D J 1964 Statistical methods in biological assay. Hafner, New York, p 668

Hewitt H B 1953 Studies of the quantitative transplantation of mouse sarcoma. British Journal of Cancer 7:367

Hewitt H B 1978 The choice of animal tumours for experimental studies of cancer therapy. Advances in Cancer Research 27:149

Hewitt H B, Blake E, Porter E H 1973 The effect of lethally irradiated cells on the transplantability of murine tumours. British Journal of Cancer 28:123

Hewitt H B, Blake E R, Walder A S 1976 A critique of the evidence for active host defence against cancer, based on personal studies of 27 murine tumours of spontaneous origin. British Journal of Cancer 33:241

Hewitt H B, Chan D P-S, Blake E R 1967 Survival curves for clonogenic cells of a murine keratinizing squamous carcinoma irradiated in vivo or under hypoxic conditions.

International Journal of Radiation Biology 12:535
Hewitt H B, Wilson C W 1959 A survival curve for mammalian leukaemia cells irradiated in vivo (implications for the treatment of mouse leukaemia by whole-body irradiation). British Journal of Cancer 13:69
Hill R P 1980 An appraisal of in vivo assays of excised tumours. British Journal of Cancer 41 (Suppl. IV):230
Hill R P, Bush R S 1969 A lung-colony assay to determine the radiosensitivity of the cells of a solid tumour. International Journal of Radiation Biology 15:435
Hill R P, Bush R S 1973 The effect of continuous or fractionated irradiation on a murine sarcoma. British Journal of Radiology 46:167
Jurin M, Suit H D 1972 In vivo and in vitro studies of the influence of the immune status of $C_3Hf$/Bu mice on the effectiveness of local irradiation of a methylcholanthrene-induced fibrosarcoma. Cancer Research 32:2201
Kallman R F 1968 Methods for the study of radiation effects on cancer cells. In: Busch H (ed) Methods in cancer research 4. Academic Press, New York, p 309
Kallman R F, Silini G, Van Putten L M 1967 Factors influencing the estimation of the in vivo survival of cells from solid tumours. Journal of the National Cancer Institute 39:539
Kovnat A, Armitage M, Tannock I 1982 Xenografts of human bladder cancer in immune-deprived mice. Cancer Research 42:3696
Peters L J, Hewitt H B 1974 The influence of fibrin formation on the transplantability of murine tumour cells: Implications for the mechanism of the Revesz effect. British Journal of Cancer 29:279
Porteous D D, Porteous K M, Hughes M J 1979 Tumour-cell killing by x-rays and immunity quantitated in a mouse model system. British Journal of Cancer 39:603
Porter E H 1980 The statistics of dose/cure relationships for irradiated tumours, parts I, II. British Journal of Radiology 53:210 and 336
Porter E H, Berry R J 1963 The efficient design of transplantable tumour assays. British Journal of Cancer 17:583
Porter E H, Hewitt H B, Blake E R 1973 The trasnplantation kinetics of tumour cells. British Journal of Cancer 27:55
Powers W E, Tolmach L J 1963 A multi-component x-ray survival curve for mouse lymphosarcoma cells irradiated in vivo. Nature 197:710
Powers W E, Tolmach L J 1964 Demonstration of an anoxic component in a mouse tumour-cell population by in vivo assay of survival following irradiation. Radiology 83:328
Reinhold H S 1965 A cell dispersion technique for use in quantitative transplantation studies with solid tumours. European Journal of Cancer 1:67
Reinhold H S 1966 Quantitative evaluation of the radiosensitivity of cells of a transplantable Rhabdomyosarcoma in the rat. European Journal of Cancer 2:33
Rice L, Urano M, Suit H D 1980 The radiosensitivity of a murine fibrosarcoma as measured by three cell survival assays. British Journal of Cancer 41 (Suppl. IV):240
Revesz L 1956 Effect of tumour cells killed by x-rays upon the growth of admixed viable cells. Nature 178:1391
Revesz L 1958 Effect of lethally damage tumour cells upon the development of admixed viable cells. Journal of the National Cancer Institute 20:1157
Silobrcic V, Suit H D 1967 Tumour-specific antigen(s) in a spontaneous mammary carcinoma of $C_3H$ mice, 1. Quantitative cell transplants into mammary-tumour-agent-positive and -free mice. Journal of the National Cancer Institute 39:1113
Steel G G, Adams K 1975 Stem-cell survival and tumour control in the Lewis lung carcinoma. Cancer Research 35:1530
Steel G G, Adams K, Stephens T C 1977 Clonogenic assays in the B16 melanoma: Response to cyclophosphamide. British Journal of Cancer 36:618
Thomassen D G, DeMars R 1982 Clonal analysis of the stepwise appearance of anchorage independence and tumorigenicity in CAK, a permanent line of mouse cells. Cancer Research 42:4054
Urano M, Fukuda N, Koike S 1973 The effect of bleomycin on survival and tumour growth in $C_3H$ mouse mammary carcinoma. Cancer Research 33:2849

# Index